Exponential Families in Theory and Practice

During the past half-century, exponential families have attained a position at the center of parametric statistical inference. Theoretical advances have been matched, and more than matched, in the world of applications, where logistic regression by itself has become the go-to methodology in medical statistics, computer-based prediction algorithms, and the social sciences. This book is based on a one-semester graduate course for first year Ph.D. and advanced master's students. After presenting the basic structure of univariate and multivariate exponential families, their application to generalized linear models including logistic and Poisson regression is described in detail, emphasizing geometrical ideas, computational practice, and the analogy with ordinary linear regression. Connections are made with a variety of current statistical methodologies: missing data, survival analysis and proportional hazards, false discovery rates, bootstrapping, and empirical Bayes analysis. The book connects exponential family theory with its applications in a way that doesn't require advanced mathematical preparation.

BRADLEY EFRON is Professor Emeritus of Statistics and Biomedical Data Science at Stanford University. He is the inventor of the bootstrap method for assessing statistical accuracy. He has published extensively on statistical theory and its applications, with particular attention to exponential families. A MacArthur fellow, he is a member of the National Academy of Sciences. He received the National Medal of Science in 2007.

INSTITUTE OF MATHEMATICAL STATISTICS TEXTBOOKS

IMS Textbooks give introductory accounts of topics of current concern suitable for advanced courses at master's level, for doctoral students and for individual study. They are typically shorter than a fully developed textbook, often arising from material created for a topical course. Lengths of 100–290 pages are envisaged. The books typically contain exercises.

In collaboration with the International Society for Bayesian Analysis (ISBA), selected volumes in the IMS Textbooks series carry the "with ISBA" designation at the recommendation of the ISBA editorial representative.

Other Books in the Series (*with ISBA)

1. *Probability on Graphs*, by Geoffrey Grilmmett
2. *Stochastic Networks*, by Frank Kelly and Elena Yudovina
3. *Bayesian Filtering and Smoothing*, by Simo Särkkä
4. *The Surprising Mathematics of Longest Increasing Subsequences*, by Dan Romik
5. *Noise Sensitivity of Boolean Functions and Percolation*, by Christophe Garban and Jeffrey E. Steif
6. *Core Statistics*, by Simon N. Wood
7. *Lectures on the Poisson Process*, by Günter Last and Mathew Penrose
8. *Probability on Graphs (Second Edition)*, by Geoffrey Grimmett
9. *Introduction to Malliavin Calculus*, by David Nualart and Eulália Nualart
10. *Applied Stochastic Differential Equations*, by Simo Särkkä and Arno Solin
11. **Computational Bayesian Statistics*, by M. Antónia Amaral Turkman, Carlos Daniel Paulino, and Peter Müller
12. *Statistical Modelling by Exponential Families*, by Rolf Sundberg
13. *Two-Dimensional Random Walk: From Path Counting to Random Interlacements*, by Serguei Popov
14. *Scheduling and Control of Queueing Networks*, by Gideon Weiss
15. *Principles of Statistical Analysis: Learning from Randomized Experiments*, by Ery Arias-Castro
16. *Exponential Families in Theory and Practice*, by Bradley Efron

Exponential Families in Theory and Practice

BRADLEY EFRON
Stanford University

CAMBRIDGE
UNIVERSITY PRESS

CAMBRIDGE
UNIVERSITY PRESS

Shaftesbury Road, Cambridge CB2 8EA, United Kingdom

One Liberty Plaza, 20th Floor, New York, NY 10006, USA

477 Williamstown Road, Port Melbourne, VIC 3207, Australia

314–321, 3rd Floor, Plot 3, Splendor Forum, Jasola District Centre, New Delhi – 110025, India

103 Penang Road, #05–06/07, Visioncrest Commercial, Singapore 238467

Cambridge University Press is part of Cambridge University Press & Assessment,
a department of the University of Cambridge.

We share the University's mission to contribute to society through the pursuit of
education, learning and research at the highest international levels of excellence.

www.cambridge.org
Information on this title: www.cambridge.org/9781108488907

DOI: 10.1017/9781108773157

First published 2023

A catalogue record for this publication is available from the British Library

ISBN 978-1-108-48890-7 Hardback
ISBN 978-1-108-71566-9 Paperback

Contents

Preface

Exponential Families in Theory and Practice is based on my notes for a graduate course designed for first-year Ph.D. and advanced master's degree students in the Statistics Department at Stanford. The course and the book focus on the elegant structure of exponential families, and how exponential family methods have transformed statistical applications in the age of high-speed computing.

Parts 1, 2, and 3 concern the basic ideas of univariate and multivariate exponential families, and their use in generalized linear models, particularly logistic and Poisson regression, the mainstays of modern applications in a variety of fields. The three parts can be covered in about twenty 50-minute lectures, leaving ten lectures for selections from Parts 4 and 5 in a one-quarter course, or fifteen in a semester. Applied topics touch on several statistical success stories: survival analysis and proportional hazards, empirical Bayes, missing data, and false discovery rates.

Homework problems, integrated into the text rather than gathered at the end, play an important role in getting the material across. For the most part the problems aren't very difficult, with the majority chosen to augment points raised in the lecture. Their main role is to help students incorporate the ideas rather than just hear them. Each week I usually assigned four or five homework problems to be turned in, and allowed students to work together on them.

Computational exercises utilize the programming language R, which is also used occasionally in the text to convey specific algorithmic details. Data sets appearing in the text are available from the author's website.

About the mathematical level: I have tried to keep this as low as possible consonant with the subject's needs. Asymptotic arguments are mostly absent, and there are almost no proofs except those that are vital to understanding the statistical points being made. A good background in multivariable calculus, linear algebra, and probability is sufficient mathematical background for the book. Exponential family theory has a strong geometrical aspect, and, whenever possible, I have substituted geometry for algebra and figures for equations.

The physics profession has an honored cohort of practitioners called "phenomenologists" who work to connect theory with applications. In that spirit, the title *Exponential Families in Theory and Practice* could be better amended to *Exponential Families* Between *Theory and Practice*. My goal was to link the powerful theory of exponential families with the modern world of statistical applications, and I hope the book will be successful in that role for both teachers and students.

Acknowledgments

The material in this book accrued over fifty years of teaching, during which time the Stanford Statistics graduate students were almost always good sports and keen critics. My associate Cindy Kirby did heroic work as editor, compositor, and occasional artist in turning messy notes into the volume you are holding. My thanks also to my Cambridge University Press editors Lauren Cowles and Diana Gilooly for their kind support during the long process of publication.

Introduction

Some great ideas are born in a flash of inspiration, perhaps announced to the world by a pathbreaking paper. R. A. Fisher's 1925 article on maximum likelihood estimation is a classic example. Nothing at all like that happened with exponential families. The theory accrued slowly over a period extending roughly between 1932 and 1970. Applications lagged behind, a turning point being the advent of logistic regression and McCullagh and Nelder's 1983 book on generalized linear models.

A salient fact is that no one person is credited with the development of exponential families, though it will be clear from these notes that Fisher's work was instrumental. The name "exponential familes" is comparatively recent. Until the late 1950s they were often referred to as "Koopman–Darmois–Pitman" families (crediting three prominent statisticians working separately in three different countries); the awkward nomenclature suggests only minor importance being attached to the ideas.

Figure 1 gives a rough schematic history of Twentieth Century statistics. The inner circle represents normal theory, the preferred venue of classical methodology. Exact inference – t tests, F tests, chi-squared statistics, ANOVA, multivariate analysis – was feasible inside the circle. Outside the circle was a general theory based on large-sample asymptotic approximations involving Taylor series, Edgeworth expansions, and the central limit theorem. A few special exact results lay outside the normal circle, relating to especially tractable distributions such as the binomial, Poisson, gamma and beta families. These are the figure's green stars. A happy surprise, though a slowly emerging one beginning in the 1930s, was that the special cases were all examples of a powerful general construction, *exponential families*, the intermediate circle in Figure 1. Within this circle, "almost exact" inferential calculations are possible, where any necessary approximations can be pictured in simple geometric diagrams. Such diagrams play a major role in what follows.

Two complementary types of mathematical development can be labeled

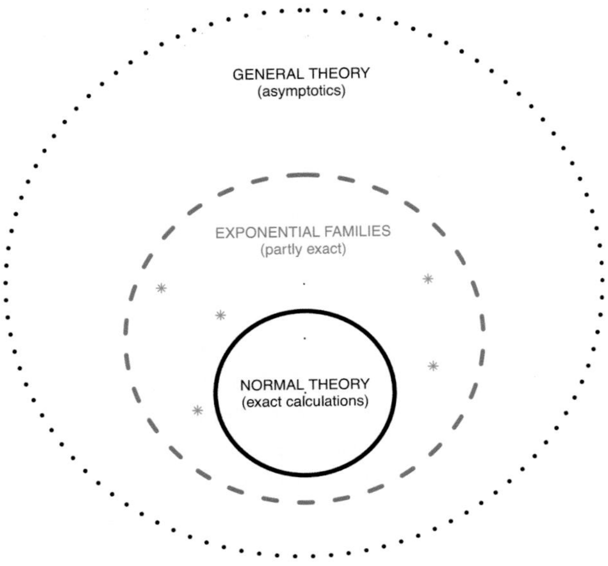

Figure 1 Three levels of statistical modeling.

as "theorem-proof" and "descriptive". The former has a worst-case aspect: "What I stated remains true even under the worst possibility of what I've allowed." Descriptive mathematics, of the kind encountered in introductions to calculus or linear algebra, is less pessimistic: disregarding pathologies, interest centers on the broad central run of useful results. This book pursues the theory of exponential families from a descriptive point of view, aiming at the parts of the theory most useful for applications.

Exponential families, used flexibly, can gracefully bridge the gap between statistical theory and its practice. These notes collect a large amount of material useful in statistical applications, but also of value to the theoretician trying to frame a new situation without immediate recourse to asymptotics. My own experience has been that when I can put a problem, applied or theoretical, into an exponential family framework, a solution is often imminent. There are almost no proofs in what follows, but hopefully enough motivation and heuristics to make the results believable if not obvious. References are given when this doesn't seem to be the case. Readers who desire a more thorough approach can look in Sundberg's excellent text, *Statistical Modelling by Exponential Families* (2019), or Brown's IMS monograph, *Fundamentals of Statistical Exponential Families* (1986).

1

One-parameter Exponential Families

The basic unit of probability theory is a probability distribution. The basic unit of statistical inference is a *family* of probability distributions. Dating from the time of Laplace and Gauss, the one-dimensional normal family[1]

$$x \sim \mathcal{N}(\mu, \sigma^2), \tag{1.1}$$

[1] Equation (1.1) means that the real-valued random variable x has density $\exp\{-(x-\mu)^2/\sigma^2\} \cdot (2\pi\sigma^2)^{-1/2}$ on the real line.

with $\mu \in (-\infty, \infty)$ and σ^2 positive, has played a dominant role in both theory and practice. A strong desire to go beyond normal models fueled the development of exponential family theory. One-parameter exponential families are useful in their own right, and crucial to understanding the multiparameter exponential families of Parts 2 through 5. Here we will present the general one-parameter family theory, and show how it plays out in familiar contexts such as the Poisson, binomial, normal, and gamma distributions.

1.1 Definitions, Notation, and Terminology

This section reviews the basic definitions for exponential families. An exponential family is a set of probability densities $\mathcal{G}$, "density" here including the possibility of discrete atoms (as in the family of binomial densities). A *one-parameter exponential family* has densities $g_\eta(y)$ of the form

$$\mathcal{G} = \left\{ g_\eta(y) = e^{\eta y - \psi(\eta)} g_0(y) m(dy),\ \eta \in A,\ y \in \mathcal{Y} \right\}, \tag{1.2}$$

where A and $\mathcal{Y}$ are subsets of the real line $\mathcal{R}^1$.

There is a more-or-less standard terminology for the elements of (1.2):

- η is the *natural* or *canonical* parameter; in familiar families like the Poisson and binomial, it often isn't the parameter we are used to working with.
- y is the *sufficient* or *natural* statistic, a name that will be more meaningful when we discuss repeated sampling situations; in many cases (the more interesting ones) $y = y(x)$ is a function of an observed data set x (as in the binomial example below); y takes values in its sample space $\mathcal{Y}$.
- The densities in $\mathcal{G}$ are defined with respect to some *carrying measure* $m(dy)$, such as the uniform measure on $[-\infty, \infty]$ for the normal family, or the discrete measure putting weight 1 on the non-negative integers ("counting measure") for the Poisson family. Usually $m(dy)$ won't be indicated in our notation. We will call $g_0(y)$ the *carrying density*.
- $\psi(\eta)$ in (1.2) is the *normalizing function* or *cumulant generating function*; it scales the densities $g_\eta(y)$ to integrate to 1 over sample space $\mathcal{Y}$,

$$\int_{\mathcal{Y}} g_\eta(y) m(dy) = \int_{\mathcal{Y}} e^{\eta y} g_0(y) m(dy) / e^{\psi(\eta)} = 1. \tag{1.3}$$

- The *natural parameter space* A consists of all η for which the integral on the right is finite,

$$A = \left\{ \eta : \int_{\mathcal{Y}} e^{\eta y} g_0(y) m(dy) < \infty \right\}. \tag{1.4}$$

Homework 1.1 Use convexity to prove that if η_1 and $\eta_2 \in A$ then so does any point in the interval $[\eta_1, \eta_2]$ (implying that A is a possibly infinite interval in $\mathcal{R}^1$).

Homework 1.2 We can reparameterize $\mathcal{G}$ in terms of $\tilde{\eta} = c\eta$ and $\tilde{y} = y/c$. Explicitly describe the reparameterized densities $\tilde{g}_{\tilde{\eta}}(\tilde{y})$.

Suppose $g_0(y)$ is any given positive function on a subset $\mathcal{Y}$ of the real line. We can construct an exponential family $\mathcal{G}$ through $g_0(y)$ by "tilting" it exponentially,

$$g_\eta(y) \propto e^{\eta y} g_0(y), \tag{1.5}$$

and then renormalizing $g_\eta(y)$ to integrate to 1,

$$g_\eta(y) = e^{\eta y - \psi(\eta)} g_0(y), \qquad \text{where } e^{\psi(\eta)} = \int_{\mathcal{Y}} e^{\eta y} g_0(y) m(dy). \tag{1.6}$$

The space A is all values of η such that the integral is finite. It seems like we might employ other tilting functions, say

$$g_\eta(y) \propto \frac{1}{1 + \eta |y|} g_0(y), \tag{1.7}$$

but only exponential tilting gives convenient properties under independent sampling.

If η_0 is any point in A we can write

$$g_\eta(y) = \frac{g_\eta(y)}{g_{\eta_0}(y)} g_{\eta_0}(y) = e^{(\eta - \eta_0) y - (\psi(\eta) - \psi(\eta_0))} g_{\eta_0}(y). \tag{1.8}$$

This is the same exponential family, now represented with

$$\eta \longrightarrow \eta - \eta_0, \quad \psi \longrightarrow \psi(\eta) - \psi(\eta_0), \quad \text{and} \quad g_0 \longrightarrow g_{\eta_0}. \tag{1.9}$$

Any member $g_{\eta_0}(y)$ of $\mathcal{G}$ can be chosen as the carrier density, with all the other members as exponential tilts of g_{η_0}. *Important*: the sample space $\mathcal{Y}$ is the *same* for all members of $\mathcal{G}$, and all put positive probability on every point in $\mathcal{Y}$. The members of $\mathcal{G}$ are absolutely continuous with respect to each other, which greatly reduces the opportunities for pathologies in exponential families.

The Poisson Family

As an important first example we consider the Poisson family. A Poisson random variable Y having expectation $\mu > 0$ takes values on the non-negative integers $\mathcal{Z}_+ = \{0, 1, \dots\}$,

$$\Pr_\mu\{Y = y\} = e^{-\mu}\mu^y/y!, \qquad \text{for } y \in \mathcal{Z}_+. \tag{1.10}$$

The densities $e^{-\mu}\mu^y/y!$, taken with respect to counting measure on $\mathcal{Y} = \mathcal{Z}_+$, can be written in exponential family form as

$$g_\eta(y) = e^{\eta y - \psi(\eta)} g_0(y) \begin{cases} \eta = \log \mu & (\mu = e^\eta) \\ \psi(\eta) = e^\eta & (= \mu) \\ g_0(y) = 1/y!. \end{cases} \tag{1.11}$$

(Here $g_0(y)$ is not a member of $\mathcal{G}$, and is not even a proper density.)

Homework 1.3 (a) Rewrite $\mathcal{G}$ so that $g_0(y)$ corresponds to the Poisson distribution with $\mu = 1$.

(b) Carry out the numerical calculations that tilt Poi(12), seen in Figure 1.1, into Poi(6).

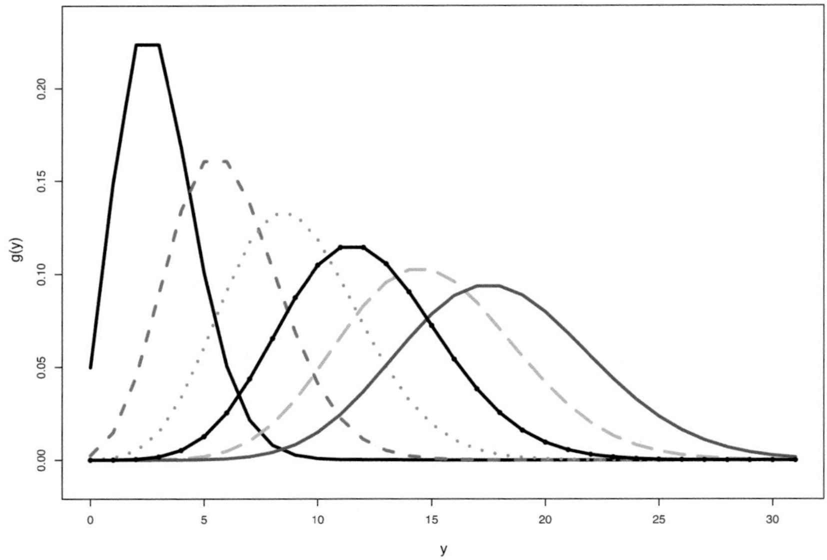

Figure 1.1 Poisson densities for $\mu = 3, 6, 9, 12, 15, 18$; heavy curve with dots for $\mu = 12$.

Even though the mathematics in (1.11) is straightforward, it is still a little surprising to see that any Poisson density is a simple exponential tilt of any other.

1.2 Moment Relationships

The name *cumulant generating function* for the normalizer $\psi(\eta)$ reflects an older methodology for finding expectations, variances, and higher-order moments. The methodology is particularly useful and easy to apply within exponential families.

Expectation and Variance

Differentiating $\exp\{\psi(\eta)\} = \int_{\mathcal{Y}} e^{\eta y} g_0(y) m(dy)$ with respect to η, and indicating differentiation by dots, gives

$$\dot{\psi}(\eta) e^{\psi(\eta)} = \int_{\mathcal{Y}} y e^{\eta y} g_0(y) m(dy) \tag{1.12}$$

and

$$\left(\ddot{\psi}(\eta) + \dot{\psi}(\eta)^2\right) e^{\psi(\eta)} = \int_{\mathcal{Y}} y^2 e^{\eta y} g_0(y) m(dy). \tag{1.13}$$

(The dominated convergence conditions for differentiating inside the integral are always satisfied inside exponential families; see Theorem 2.2 of Brown, 1986.) Multiplying by $\exp\{-\psi(\eta)\}$ gives expressions for the expectation μ_η and variance V_η of Y,

$$\dot{\psi}(\eta) = \mu_\eta = E_\eta\{Y\}, \tag{1.14}$$
$$\ddot{\psi}(\eta) = V_\eta = \mathrm{Var}_\eta\{Y\}, \tag{1.15}$$

where E_η and Var_η indicate expectation and variance under density g_η. V_η is greater than 0, implying that $\psi(\eta)$ has a positive second derivative everywhere, in other words, that $\psi(\eta)$ is convex. Except in trivial cases, the variance V_η is positive for all $\eta \in A$.

Notice that

$$\dot{\mu} = \frac{d\mu}{d\eta} = V_\eta > 0.$$

The mapping from η to μ is 1:1 increasing and infinitely differentiable. We can index the family $\mathcal{G}$ just as well with μ, the *expectation parameter*, as with η. Functions like $\psi(\eta)$, E_η, and V_η can just as well be thought of as functions of μ. We will sometimes write ψ, V, etc. when it's not necessary to specify the argument. Notations such as V_μ formally mean $V_{\eta(\mu)}$.

Note Suppose that ζ is a parameter that can be defined as a function of either η or μ,

$$\zeta = h(\eta) = H(\mu).$$

Let $\dot{h} = dh/d\eta$ and $H' = dH/d\mu$. Then

$$H' = \dot{h}\frac{d\eta}{d\mu} = \frac{\dot{h}}{V}. \tag{1.16}$$

Skewness and Kurtosis

The first two moments of a random variable Y describe its expectation and variance. The third and fourth moments give its *skewness* and *kurtosis*, valuable for higher-order asymptotic approximations. For instance, a first-order Edgeworth expansion says that

$$\Pr\{Y \leq \text{ median } (Y)\} \doteq 0.5 + \frac{1}{6\sqrt{2\pi}} \text{ SKEWNESS } (Y),$$

while the second-order approximation also involves Y's kurtosis.

A pre-computer technology, *cumulants*[2] are certain linear combinations of moments that are easy to deal with in repeated sampling situations (Section 1.3). $\psi(\eta)$ is the *cumulant generating function* for g_0 and $\psi(\eta) - \psi(\eta_0)$ is the CGF for $g_{\eta_0}(y)$, that is,

$$e^{\psi(\eta)-\psi(\eta_0)} = \int_{\mathcal{Y}} e^{(\eta-\eta_0)y} g_{\eta_0}(y) m(dy).$$

By definition, the Taylor series for $\psi(\eta) - \psi(\eta_0)$ has the cumulants k_j of $g_{\eta_0}(y)$ as its coefficients,

$$\psi(\eta) - \psi(\eta_0) = k_1(\eta - \eta_0) + \frac{k_2}{2}(\eta - \eta_0)^2 + \frac{k_3}{6}(\eta - \eta_0)^3 + \cdots .$$

[2] Cumulants add correctly under independent sampling: if X and Y are independent then the jth cumulant of $X + Y$ is the sum of their jth cumulants, this holding for all j. This isn't true for central jth moments $E_0\{Y - \mu_0\}^j$ for $j > 3$. Cumulants are an algebraic computational tool for simplifying higher-order moment relationships, but here we will never go beyond $j = 4$. Older texts, such as Kendall and Stuart (1958), tabulate the relations of cumulants and moments up to $j = 10$.

Equivalently, letting dots indicate derivatives,

$$\dot{\psi}(\eta_0) = k_1 \quad (= \mu_0), \qquad \ddot{\psi}(\eta_0) = k_2 \quad (= V_0),$$
$$\dddot{\psi}(\eta_0) = k_3 \quad \left(= E_0\{Y - \mu_0\}^3\right),$$
$$\ddddot{\psi}(\eta_0) = k_4 \quad \left(= E_0\{Y - \mu_0\}^4 - 3V_0^2\right),$$

etc., where $k_1, k_2, k_3, k_4, \ldots$ are the cumulants of g_{η_0}.

A real-valued random variable Y has skewness and kurtosis defined by

$$\text{SKEWNESS}(Y) = \frac{E(Y - EY)^3}{(\text{Var}(Y))^{3/2}} \equiv \text{“}\gamma\text{”} = \frac{k_3}{k_2^{3/2}}$$

and

$$\text{KURTOSIS}(Y) = \frac{E(Y - EY)^4}{(\text{Var}(Y))^2} - 3 \equiv \text{“}\delta\text{”} = \frac{k_4}{k_2^2}.$$

Putting this together, if $Y \sim g_\eta(\cdot)$ is an exponential family, then

$$Y \sim \Big[\underset{\text{expectation}}{\underset{\uparrow}{\dot{\psi},}} \quad \underset{\substack{\text{standard}\\ \text{deviation}}}{\underset{\uparrow}{\ddot{\psi}^{1/2},}} \quad \underset{\text{skewness}}{\underset{\uparrow}{\dddot{\psi}/\ddot{\psi}^{3/2},}} \quad \underset{\text{kurtosis}}{\underset{\uparrow}{\ddddot{\psi}/\ddot{\psi}^{2}}} \Big], \tag{1.17}$$

where the derivatives are taken at η.

For the Poisson family

$$\psi = e^\eta = \mu,$$

so all the cumulants equal μ

$$\dot{\psi} = \ddot{\psi} = \dddot{\psi} = \ddddot{\psi} = \mu,$$

giving

$$Y \sim \Big[\underset{\text{exp}}{\underset{\uparrow}{\mu,}} \quad \underset{\text{st dev}}{\underset{\uparrow}{\sqrt{\mu},}} \quad \underset{\text{skew}}{\underset{\uparrow}{1/\sqrt{\mu},}} \quad \underset{\text{kurt}}{\underset{\uparrow}{1/\mu}} \Big]. \tag{1.18}$$

A Useful Result

Continuing to use dots for derivatives with respect to η and primes for derivatives with μ, notice that

$$\gamma = \frac{\dddot{\psi}}{\ddot{\psi}^{3/2}} = \frac{\dot{V}}{V^{3/2}} = \frac{V'}{V^{1/2}} \tag{1.19}$$

(using $H' = \dot{h}/V$). Therefore

$$\gamma = 2\left(\sqrt{V}\right)' = 2\frac{d}{d\mu}\,\text{sd}_\mu, \tag{1.20}$$

where $\text{sd}_\mu = V_\mu^{1/2}$ is the standard deviation of y. In other words, $\gamma/2$ is the rate of change of sd_μ with respect to μ; this plays a role in the theory of bootstrap confidence intervals (Part 5).

Homework 1.4 Show that

$$\text{(a)}\ \delta = V'' + \gamma^2 \quad \text{and} \quad \text{(b)}\ \gamma' = \frac{\delta - {}^3/_2\gamma^2}{\text{sd}}.$$

Note The classical exponential families – binomial, Poisson, normal, etc. – are those with closed-form CGFs ψ, yielding neat expressions for means, variances, skewnesses, and kurtoses.

Modern computing power lets us work with general exponential families where results like (1.17) can be exploited numerically, no matter what the form of $\psi(\eta)$.

Unbiased Estimate of η

By definition y is an unbiased estimate of μ (and, in fact, by completeness the only unbiased estimate of form $t(y)$). What about η?

- Let $l_0(y) = \log g_0(y)$ and $l_0'(y) = dl_0(y)/dy$.
- Suppose $\mathcal{Y} = [y_0, y_1]$ (a possibly infinite interval) and that $m(y) = 1$ for all $y \in \mathcal{Y}$.

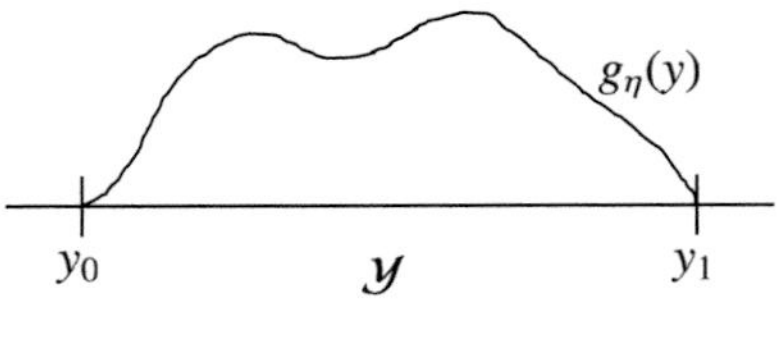

Lemma 1.1

$$E_\eta\left\{-l_0'(y)\right\} = \eta - \left(g_\eta(y_1) - g_\eta(y_0)\right).$$

Homework 1.5 Prove Lemma 1.1. (*Hint*: Integration by parts.)

So, if $g_\eta(y) = 0$ (or $\to 0$) at the extremes of $\mathcal{Y}$, then $-l_0'(y)$ is a unbiased estimate of η.

Homework 1.6 Numerically calculate values of $-l_0'(y)$ to estimate η using Lemma 1.1 for $y \sim \text{Poi}(\mu)$. Does it work?

1.3 Repeated Sampling

One-parameter exponential families have a crucial property that makes them simple to deal with, both in theory and practice: in repeated sampling situations, they retain one-parameter exponential family structure.[3]

Suppose $y_1, \ldots, y_n$ is an independent and identically distributed (i.i.d.) sample from an exponential family $\mathcal{G}$:

$$y_1, \ldots, y_n \overset{\text{iid}}{\sim} g_\eta(\cdot), \tag{1.21}$$

for an unknown value of the parameter $\eta \in A$. The density of $\boldsymbol{y} = (y_1, \ldots, y_n)$ is

$$\begin{aligned}\prod_{i=1}^{n} g_\eta(y_i) &= e^{\sum_1^n(\eta y_i - \psi)} \prod_{i=1}^{n} g_0(y_i) \\ &= e^{n(\eta\bar{y} - \psi)} \prod_{i=1}^{n} g_0(y_i),\end{aligned}$$

where $\bar{y} = \sum_{i=1}^n y_i/n$. Letting $g_\eta^{(n)}(\boldsymbol{y})$ indicate the density of $\boldsymbol{y}$ with respect to $\prod_{i=1}^n m(dy_i)$,

$$g_\eta^{(n)}(\boldsymbol{y}) = e^{n(\eta\bar{y} - \psi(\eta))} \prod_{i=1}^{n} g_0(y_i). \tag{1.22}$$

This is a one-parameter exponential family, with:

- natural parameter $\eta^{(n)} = n\eta$ (so $\eta = \eta^{(n)}/n$);
- sufficient statistic $\bar{y} = \sum_1^n y_i/n$ ($\bar{\mu} = E_{\eta^{(n)}}\{\bar{y}\} = \mu$);
- normalizing function $\psi^{(n)}(\eta^{(n)}) = n\psi(\eta^{(n)}/n)$;
- carrier density $\prod_{i=1}^n g_0(y_i)$ (with respect to $\prod m(dy_i)$).

Homework 1.7 Show that, in the bracket notation of (1.17),

$$\bar{y} \sim \left[\mu, \sqrt{\frac{V}{n}}, \frac{\gamma}{\sqrt{n}}, \frac{\delta}{n}\right].$$

Note In the following, we usually index the parameter space by η rather than $\eta^{(n)}$.

[3] The older name, "Koopman–Darmois–Pitman" families, came from the separate efforts of the three authors to show that, under mild conditions, *only* definition (1.2) allowed this kind of sufficiency property.

Notice that $\boldsymbol{y}$ is now a vector, and that the tilting factor $e^{\eta^{(n)}\bar{y}}$ is tilting the *multivariate* carrier density $\prod_1^n g_0(y_i)$. This is still a one-parameter exponential family because the tilting is in a single direction, along $\mathbf{1} = (1, \ldots, 1)$.

The sufficient statistic $\bar{y}$ also has a one-parameter exponential family of densities,

$$g_\eta^{(n)}(\bar{y}) = e^{n(\eta\bar{y}-\psi)} g_0^{(n)}(\bar{y}),$$

where $g_0^{(n)}(\bar{y})$ is the g_0 density of $\bar{y}$ with respect to $m^{(n)}(d\bar{y})$, the induced carrying measure.

The density (1.22) can also be written (ignoring the carrier) as

$$e^{\eta S - n\psi}, \qquad \text{where } S = \sum_{i=1}^n y_i.$$

This moves a factor of n from the definition of the natural parameter to the definition of the sufficient statistic. For any constant c we can re-express an exponential family $\{g_\eta(y) = \exp(\eta y - \psi) g_0(y)\}$ by mapping η to η/c and y to cy. This tactic will be useful when we consider multiparameter exponential families.

Homework 1.8 $y_1, \ldots, y_n \overset{\text{iid}}{\sim} \text{Poi}(\mu)$. Describe the distributions of $\bar{y}$ and S, and say what are the exponential family quantities $(\eta, y, \psi, g_0, m, \mu, V)$ in both cases.

1.4 Maximum Likelihood Estimation in Exponential Familes

This section briefly reviews some basic results on maximum likelihood estimation (also with a few words about testing). The methodology is particularly simple in exponential families, as we will see. A good reference is Lehmann and Casella (1998), *Theory of Point Estimation.*

Suppose we observe a random sample $\boldsymbol{y} = (y_1, \ldots, y_n)$ from a member $g_\eta(y)$ of an exponential family $\mathcal{G}$,

$$y_i \overset{\text{iid}}{\sim} g_\eta(y), \qquad i = 1, \ldots, n,$$

and wish to estimate η. According to (1.22) in Section 1.3, the density of $\boldsymbol{y}$ is

$$g_\eta^{(n)}(\boldsymbol{y}) = e^{n[\eta\bar{y}-\psi(\eta)]} \prod_{i=1}^n g_0(y_i), \tag{1.23}$$

where $\bar{y} = \sum_1^n y_i/n$. The log likelihood function $l_\eta(\boldsymbol{y}) = \log g_\eta^{(n)}(\boldsymbol{y})$, with $\boldsymbol{y}$ fixed and η varying, is

$$l_\eta(\boldsymbol{y}) = n\left[\eta\bar{y} - \psi(\eta)\right],$$

giving *score function* $\dot{l}_\eta(\boldsymbol{y}) = \partial/\partial\eta\, l_\eta(y)$ equaling

$$\dot{l}_\eta(\boldsymbol{y}) = n(\bar{y} - \mu) \tag{1.24}$$

(remembering that $\dot{\psi}(\eta) = \partial/\partial\eta\,\psi(\eta)$ equals μ, the expectation parameter).

The maximum likelihood estimate (MLE) of η is the value $\hat{\eta}$ satisfying

$$\dot{l}_{\hat{\eta}}(\boldsymbol{y}) = 0.$$

Looking at (1.24), $\hat{\eta}$ is that η such that $\mu = \dot{\psi}(\eta)$ equals $\bar{y}$, that is,

$$\hat{\eta} : E_{\eta=\hat{\eta}}\left\{\bar{Y}\right\} = \bar{y}.$$

In other words, the MLE matches the theoretical expectation of $\bar{Y}$ to the observed mean $\bar{y}$.

We can also take the score function with respect to μ,

$$\frac{\partial}{\partial\mu} l_\eta(\boldsymbol{y}) = \dot{l}_\eta(\boldsymbol{y})\frac{\partial\eta}{\partial\mu} = \frac{\dot{l}_\eta(\boldsymbol{y})}{V} = \frac{n(\bar{y}-\mu)}{V}. \tag{1.25}$$

This gives

$$\left.\frac{\partial}{\partial\mu} l_\eta(\boldsymbol{y})\right|_{\mu=\bar{y}} = 0,$$

which shows that the MLE of μ is

$$\hat{\mu} = \bar{y}.$$

But $\mu = \dot{\psi}(\eta)$, a monotone one-to-one function; since MLEs map in the obvious way, we get

$$\hat{\eta} = \dot{\psi}^{-1}(\bar{y}).$$

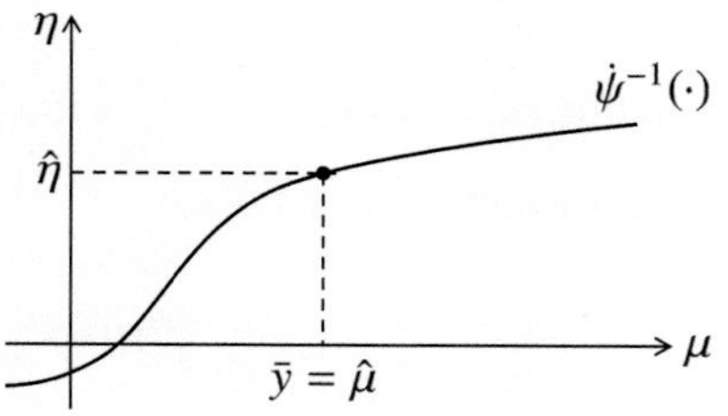

For the Poisson $\hat{\eta} = \log\bar{y}$, and for the binomial, according to what we will see in Section 1.5,

$$\hat{\eta} = \log\left(\frac{\hat{\pi}}{1-\hat{\pi}}\right), \qquad \text{where } \hat{\pi} = \frac{y}{N}.$$

Fisher information is the expected square of the score function – which, since the expected score is always zero, is also its variance – denoted

$$i_\eta^{(n)} = nV$$

for the information for η. We write simply i_η for the case $n = 1$. The information for μ is

$$i_\eta^{(n)}(\mu) = n/V,$$

using (1.25), the notation being understood as the information for μ in a sample of size n, evaluated for $g_\eta(\mathbf{y})$. As always, V stands for V_η, the variance of a single observation y from $g_\eta(\cdot)$.

Let $\zeta = h(\eta)$ be any smooth function of η, also expressed as, say,

$$\zeta = H(\mu) = h\left(\dot{\psi}^{-1}(\mu)\right).$$

Then ζ has MLE $\hat{\zeta} = h(\hat{\eta}) = H(\hat{\mu})$ and score

$$\frac{\partial}{\partial \zeta} l_\eta(\mathbf{y}) = \frac{\dot{l}_\eta(\mathbf{y})}{\dot{h}(\eta)}.$$

Figure 1.2 and Table 1.1 show the MLE and information relationships.

In general, the Fisher information i_θ for a one-parameter family $f_\theta(x)$ has two expressions in terms of the first and second derivatives of the log likelihood,

$$i_\theta = E\left\{\left(\frac{\partial l_\theta}{\partial \theta}\right)^2\right\} = -E\left\{\frac{\partial^2 l_\theta}{\partial \theta^2}\right\}. \tag{1.26}$$

For $i_\eta^{(n)}$, the Fisher information for η in $\mathbf{y} = (y_1, \ldots, y_n)$, we have

$$-\ddot{l}_\eta(\mathbf{y}) = -\frac{\partial^2}{\partial \eta^2} n(\eta \bar{y} - \psi) = -\frac{\partial}{\partial \eta} n(\bar{y} - \mu) = nV_\eta = i_\eta^{(n)}, \tag{1.27}$$

so in this case $-\ddot{l}_\eta(\mathbf{y})$ gives $i_\eta^{(n)}$ without requiring an expectation over $\mathbf{y}$.

Homework 1.9 (a) Does

$$i_\eta^{(n)}(\mu) = -\frac{\partial^2}{\partial \mu^2} l_\eta(\mathbf{y}) \text{ ?}$$

(b) Does

$$i_{\eta=\hat{\eta}}^{(n)}(\mu) = -\frac{\partial}{\partial \mu^2} l_\eta(\mathbf{y})\Big|_{\eta=\hat{\eta}} \qquad (\hat{\eta} \text{ the MLE}) \text{ ?}$$

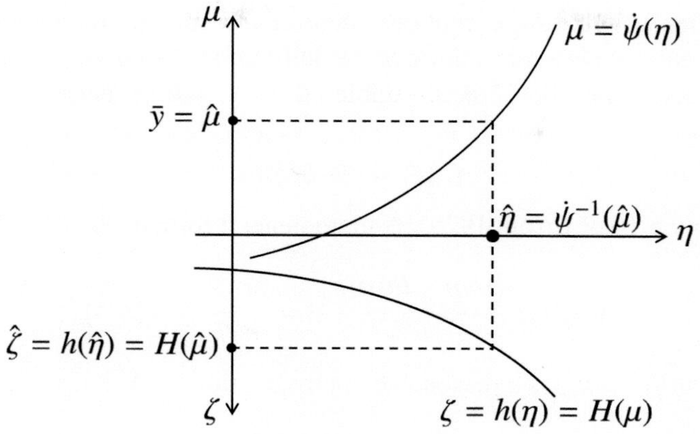

Figure 1.2 Maximum likelihood estimates.

Table 1.1 *Score functions and Fisher information.*

	Score functions	Fisher information
$\eta:$	$\dot{l}_\eta(\mathbf{y}) = n(\bar{y} - \mu)$	$i_\eta^{(n)} = \mathrm{Var}_\eta\left[\dot{l}_\eta(\mathbf{y})\right] = nV = ni_\eta$
$\mu:$	$\dfrac{\partial l_\eta(\mathbf{y})}{\partial \mu} = \dfrac{n(\bar{y} - \mu)}{V}$	$i_\eta^{(n)}(\mu) = \dfrac{n}{V} = ni_\eta(\mu)$
$\zeta:$	$\dfrac{\partial l_\eta(\mathbf{y})}{\partial \zeta} = \dfrac{n(\bar{y} - \mu)}{\dot{h}(\eta)}$	$i_\eta^{(n)}(\zeta) = \dfrac{nV}{\dot{h}(\eta)^2} = ni_\eta(\zeta)$

Cramér–Rao Lower Bound

The Cramér–Rao lower bound (CRLB) for an unbiased estimator $\bar{\zeta}$ of a general parameter $\zeta = h(\eta)$ is

$$\mathrm{Var}_\eta(\bar{\zeta}) \geq \frac{1}{i_\eta^{(n)}(\zeta)} = \frac{\dot{h}(\eta)^2}{nV_\eta}. \tag{1.28}$$

For $\zeta \equiv$ the expectation parameter μ we get

$$\mathrm{Var}(\bar{\mu}) \geq \frac{V_\eta^2}{nV_\eta} = \frac{V_\eta}{n}. \tag{1.29}$$

In this case the MLE $\hat{\mu} = \bar{y}$ is unbiased and achieves the CRLB. This happens only for μ or linear functions of μ, and not for η, for instance. The regularity conditions necessary for the CRLB are almost always satisfied in exponential families, exceptions occuring at boundary points η in those unusual cases where A is a finite or partially finite closed set.

In general, the MLE $\hat{\zeta}$ is *not* unbiased for $\zeta = h(\eta)$, but the bias is of order $1/n$,

$$E_\eta\{\hat{\zeta}\} = \zeta + B(\eta)/n$$

(see Section 10 of Efron, 1975). A more general form of the CRLB gives

$$\mathrm{Var}_\eta(\hat{\zeta}) \geq \frac{\left(\dot{h}(\eta) + \dot{B}(\eta)/n\right)^2}{nV_\eta} = \frac{\dot{h}(\eta)^2}{nV_\eta} + O\left(n^{-2}\right).$$

Usually $\dot{h}(\eta)^2/(nV_\eta)$ is a reasonable approximation for $\mathrm{Var}_\eta(\hat{\zeta})$.

Delta Method

The *delta method* uses a first-order Taylor series expansion to calculate approximate variances. If X has mean μ and variance σ^2, then $Y = H(X) \doteq H(\mu) + H'(\mu)(X - \mu)$ has approximate mean and variance

$$Y \mathrel{\dot\sim} \left[H(\mu), \sigma^2 \left(H'(\mu)\right)^2\right].$$

Homework 1.10 Show that if $\zeta = h(\eta)$, then the MLE $\hat{\zeta}$ has delta method approximate variance

$$\mathrm{Var}_\eta(\hat{\zeta}) \doteq \frac{\dot{h}(\eta)^2}{nV_\eta},$$

in accordance with the CRLB $(i_\eta^{(n)}(\zeta))^{-1}$. (In practice one must substitute $\hat{\eta}$ for η in order to estimate $\mathrm{Var}_\eta(\hat{\zeta})$.)

The simple exponential form of exponential family densities has happy consequences for hypothesis testing as well as estimation. Suppose we wish to test the null hypothesis

$$H_0\colon \eta = \eta_0 \quad \text{versus} \quad H_1\colon \eta = \eta_1$$

for values $\eta_1 > \eta_0$. From (1.23) we get

$$\log \frac{g_{\eta_1}^{(n)}(\bar{y})}{g_{\eta_0}^{(n)}(\bar{y})} = n\left[(\eta_1 - \eta_0)\bar{y} - (\psi(\eta_1) - \psi(\eta_0))\right],$$

which is an increasing function of $\bar{y}$. By the Neyman–Pearson lemma, the most powerful level α test of H_0 ("MP$_\alpha$") rejects for $\bar{y} \geq \overline{Y}_0^{(1-\alpha)}$, where $\overline{Y}_0^{(1-\alpha)}$ is the $(1-\alpha)$th quantile of $\overline{Y}$ under H_0. But this doesn't depend on η_1, so the test is uniformly most powerful level α ("UMP$_\alpha$").

For nonexponential familes such as the Cauchy translation family

$$g_\eta(y) = \frac{1}{\pi}\frac{1}{1+(y-\eta)^2},$$

the MP_α test depends on η_1. Efron (1975) shows that in a certain geometric sense a one-parameter exponential family is a straight line through the space of probability distributions, and this accounts for the UMP property.

1.5 Some Important One-parameter Exponential Families

A good first course in statistics will include various distribution families – normal, Poisson, binomial, gamma – all of which turn out to be one-parameter exponential families. We introduced the Poisson in Section 1.1. This section examines other well-known families, and some that are not so well known. All of these have one important thing in common: their normalizing function $\psi(\eta)$, the CGF, has a closed-form expression. Modern computing ability lets us construct useful exponential families where $\psi(\eta)$ is *not* closed-form, a first example appearing in Section 1.7.

Normal with Variance 1

Normal distributions have played a central role in the evolution of inferential statistics. We begin with the simplest case, where the observed variable Y is normal with unknown expectation μ and fixed variance 1, indicated $Y \sim \mathcal{N}(\mu, 1)$. The family $\mathcal{G}$ has densities

$$g_\mu(y) = \frac{1}{\sqrt{2\pi}} e^{-\frac{1}{2}(y-\mu)^2} \qquad (\mu, y \in \mathcal{R}^1) \tag{1.30}$$

with respect to Lebesque measure $m(dy) = dy$. This can be written in exponential family form as

$$g_\mu(y) = e^{\mu y - \mu^2/2} \cdot \frac{1}{\sqrt{2\pi}} e^{-y^2/2}.$$

In terms of the definitions following (1.1), μ is the expectation parameter $E_\eta\{Y\}$, and

$$\eta = \mu, \quad y = y, \quad \psi = \frac{1}{2}\eta^2, \quad g_0(y) = \frac{1}{\sqrt{2\pi}} e^{-\frac{1}{2}y^2}$$

($g_0(y)$, the *standard normal density*, is often denoted as $\phi(y)$). The variance function is $V_\eta = 1$.

Homework 1.11 Suppose $Y \sim \mathcal{N}(\mu, \sigma^2)$ with σ^2 fixed and known. Derive η, y, ψ, and g_0.

Binomial

$Y \sim \text{Bi}(N, \pi)$, N known, indicates the number of successes in N independent flips of a coin with probability π of success. The density function is

$$g(y) = \binom{N}{y} \pi^y (1-\pi)^{N-y} \qquad (y = 0, 1, \dots, N) \tag{1.31}$$

with respect to counting measure on $\{0, 1, \dots, N\}$. This can be written as

$$\binom{N}{y} e^{\left(\log \frac{\pi}{1-\pi}\right) y + N \log(1-\pi)},$$

a one-parameter exponential family, with:

- $\eta = \log[\pi/(1-\pi)]$ (so $\pi = (1+e^{-\eta})^{-1}$, $1-\pi = (1+e^{\eta})^{-1}$);
- $A = (-\infty, \infty)$, $\mathcal{Y} = \{0, 1, \dots, N\}$;
- $y = y$;
- expectation parameter $\mu = N\pi = N(1+e^{-\eta})^{-1}$;
- $\psi(\eta) = N \log(1+e^{\eta})$;
- variance function $V = N\pi(1-\pi)$ $(= \mu(1-\mu/N))$;
- $g_0(y) = \binom{N}{y}$.

Homework 1.12 Show that the binomial has skewness and kurtosis

$$\gamma = \frac{1-2\pi}{\sqrt{N\pi(1-\pi)}} \quad \text{and} \quad \delta = \frac{1-6\pi(1-\pi)}{N\pi(1-\pi)}.$$

Homework 1.13 Notice that $A = (-\infty, \infty)$ does *not* include the cases $\pi = 0$ or $\pi = 1$. Why not?

Gamma and Chi-squared

The notation

$$Y \sim \lambda G_N \tag{1.32}$$

indicates that Y is a scaled gamma variable having positive scale factor λ, density

$$g(y) = \frac{y^{N-1} e^{-y/\lambda}}{\lambda^N \Gamma(N)}, \qquad \text{for } \mathcal{Y} = (0, \infty), \tag{1.33}$$

and N positive, fixed, and known. With λ the unknown parameter, this is a one-parameter exponential family,

$$\begin{aligned} \eta &= -\frac{1}{\lambda}; \quad \mu = N\lambda = -\frac{N}{\eta}; \\ V &= \frac{N}{\eta^2} = \frac{\mu^2}{N} = N\lambda^2; \\ \psi &= -N\log(-\eta); \quad \gamma = \frac{2}{\sqrt{N}}; \quad \delta = \frac{6}{N}. \end{aligned} \tag{1.34}$$

Situation (1.32) is denoted $Y \sim \text{Gamma}(N, \lambda)$, or sometimes $Y \sim \text{Gamma}(N, {}^1/_\lambda)$, where $r = {}^1/_\lambda$ is the *rate* parameter. *Additivity*: if $Y_i \overset{\text{ind}}{\sim} \lambda G_{N_i}$ for $i = 1, \ldots, K$, then $\sum_1^K Y_i \sim \lambda G_{\sum N_i}$.

Homework 1.14 Derive the skewness and kurtosis γ and δ.

By definition, a chi-squared random variable with m degrees of freedom is twice a gamma with $N = m/2$, written as

$$Y \sim \chi_m^2 = 2G_{m/2}.$$

Chi-squared distributions apply to estimates of variance from normal observations. If

$$x_i \overset{\text{ind}}{\sim} \mathcal{N}(0, \sigma^2), \qquad \text{for } i = 1, \ldots, m,$$

then

$$\hat{\sigma}^2 = \sum_1^m \frac{x_i^2}{m} \sim \frac{\sigma^2 \chi_m^2}{m}. \tag{1.35}$$

In this case $\hat{\sigma}^2$ is an unbiased estimate of σ^2, having mean, standard deviation, skewness, and kurtosis, in the notation of (1.17),

$$\hat{\sigma}^2 \sim \left[\sigma^2, \frac{\sigma^2}{\sqrt{m/2}}, \frac{2}{\sqrt{m/2}}, \frac{12}{m}\right]. \tag{1.36}$$

Chi-squared distributions entered statistics from the world of 19th century physics. The *Maxwell–Boltzmann* distribution describes the velocity of gas molecules. Individual molecules are assumed to have independent, normal signed speeds in three dimensions,

$$v_1, v_2, v_3 \overset{\text{ind}}{\sim} \mathcal{N}(0, \sigma^2),$$

in which case their velocity $v = (v_1^2 + v_2^2 + v_3^2)^{1/2}$ is distributed as

$$v \sim \sigma\sqrt{\chi_3^2},$$

a scaled "chi" distribution with three degrees of freedom. Boltzmann's theory of kinetic gases says that $\sigma = k_B T/\text{mass}$, where T is temperature, mass is the molecule's mass, and k_B is Boltzmann's constant. A surprising fact says that v is typically in the range of a thousand miles per hour at room temperature, something to contemplate as you relax on a nice summer's day.

Homework 1.15 Derive the Maxwell–Boltzmann density $f(v)$.

The Negative Binomial Distribution

A coin with probability of heads θ is flipped until exactly k heads are observed. Let $Y = \{\#\,\text{tails observed}\}$. It has density

$$\begin{aligned} g(y) &= \binom{y+k-1}{k-1}(1-\theta)^y\theta^k \\ &= \binom{y+k-1}{k-1}e^{[\log(1-\theta)]y+k\log\theta}, \end{aligned} \tag{1.37}$$

and sample space $\mathcal{Y} = \{0, 1, 2, \dots\}$. This is a one-parameter exponential family with

$$\begin{aligned} \eta = \log(1-\theta), \quad \psi(\eta) = -k\log(1-e^\eta), \\ \mu = k\frac{1-\theta}{\theta}, \quad V = \frac{\mu}{\theta}. \end{aligned} \tag{1.38}$$

The negative binomial can be thought of as an overdispersed version of the Poisson. For a given value of μ, its variance V is always greater than the Poisson variance μ – as illustrated in Figure 1.3 – making the negative binomial useful for situations with count data where there are reasons to suspect overdispersion.[4] If $k \to \infty$ with θ changing to keep μ fixed, then $Y \to \text{Poi}(\mu)$.

For a given value of μ, the ratio $V/\mu = 1 + \mu/k$ decreases to 1 as

$$\phi \equiv 1/k,$$

with ϕ often called the *dispersion parameter*; distribution (1.37) is denoted by $\text{NB}(\mu, \phi)$. A minor disadvantage of the negative binomial family is that it is not exponential in the dispersion parameter ϕ (or in k). Section 3.9 discusses the "double Poisson" distribution, which is a full two-parameter exponential family of overdispersed Poisson distributions.

[4] As used in the popular algorithm DESeq2 for the analysis of genetic sequence data (Love et al., 2014).

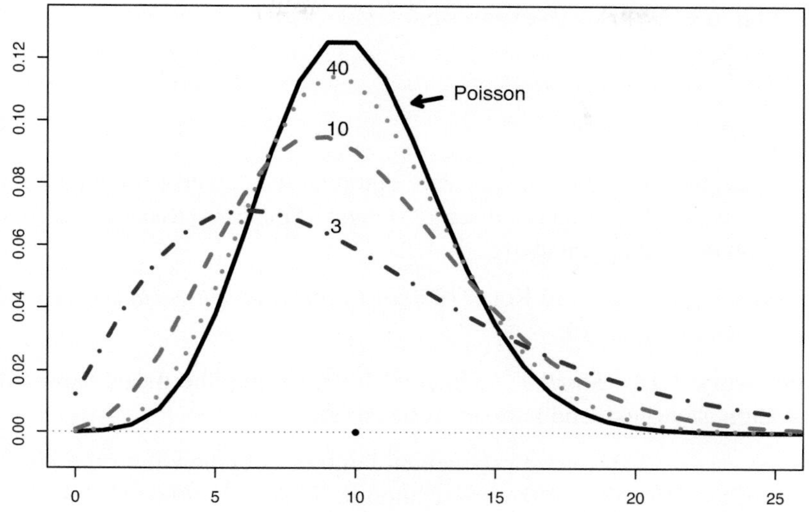

Figure 1.3 Poisson(10) and negative binomials with mean 10, $k = 40, 10$, and 3.

Homework 1.16 Notice that $\psi = k\psi_0$ where ψ_0 is the CGF for $k = 1$. Give a simple explanation for this. How does this affect the expressions for μ and V? What about the expression for the skewness γ?

Homework 1.17 Show that if μ is drawn from a gamma distribution and $y \sim \text{Poi}(\mu)$ is observed, then, marginally, y has a negative binomial distribution. (Another name for the negative binomial is "gamma-Poisson".)

Inverse Gaussian

Let $W(t)$ be a *Wiener process* with drift $1/\mu$, that is, $W(t) \sim \mathcal{N}(t/\mu, t)$ with $\text{Cov}[W(t), W(t+d)] = t$. Define Y as the first passage time to $W(t) = 1$. Then it turns out that Y has the "inverse Gaussian" or Wald density

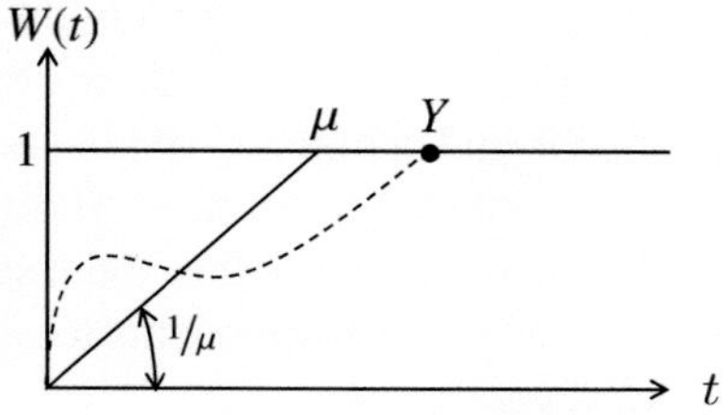

$$g(y) = \frac{1}{\sqrt{2\pi y^3}} e^{-\frac{(y-\mu)^2}{2\mu^2 y}}$$

(y and μ in $\mathcal{R}'$). This is an exponential family with:

- $\eta = -(2\mu^2)^{-1}$;
- $\psi = -\sqrt{-2\eta}$;
- $V = \mu^3$.

We might have called the negative binomial the "inverse binomial" instead, since its definition is a discrete version of the first-passage time construction in the diagram above.

REFERENCE Johnson and Kotz (1970a), *Continuous Univariate Distributions Vol. 1*, Chapter 15.

Homework 1.18 Show $Y \sim [\mu, \mu^{3/2}, 3\mu^{1/2}, 15\mu]$ as the mean, standard deviation, skewness, and kurtosis, respectively.

One way to characterize exponential families is by how the variance V behaves as a function of the expectation μ. Table 1.2 shows V_μ equaling various powers of μ. There is no family with $V_\mu \propto \mu^{1.5}$, say, but quasi-likelihood methods (Section 3.9) let us act as if there is. This was a tactic used in the early history of generalized linear models but will not be explored here.

Table 1.2 *V as a function of μ in some one-parameter exponential families.*

	Normal	Poisson	Scaled Gamma	Inverse Normal
$V_\mu \propto$	μ^0	μ	μ^2	μ^3

2×2 *Tables (Tilted Hypergeometric Family)*

The diagram at right shows a hypothetical two-by-two table comparing men's and women's responses to a yes/no question, perhaps "Have you attended a ballet within the last five years?" The N respondents have provided counts

$$X = (X_1, X_2, X_3, X_4)$$

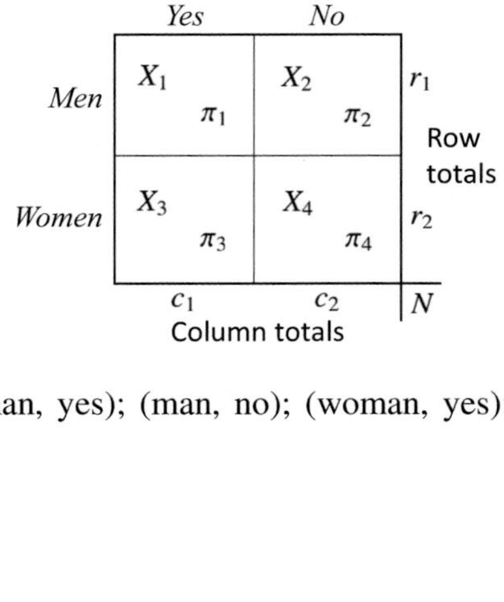

for the four possible categories – (man, yes); (man, no); (woman, yes);

(woman, no) – labeled 1, 2, 3, 4 as shown. As discussed in Section 2.9, X has a four-category multinomial distribution, with true probabilities

$$\pi = (\pi_1, \pi_2, \pi_3, \pi_4)$$

for the four categories.

Perhaps we would like to answer the question, "Do men and women differ in their ballet attendance?" The question can be symmetrically stated in terms of the *log odds* parameter

$$\theta = \log\left(\frac{\pi_1/\pi_2}{\pi_3/\pi_4}\right), \tag{1.39}$$

as a test of the null hypothesis H_0: $\theta = 0$ (i.e., men and women have the same probability of answering yes). Karl Pearson suggested the chi-squared test of H_0 in the early 1900s. *Fisher's exact test* of H_0 (Fisher, 1925) leads to a one-parameter exponential family, the "tilted hypergeometric".

Testing H_0 in terms of X seems awkward since X is four-dimensional. Fisher suggested conditioning the 2×2 table on its marginal sums (r_1, r_2, c_1, c_2) or equivalently conditioning on (N, r_1, c_1), since $r_2 = N - r_1$ and $c_2 = N - c_1$. With the marginals fixed, we need only know x_1 to fill in the 2×2 table.

Fisher's suggestion was to base the test of H_0: $\theta = 0$ on the conditional distribution of x_1 given (r_1, r_2, c_1, c_2). Under H_0, x_1 has the *hypergeometric distribution*

$$g_0(x_1 \mid r_1, r_2, c_1, c_2) = \binom{r_1}{x_1}\binom{r_2}{c_1 - x_1}\Big/\binom{N}{c_1}, \tag{1.40}$$

for x_1 in the set of possible integer values consistent with the marginal constraints

$$\max(0, c_1 - r_2) \leq x_1 \leq \min(c_1, r_1); \tag{1.41}$$

x_1 has

$$\text{expectation} = \frac{r_1 c_1}{N} \quad \text{and} \quad \text{variance} = \frac{r_1 r_2 c_1 c_2}{N^2(N-1)}.$$

Conditioning has the effect of reducing a four-dimensional testing problem to one dimension, at the expense of losing whatever information the marginal totals have on H_0, not much in most situations.

What happens if H_0 is *not* true? Beginning with the probabilities $(\pi_1, \pi_2, \pi_3, \pi_4)$ for the four-category multinomial pictured above, Section 2.8 and

Section 2.9 show that the conditional distribution of x_1 given (r_1, r_2, c_1, c_2) forms a one-parameter exponential family

$$g_\theta(x_1) = \frac{g_0(x_1)e^{\theta x_1}\binom{N}{c_1}}{C(\theta)}, \tag{1.42}$$

where

$$C(\theta) = \sum_{x_1} \binom{r_1}{x_1}\binom{r_2}{c_1 - x_1} e^{\theta x_1},$$

x_1 as in (1.41). That is, we tilt the hypergeometric distribution (1.40) according to $e^{\theta x_1}$, θ the log odds parameter (1.39), and then renormalize to make g_θ sum to 1.

Reference Lehmann and Romano (2005), *Testing Statistical Hypotheses*, Section 4.5.

The Ulcer Data

A clinical trial was held in 41 cities comparing a new ulcer surgery, the Treatment, with the standard surgery, or Control. The 2×2 table at right shows the outcomes for city 14.[5] The obvious estimate of θ is

$$\hat\theta = \log\left(\frac{9/12}{7/17}\right) = 0.600.$$

	Success	*Failure*	
Treatment	**9**	**12**	**21**
Control	**7**	**17**	**24**
	16	**29**	**45**

Figure 1.4 graphs the likelihood, i.e., expression (1.42) as a function of θ, with the data held fixed (normalized so that $\max\{L(\theta)\} = 1$).

Homework 1.19 (a) Compute the likelihood $L(\theta) = g_\theta(\hat\theta)$ numerically and verify that it is maximized at $\hat\theta = 0.600$.

(b) Verify numerically that

$$-\left.\frac{d^2 \log L(\theta)}{d\theta^2}\right|_{\hat\theta} = 2.56.$$

(Note that the quantity on the left is sometimes called "the observed Fisher information", as discussed in Part 4.)

(c) Using this result, guess the variance of $\hat\theta$. *Hint*: Think of the same calculation if $\hat\theta \sim \mathcal{N}(\theta, \sigma^2)$.

[5] The results for all 41 cities are in the file named `ulcdata`.

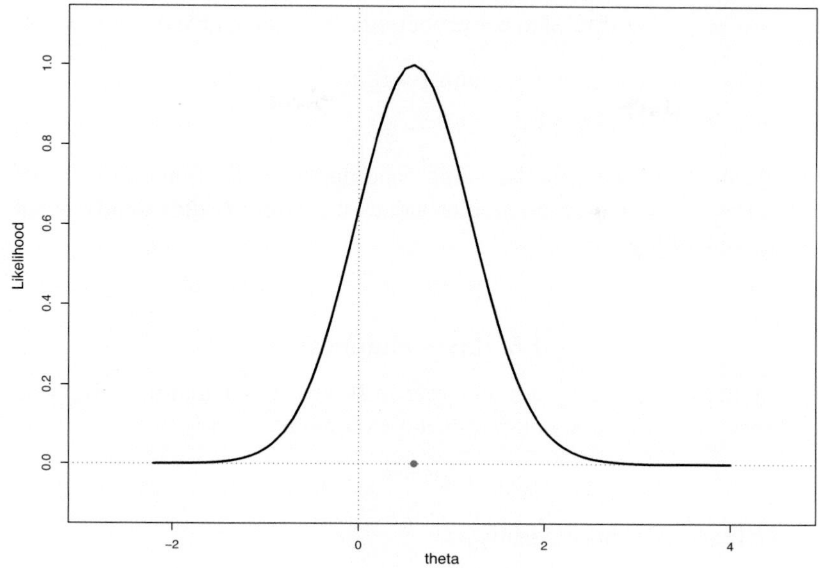

Figure 1.4 `ulcdata` #14; likelihood function for log odds ratio θ; max at $\theta = 0.600$; $-\ddot{l} = 2.56$.

The Structure of One-parameter Exponential Families

Suppose $f_\theta(x)$, θ and x possibly vectors, is a family of densities satisfying

$$\log f_\theta(x) = A(\theta)B(x) + C(\theta) + D(x),$$

with A, B, C, D real. Then $\{f_\theta(x)\}$ is a one-parameter exponential family with:

- $\eta = A(\theta)$;
- $y = B(x)$;
- $\psi = -C(\theta)$;
- $\log g_0 = D(x)$.

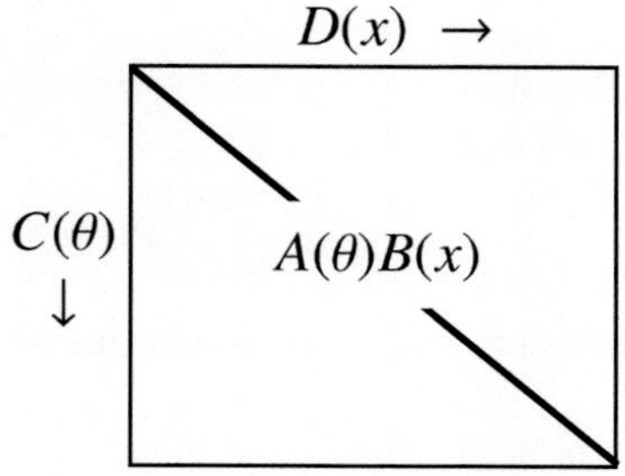

A two-way table of $\log f_\theta(x)$ would have additive components $C(\theta) + D(x)$, and an interaction term $A(\theta)B(x)$.

Homework 1.20 I constructed a 14×9 matrix P with ijth element

$$p_{ij} = \text{Bi}(x_i, \theta_j, 13),$$

the binomial probability of x_i for probability θ_j, sample size $n = 13$, where

$$x_i = i, \qquad \text{for } i = 0, 1, \dots, 13,$$
$$\theta_j = 0.1, \dots, 0.9.$$

Then I calculated the singular value decomposition (R function `svd`) of $\log P$. How many non-zero singular values did I see? (Equivalently, what was the rank of $\log P$?)

1.6 Bayes Families

Exponential families play a major role in Bayesian calculations. Suppose we observe $Y = y$ from a one-parameter exponential family

$$g_\eta(y) = e^{\eta y - \psi(\eta)} g_0(y),$$

where η itself has a prior density

$$\eta \sim \pi(\eta)$$

with respect to Lebesgue measure on a set $\mathcal{A}$. Bayes rule

$$\pi(\eta \mid y) = \pi(\eta) g_\eta(y)/g(y), \tag{1.43}$$

yields the posterior density of η given y, $\pi(\eta \mid y)$, where $g(y)$ is the *marginal density*

$$g(y) = \int_{\mathcal{A}} \pi(\eta) g_\eta(y)\, d\eta.$$

(Note that here $g_\eta(y)$ is the *likelihood function*, with y fixed and η varying.) Putting all this together gives

$$\pi(\eta \mid y) = e^{y\eta - \log(g(y)/g_0(y))} \left(\pi(\eta) e^{-\psi(\eta)}\right). \tag{1.44}$$

We recognize this as a one-parameter exponential family with:

- natural parameter "η" $= y$;
- sufficient statistic "y" $= \eta$;
- CGF $\psi = \log(g(y)/g_0(y))$;
- carrier $g_0 = \pi(\eta) e^{-\psi(\eta)}$.

Homework 1.21 (a) Show that prior $\pi(\eta)$ for η corresponds to prior $\pi(\eta)/V_\eta$ for μ. (b) What is the posterior density $\pi(\mu \mid y)$ for μ?

Conjugate Priors

Certain choices of $\pi(\eta)$ yield particularly simple forms for $\pi(\eta \mid y)$ or $\pi(\mu \mid y)$, and these are called *conjugate priors*. They play an important role in modern Bayesian applications. As an example, the conjugate prior for Poisson is the gamma.

Homework 1.22 (a) Suppose $y \sim \text{Poi}(\mu)$ and $\mu \sim mG_\nu$, a scale multiple of a gamma with ν degrees of freedom. Show that

$$\mu \mid y \sim \frac{m}{m+1} G_{y+\nu}.$$

(b) Also show that

$$E\{\mu \mid y\} = \frac{m}{m+1} y + \frac{1}{m+1}(m\nu)$$

(compared to $E\{\mu\} = m\nu$ *a priori*, so $E\{\mu \mid y\}$ is a linear combination of y and $E\{\mu\}$).

(c) What is the posterior distribution of μ having observed $y_1, y_2, \dots, y_n \overset{\text{iid}}{\sim} \text{Poi}(\mu)$?

Diaconis and Ylvisaker (1979) provide a general formulation of conjugacy: if we observe

$$y_1, \dots, y_n \overset{\text{iid}}{\sim} g_\eta(y) = e^{\eta y - \psi(\eta)} g_0(y),$$

the conjugate prior for μ with respect to Lebesgue measure is

$$\pi_{n_0, y_0}(\mu) = c_0 e^{n_0[\eta y_0 - \psi(\eta)]} \big/ V_\eta, \tag{1.45}$$

where y_0 is notionally the average of n_0 hypothetical prior observations of y (c_0 is the constant making $\pi_{n_0,y_0}(\mu)$ integrate to 1). Prior (1.45) yields a particularly convenient posterior density for μ:

Theorem 1.2

$$\pi(\mu \mid y_1, \dots, y_n) = \pi_{n_+, y_+}(\mu),$$

where

$$n_+ = n_0 + n \quad \textit{and} \quad y_+ = \frac{(n_0 y_0 + \sum_1^n y_i)}{n_+}.$$

Moreover,

$$E\{\mu \mid y_1, \dots, y_n\} = y_+.$$

The first result, which justifies the notional interpretation of n_0 and y_0, is almost immediate, but the second is more involved and won't be verified here.

Homework 1.23 Make the explicit connections between Theorem 1.2 and Homework 1.22.

Binomial Case

Suppose $y_1, \dots, y_n$ are independent *Bernoulli* observations,

$$y_i = \begin{cases} 0 & \text{with probability } 1 - \pi \\ 1 & \text{with probability } \pi, \end{cases}$$

so $y = \sum_{i=1}^{n} y_i$ is binomial, $y \sim \text{Bi}(n, \pi)$. As in Section 1.5, y is the sufficient statistic of a one-parameter exponential family having $\eta = \log[\pi/(1 - \pi)]$ and $\mu = n\pi$.

Homework 1.24 Remembering that μ equals π in the binomial case, show that the conjugate family (1.45) is

$$\pi_{n_0, y_0}(\pi) = c_0 \pi^{s_1 - 1}(1 - \pi)^{s_0 - 1}, \tag{1.46}$$

where (s_1, s_0) are the number of 1s and 0s in the hypothetical prior sample (a "beta" distribution; see Part 2). Theorem 1.2 gives posterior expectation

$$E\{\pi \mid y_1, \dots, y_n\} = \frac{s_1 + y}{n_0 + n}. \tag{1.47}$$

The interpretation of the prior (1.46) is as a hypothetical binomial sample of size n_0, with observed number $s_1 = n_0 y_0$ of successes. Current Bayes practice favors using small amounts of hypothetical prior information, in the binomial case maybe $s_1 = 1$ and $n_0 = 2$ (so $y_0 = 1/2$), giving

$$\hat{\theta} = \frac{1 + y}{2 + n},$$

pulling $\hat{\theta}$ a little toward $1/2$, compared to the MLE y/n.

Tweedie's Formula

Equation (1.44) gave

$$\pi(\eta \mid y) = e^{y\eta - \lambda(y)} \pi_0(y),$$

where

$$\pi_0(y) = \pi(\eta)e^{-\psi(\eta)} \quad \text{and} \quad \lambda(y) = \log \frac{g(y)}{g_0(y)},$$

with $g(y)$ the marginal density of y. Define

$$l(y) = \log g(y) \quad \text{and} \quad l_0(y) = \log g_0(y).$$

We can now differentiate $\lambda(y)$ with respect to y to get the posterior moments (and cumulants) of η given y,

$$E\{\eta \mid y\} = \lambda'(y) = l'(y) - l_0'(y),$$
$$\text{Var}\{\eta \mid y\} = \lambda''(y) = l''(y) - l_0''(y).$$

Homework 1.25 Suppose $y \sim \mathcal{N}(\mu, \sigma^2)$, σ^2 known, where μ has prior density $\pi(\mu)$. Show that the posterior mean and variance of μ given y is

$$\mu \mid y \sim \left[y + \sigma^2 l'(y), \sigma^2\left(1 + \sigma^2 l''(y)\right)\right]. \tag{1.48}$$

Reference Efron (2011), "Tweedie's formula and selection bias", *JASA* 1602–1614.

1.7 Empirical Bayes Inference

Bayes rule, when applicable, provides a wonderfully satisfying path for statistical inference. The catch is that the prior density, $\pi(\eta)$ in (1.43), is most often unknown in typical applications. A surprising development, post-World War II, was that when simultaneously dealing with many similar inference problems, the data itself may provide an estimate of the prior. This is the *empirical Bayes* concept, an approach where exponential families have played a central part.

Table 1.3 displays the *insurance data*, a summary of one year's record of claims from a European auto insurance company: 7840 of the 9461 policyholders made no claims during the year, 1317 made a single claim, 239 made two claims each, going on to the one person, possibly a very bad driver, who made seven claims. In the notation of Table 1.3,

$$y_x = \#\,\{\text{policyholders who made } x \text{ claims}\}, \tag{1.49}$$

for $x = 0, 1, \ldots, 7$.

Suppose the company wants to know how many accident claims it can expect next year from a driver with x claims this year. A commonly used

Table 1.3 *Insurance data counts and claims, and two empirical Bayes estimates of future claims per driver.*

Claims x	0	1	2	3	4	5	6	7
Counts y_x	7840	1317	239	42	14	4	4	1
Robbins' formula	0.168	0.363	0.527	1.33	1.43	6.00	1.25	
Gamma MLE	0.164	0.398	0.632	0.87	1.10	1.34	1.57	

actuarial model assumes that each driver has a Poisson distribution of annual accidents, $g_\mu(x) = e^{-\mu}\mu^x/x!$, μ varying from driver to driver, and with μ having some prior density $\pi(\mu)$,

$$\pi(\mu) \longrightarrow \mu \longrightarrow x \sim \text{Poi}(\mu). \tag{1.50}$$

The insurance company would like to know the Bayes posterior expectation of μ given x,

$$E\{\mu \mid x\} = \int_0^\infty \mu\pi(\mu \mid x)\, d\mu = \frac{\int_0^\infty [e^{-\mu}\mu^{x+1}/x!]\pi(\mu)\, d\mu}{\int_0^\infty [e^{-\mu}\mu^x/x!]\pi(\mu)\, d\mu}, \tag{1.51}$$

but unless they know the prior $\pi(\mu)$ this is out of reach.

Here is where empirical Bayes makes its appearance. Notice that we can rewrite (1.51) as

$$\begin{aligned} E\{\mu \mid x\} &= \frac{(x+1)\int_0^\infty \left[e^{-\mu}\mu^{x+1}/(x+1)!\right]\pi(\mu)\, d\mu}{\int_0^\infty [e^{-\mu}\mu^x/x!]\pi(\mu)\, d\mu} \\ &= \frac{(x+1)g(x+1)}{g(x)}, \end{aligned} \tag{1.52}$$

where $g(x)$ is the marginal density of x,

$$g(x) = \int_0^\infty \frac{e^{-\mu}\mu^x}{x!}\pi(\mu)\, d\mu.$$

Homework 1.26 Give a careful derivation of (1.51)–(1.52).

We don't know $g(\cdot)$ either but, as the marginal distribution of x, it has an obvious estimate in terms of the counts y_x,

$$\hat{g}(x) = \frac{y_x}{N} \qquad \left(N = \sum y_x\right),$$

$N = 9461$ here; (1.52) leads to *Robbins' formula* (Robbins, 1956)

$$\widehat{E}\{\mu \mid x\} = (x+1)\frac{y_{x+1}}{y_x}. \tag{1.53}$$

The third line of Table 1.3 shows $\widehat{E} = \{\mu \mid x = 0\} = 0.168$, so last year's perfect driver can expect about one-sixth of a claim this year, and so on up the table.

Small values of y_x make $\widehat{E}\{\mu \mid x\}$ erratic near the right end of Table 1.3. We can get a less variable estimate of $E\{\mu \mid x\}$ by assuming a parametric model for $\pi(\mu)$ in (1.52). A natural choice is the conjugate prior, the scaled Gamma,

$$\pi(\mu) = \frac{\mu^{a-1}e^{-\mu/b}}{b^a\Gamma(a)} \qquad (\mu > 0), \tag{1.54}$$

that is, $\mu \sim bG_a$ (1.32) (making the marginal $g(x)$ negative binomial, as in Homework 1.17). Choosing (a, b) to be the maximum likelihood estimates based on the counts $y_0, y_1, \ldots, y_n$, and substituting $\hat{\pi}(\mu)$ for $\pi(\mu)$ in (1.52), gave the estimates $\widehat{E}\{\mu \mid x\}$ in the last row of Table 1.3.

Homework 1.27 What is Robbins' formula if x is binomial rather than Poisson in (1.50)?

Robbins' formula applies to Poisson sampling models (1.50). Suppose instead we have a normal model,

$$\pi(\mu) \longrightarrow \mu \longrightarrow y \sim \mathcal{N}(\mu, \sigma^2), \tag{1.55}$$

σ^2 known. Tweedie's formula (1.48) says that

$$E\{\mu \mid y\} = y + \sigma^2 l'(y), \tag{1.56}$$

with $l'(y)$ the derivative of the log marginal density $g(y)$; y is the MLE of μ, the usual frequentist (non-Bayesian) estimate of μ, so (1.56) amounts to

$$E\{\mu \mid y\} = \text{MLE} + \text{Bayes correction}. \tag{1.57}$$

Empirical Bayes methods come into play when we have many realizations of (1.55) to deal with at once, say $y_i \sim \mathcal{N}(\mu_i, \sigma^2)$ for $i = 1, \ldots, N$, where we can use all the data to estimate the Bayes correction for each case. In other words, we can enjoy the advantages of Bayesian estimation without the requisite prior knowledge. A microarray example follows next.

In a study of prostate cancer, $n = 102$ men each had his genetic expression level x_{ij} measured on $N = 6033$ genes,

$$x_{ij} = \begin{cases} i = 1, \ldots, N & \text{genes} \\ j = 1, \ldots, n & \text{men.} \end{cases}$$

There were:

- $n_1 = 50$ healthy controls;
- $n_2 = 52$ prostate cancer patients.

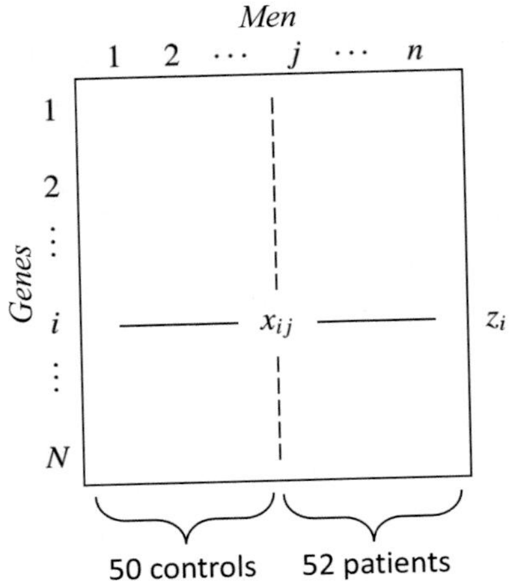

For gene$_i$, let t_i equal a two-sample t statistic comparing patients with controls and

$$z_i = \Phi^{-1}\left[F_{100}(t_i)\right] \qquad (F_{100} \text{ CDF of a } t_{100} \text{ distribution});$$

z_i is a *z-value*, i.e., a statistic having a $\mathcal{N}(0, 1)$ distribution under the null hypothesis that there is no difference in gene$_i$ expression between patients and controls.[6]

A reasonable model for the z_is is

$$z_i \sim \mathcal{N}(\delta_i, 1),$$

where δ_i is the *effect size* for gene i. (In terms of our previous notation, z and δ are playing the roles of y and μ.) The investigators were looking for genes with large values of δ_i, either positive or negative. Figure 1.5 shows the histogram of the 6033 z_i values. It is a little wider than a $\mathcal{N}(0, 1)$ density, suggesting some non-null ($\delta_i \neq 0$) genes. Which ones and how much?

An empirical Bayes analysis proceeds in four steps:

1. Compute $z_1, \ldots, z_N$; $N = 6033$.
2. Fit a smooth parametric estimate $\hat{g}(z)$ to histogram (details in Part 2).
3. Numerically differentiate $\hat{l}(z) = \log \hat{g}(z)$ to get $\hat{l}'(x)$.
4. Estimate $E\{\delta_i \mid z_i\}$ by

$$E\{\delta_i \mid z_i\} = z_i + \hat{l}'(z_i), \tag{1.58}$$

for $i = 1, \ldots, N$. Notice that *all* of the z-values play a role in the estimation of any one δ_i, through their part in estimating $\hat{g}(\cdot)$.

6 Since $t_i \sim F_{100}$ under the null hypothesis that $\delta_i = 0$, the "probability integral transformation" $F_{100}(t_i)$ has a uniform distribution over $[0, 1]$; then the inverse transformation $Z = \Phi^{-1}(F_{100}(t_i)) \sim \mathcal{N}(0, 1)$.

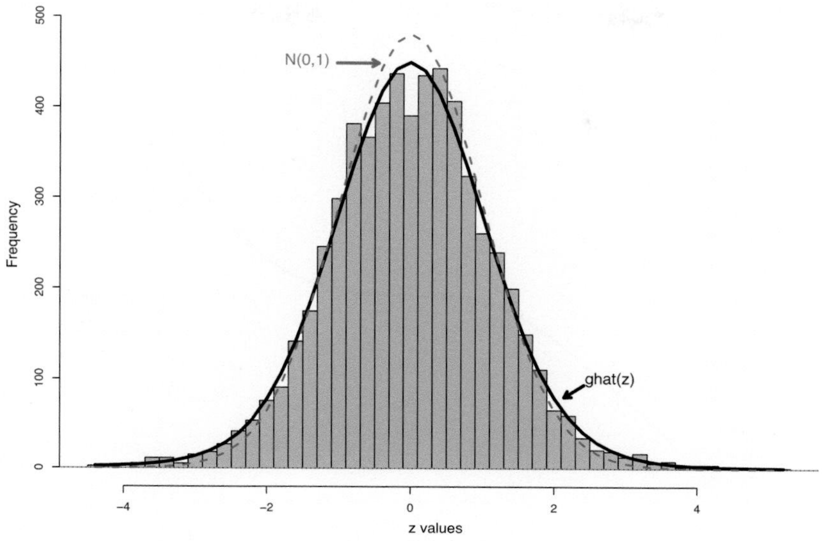

Figure 1.5 *Prostate data* microarray study. 6033 z-values; heavy curve is $\hat{g}(z)$ from GLM fit; dashed line is $\mathcal{N}(0, 1)$.

Figure 1.6 shows $\widehat{E}\{\delta \mid z\}$. It is near zero ("nullness") for $|z| \leq 2$. At $z = 3$, $\widehat{E}\{\delta \mid z\} = 1.31$. At $z = 5.29$, the largest observed z_i value (gene #610), $E\{\delta \mid z\} = 3.94$. Even though each z_i is unbiased for its δ_i it isn't true that $z_{i_{\max}}$, $i_{\max} = 610$, is unbiased for $\delta_{i_{\max}}$. The Bayes correction in (1.57) is quite negative, an example of *selection bias* or *the winner's curse*: being largest in a group of unbiased estimates involves luck as well as a genuinely large value of δ, and that's what empirical Bayes is accounting for in Figure 1.6.

The purpose of a large-scale study like that for the prostate data is to weed out the great proportion, say π_0, of null genes, those having $\delta_i = 0$, in order to focus attention on those with large effect sizes. The *local false discovery rate* "fdr(z_i)" is the posterior probability of nullness,

$$\text{fdr}(z_i) = \Pr\{\delta_i = 0 \mid z_i\}.$$

Homework 1.28 (a) Show that

$$\text{fdr}(z) = \pi_0 g_0(z)/g(z), \tag{1.59}$$

with π_0 the prior probability of nullness, $g(z)$ the marginal density of z,

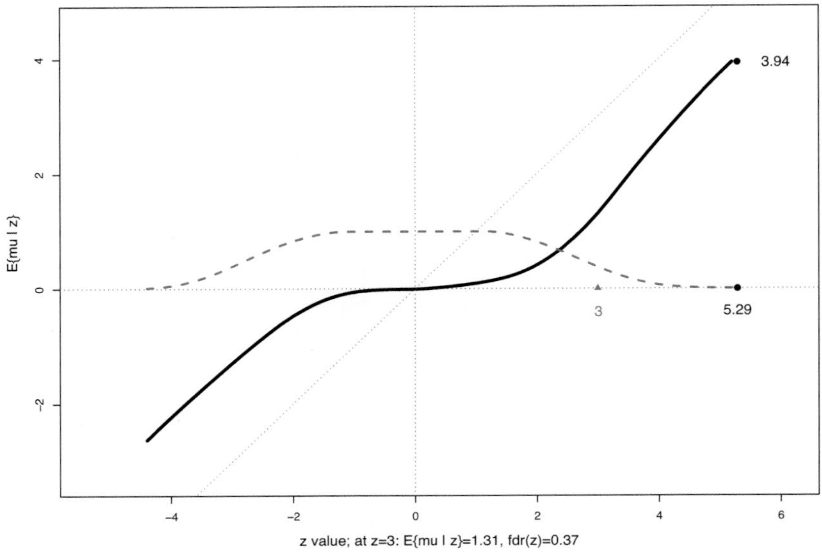

Figure 1.6 Tweedie estimate of $E\{\mu \mid z\}$, prostate study. Dashed curve is estimated local false discovery rate fdr(z).

and

$$g_0(z) = \frac{e^{-z^2/2\sigma^2}}{\sqrt{2\pi\sigma^2}}.$$

(b) If $\pi(\delta)$ is the prior density of δ, as in (1.45), show that

$$E\{\delta \mid z\} = \frac{d}{dz} \log \text{fdr}(z).$$

The red dashed curve in Figure 1.6 is $\widehat{\text{fdr}}(z) = g_0(z)/\hat{g}(z)$, taking $\pi_0 = 1$ as an upper bound, usually a close one in practice; $\widehat{\text{fdr}}(3) = 0.37$ for the prostate data, so even $z = 3$ standard deviations away from 0, there is still substantial probability of $\delta = 0$.

1.8 Deviance and Hoeffding's Formula

Traditional normal theory methods depend on notions of Euclidean distance. *Deviance* is an analogue of Euclidean distance applying to exponential families. By definition the deviance $D(g_1, g_2)$ between g_1 and g_2 is

$$\begin{aligned} D(g_1, g_2) &= 2E_{g_1}\left\{\log \frac{g_1(y)}{g_2(y)}\right\} \\ &= 2\int_{\mathcal{Y}} g_1(y)\left(\log \frac{g_1(y)}{g_2(y)}\right) m(dy). \end{aligned} \tag{1.60}$$

We will also write $D(\eta_1, \eta_2)$, $D(\mu_1, \mu_2)$, or just $D(1, 2)$ as convenient. If $g_1(y)$ and $g_2(y)$ are densities $g_{\eta_1}(y)$ and $g_{\eta_2}(y)$ in an exponential family, then it is easy to verify that

$$D(1, 2) = 2\left[(\eta_1 - \eta_2)\mu_1 - (\psi(\eta_1) - \psi(\eta_2))\right].$$

Homework 1.29 Show that $D(1, 2) \geq 0$, with strict inequality unless the two densities are identical.

Note In general, $D(1, 2) \neq D(2, 1)$.

The "Kullback–Leibler distance", using an older name for the same idea, equals $D(\eta_1, \eta_2)/2$. Information theory uses "mutual information" for $D(f(x, y), f(x)f(y))/2$, where $f(x, y)$ is a bivariate density and $f(x)$ and $f(y)$ its marginals.

Homework 1.30 Verify these formulas for the deviance.

Normal $Y \sim \mathcal{N}(\mu, 1)$: $D(\mu_1, \mu_2) = (\mu_1 - \mu_2)^2$

(This motivates the factor 2 in (1.60).)

Poisson $Y \sim \text{Poi}(\mu)$: $D(\mu_1, \mu_2) = 2\mu_1\left[\log\left(\frac{\mu_1}{\mu_2}\right) - \left(1 - \frac{\mu_2}{\mu_1}\right)\right]$

Binomial $Y \sim \text{Bi}(N, \pi)$: $D(\pi_1, \pi_2) = 2N\left[\pi_1 \log\left(\frac{\pi_1}{\pi_2}\right) + (1 - \pi_1)\log\left(\frac{1 - \pi_1}{1 - \pi_2}\right)\right]$

Gamma $Y \sim \lambda G_N$: $D(\lambda_1, \lambda_2) = 2N\left[\log\left(\frac{\lambda_2}{\lambda_1}\right) + \left(\frac{\lambda_1}{\lambda_2} - 1\right)\right] = 2N\left[\log\left(\frac{\mu_2}{\mu_1}\right) + \left(\frac{\mu_1}{\mu_2} - 1\right)\right]$

Negative binomial (1.37): $D(\theta_1, \theta_2) = k\left[\left(\frac{1 - \theta_1}{\theta_1}\right)\log\left(\frac{1 - \theta_1}{1 - \theta_2}\right) + \log\left(\frac{\theta_1}{\theta_2}\right)\right]$

Hoeffding's Formula

An exponential family of densities $\mathcal{G} = \{g_\eta(y) = \exp(\eta y - \psi(\eta))\}$ can be rewritten in a form that is particularly helpful in discussing maximum likelihood estimation:

Lemma 1.3 (Hoeffding, 1965) *Let $\hat{\eta}$ be the MLE of η and $\hat{\mu} = y$ the MLE of μ. Then*

$$g_\eta(y) = g_{\hat{\eta}}(y) e^{-D(\hat{\eta},\eta)/2} \tag{1.61}$$

or, reparameterizing $\mathcal{G}$ in terms of μ (and recalling that $\hat{\mu} = y$),

$$g_\mu(y) = g_y(y) e^{-D(y,\mu)/2}. \tag{1.62}$$

This says that a plot of the log likelihood $\log g_\mu(y)$ declines from its maximum at $\mu = y$ according to the deviance,

$$\log g_\mu(y) = \log g_y(y) - \frac{D(y,\mu)}{2}.$$

In our applications of the deviance, the first argument will always be the data, the second a proposed value of the unknown parameter.

Proof The deviance in an exponential family is

$$\begin{aligned}\frac{D(\eta_1,\eta_2)}{2} = E_{\eta_1}\left\{\log \frac{g_{\eta_1}(y)}{g_{\eta_2}(y)}\right\} &= E_{\eta_1}\{(\eta_1 - \eta_2)y - \psi(\eta_1) + \psi(\eta_2)\} \\ &= (\eta_1 - \eta_2)\mu_1 - \psi(\eta_1) + \psi(\eta_2),\end{aligned}$$

and

$$\frac{g_\eta(y)}{g_{\hat{\eta}}(y)} = \frac{e^{\eta y - \psi(\eta)}}{e^{\hat{\eta} y - \psi(\hat{\eta})}} = e^{(\eta - \hat{\eta})y - \psi(\eta) + \psi(\hat{\eta})} = e^{(\eta - \hat{\eta})\hat{\mu} - \psi(\eta) + \psi(\hat{\eta})}.$$

Taking $\eta_1 = \hat{\eta}$ and $\eta_2 = \eta$ above, this last is $e^{-D(\hat{\eta},\eta)/2}$. ■

Repeated Sampling

If $\boldsymbol{y} = (y_1, \ldots, y_n)$ is an i.i.d. sample from $g_\eta(\cdot)$ then the deviance based on $\boldsymbol{y}$, say $D^{(n)}(\eta_1, \eta_2)$, is

$$\begin{aligned}D^{(n)}(\eta_1,\eta_2) = 2E_{\eta_1}\left\{\log \frac{g^{(n)}_{\eta_1}(\boldsymbol{y})}{g^{(n)}_{\eta_2}(\boldsymbol{y})}\right\} &= 2E_{\eta_1}\left\{\log \prod_{i=1}^n \frac{g_{\eta_1}(y_i)}{g_{\eta_2}(y_i)}\right\} \\ = 2\sum_{i=1}^n E_{\eta_1}\left\{\log \frac{g_{\eta_1}(y_i)}{g_{\eta_2}(y_i)}\right\} &= nD(\eta_1,\eta_2).\end{aligned}$$

(This shows up in the binomial, Poisson, gamma, and negative binomial cases of Homework 1.30.) Hoeffding's formula (1.61) applied to $\boldsymbol{y}$ is

$$g_{\eta}^{(n)}(\boldsymbol{y}) = g_{\hat{\eta}}^{(n)}(\boldsymbol{y})e^{-nD(\hat{\eta},\eta)/2}. \tag{1.63}$$

For η_2 near η_1, the deviance is related to the Fisher information $i_{\eta_1} = V_{\eta_1}$ (in a single observation y, for η_1 and at η_1):

$$D(\eta_1, \eta_2) = i_{\eta_1}(\eta_2 - \eta_1)^2 + O(\eta_2 - \eta_1)^3.$$

Proof

$$\begin{aligned}\frac{\partial}{\partial \eta_2} D(\eta_1, \eta_2) &= \frac{\partial}{\partial \eta_2} 2\left[(\eta_1 - \eta_2)\mu_1 - (\psi(\eta_1) - \psi(\eta_2))\right] \\ &= 2(-\mu_1 + \mu_2) \\ &= 2(\mu_2 - \mu_1).\end{aligned}$$

Also

$$\frac{\partial^2}{\partial \eta_2^2} D(\eta_1, \eta_2) = 2\frac{\partial \mu_2}{\partial \eta_2} = 2V_{\eta_2}.$$

Therefore

$$\left.\frac{\partial}{\partial \eta_2} D(\eta_1, \eta_2)\right|_{\eta_2=\eta_1} = 0 \quad \text{and} \quad \left.\frac{\partial^2}{\partial \eta_2^2} D(\eta_1, \eta_2)\right|_{\eta_2=\eta_1} = 2V_{\eta_1},$$

so a Taylor expansion gives

$$D(\eta_1, \eta_2) = 2V_{\eta_1}\frac{(\eta_2 - \eta_1)^2}{2} + O(\eta_2 - \eta_1)^3. \qquad \blacksquare$$

Homework 1.31 What is $\partial^3 D(\eta_1, \eta_2)/\partial \eta_2^3$? Give an improved version of the relationship above.

REFERENCE Efron (1978), "The geometry of exponential families", *Ann. Stat.* 362–376.

An Informative Picture

We know that $\psi(\eta)$ is a convex function of η since $\ddot{\psi}(\eta) = V_\eta > 0$. Figure 1.7 shows $\psi(\eta)$ passing through $(\eta_1, \psi(\eta_1))$ at slope $\mu_1 = \dot{\psi}(\eta_1)$. The difference between $\psi(\eta_2)$ and the linear bounding line $\psi(\eta_1) + (\eta_2 - \eta_1)\mu_1$ is $\psi(\eta_2) - \psi(\eta_1) + (\eta_1 - \eta_2)\mu_1 = D(\eta_1, \eta_2)/2$.

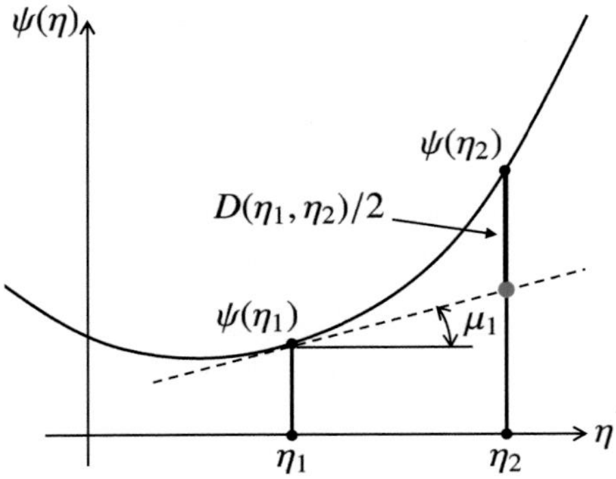

Figure 1.7 Convex function $\psi(\eta)$ passing through $(\eta_1, \psi(\eta_1))$ at slope $\mu_1 = \dot{\psi}(\eta_1)$.

Unlike our other results, Figure 1.7 depends on parameterizing the deviance as $D(\eta_1, \eta_2)$. A version that uses $D(\mu_1, \mu_2)$ depends on the *dual function* $\phi(y)$ to $\psi(y)$,

$$\phi(y) = \max_{\eta} \{\eta y - \psi(\eta)\} .$$

Homework 1.32 Show that:

(a) $\phi(\mu) = \eta\mu - \psi(\eta)$, where $\mu = \dot{\psi}(\eta)$;
(b) $\phi(\mu)$ is convex as a function of μ;
(c) $d\phi(\mu)/d\mu = \eta$.

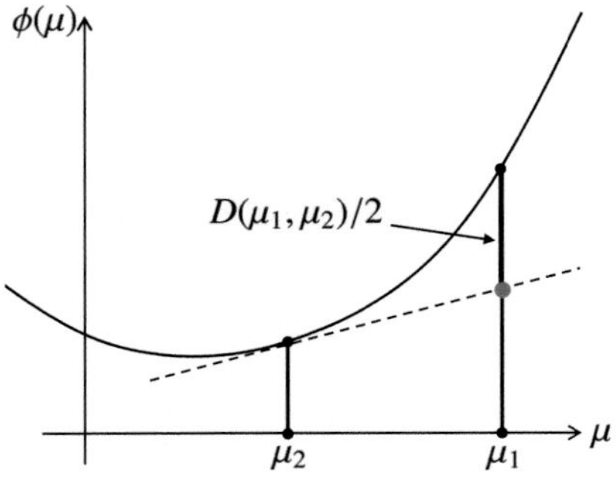

Now verify the diagram here.

Homework 1.33 *Parametric bootstrap*: We resample y^* from $g_{\hat{\eta}}(\cdot)$, $\hat{\eta}$ the

MLE based on y. Show that, given y and $\hat{\eta}$,

$$g_\eta(y^*) = g_{\hat{\eta}}(y^*)e^{(\eta-\hat{\eta})(y^*-y)-D(\hat{\eta},\eta)/2}.$$

Deviance Residuals

The idea here is that if $D(y, \mu)$ is the exponential family analogue of $(y-\mu)^2$ in a normal model, then

$$\operatorname{sign}(y-\mu)\sqrt{D(y,\mu)}$$

should be the exponential family analogue of the normal residual $y - \mu$.

We will work in the repeated sampling framework

$$y_i \overset{\text{iid}}{\sim} g_\mu(\cdot), \qquad i = 1, \ldots, n,$$

with MLE $\hat{\mu} = \bar{y}$ and total deviance $D^{(n)}(\hat{\mu}, \mu) = nD(\bar{y}, \mu)$. The *deviance residual*, of $\hat{\mu} = \bar{y}$ from true mean μ, is defined to be

$$R = \operatorname{sign}(\bar{y}-\mu)\sqrt{D^{(n)}(\bar{y},\mu)}. \tag{1.64}$$

The hope is that R will be nearly $\mathcal{N}(0, 1)$, at least closer to normal than the more obvious "Pearson residual"

$$R_P = \frac{\bar{y}-\mu}{\sqrt{V_\mu/n}}$$

(called "z_i" later). Our hope is bolstered by the following theorem, verified in Appendix C of McCullagh and Nelder (1983).

Theorem 1.4 *The asymptotic distribution of R as $n \to \infty$ is*

$$R \mathrel{\dot\sim} \mathcal{N}\left[-a_n, (1+b_n)^2\right], \tag{1.65}$$

where a_n and b_n are defined in terms of the skewness γ_μ and kurtosis δ_μ of the original ($n = 1$) exponential family,

$$a_n = \frac{\gamma_\mu/6}{\sqrt{n}} \quad \textit{and} \quad b_n = \frac{(7/36)\,\gamma_\mu^2 - \delta_\mu}{n}.$$

The normal approximation in (1.65) *is accurate through $O_p(n^{-1})$, with errors of order $O_p(n^{-3/2})$, for instance,*

$$\Pr\left\{\frac{R+a_n}{1+b_n} > 1.96\right\} = 0.025 + O\left(n^{-3/2}\right)$$

(so-called "third-order accuracy").

Corollary

$$D^{(n)}(\bar{y}, \mu) = R^2 \mathrel{\dot\sim} \left(1 + \frac{5\gamma_\mu^2 - 3\delta_\mu}{12n}\right) \cdot \chi_1^2,$$

where χ_1^2 is a chi-squared random variable with degrees of freedom 1. Since, according to Hoeffding's formula,

$$D^{(n)}(\bar{y}, \mu) = 2 \log \frac{g_{\hat\mu}^{(n)}(\mathbf{y})}{g_\mu^{(n)}(\mathbf{y})},$$

this is an improved version of Wilks' theorem: $2\log(g_{\hat\mu}/g_\mu) \to \chi_1^2$ in one-parameter situations.

The constants a_n and b_n are called "Bartlett corrections". The theorem says that

$$R \mathrel{\dot\sim} \frac{Z + a_n}{1 + b_n}, \qquad \text{where } Z \sim \mathcal{N}(0, 1).$$

Since $a_n = O(n^{-1/2})$ and $b_n = O(n^{-1})$, the expectation correction in (1.65) is more important than the variance correction.

Homework 1.34 Consider the gamma case, $y \sim \lambda G_N$ with N fixed (N can be thought of as n).

(a) Show that the deviance residual $\text{sign}(y - \lambda N)\sqrt{D(y, \lambda N)}$ has the same distribution for all choices of λ.

(b) What is the skewness of the Pearson residual $(y - \lambda N)/\lambda(N^{1/2})$?

(c) Use our previous results to show that

$$D^{(n)}(\bar{y}, \mu) \doteq R_P^2 + \frac{\gamma}{6\sqrt{n}} R_P^3 + O_P\left(n^{-1}\right).$$

As an example, Figure 1.8 is a simulation showing 2000 replications of $\bar{y} = \sum_1^5 y_i$, where the y_i are independent G_1 variates; that is, standard one-sided exponentials, as in the Gamma case of Homework 1.30 with $\lambda = N = 1$. This makes $\bar{y} \sim G_5/5$, so that the deviance residual R in (1.64) is calculated as in Homework 1.30 again, now with $N = 5$. The qq-plot shows the deviance residuals (black) much closer to $\mathcal{N}(0, 1)$ than the Pearson residuals (red).

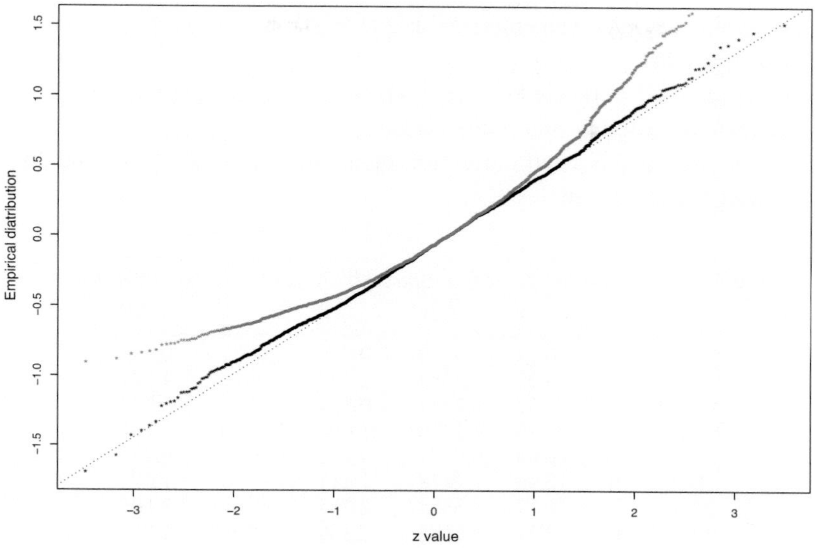

Figure 1.8 qq comparison of deviance residuals (black) with Pearson residuals (red); gamma $N = 1$, $\lambda = 1$, $n = 5$; $B = 2000$ simulations.

An Example of Deviance Analysis

Reference Thisted and Efron (1987), "Did Shakespeare write a newly discovered poem?", *Biometrika* 445–455.

On November 14, 1985, Gary Taylor, a respected Shakespearean scholar, found a short poem of 429 words in the Bodleian Library that he attributed to Shakespeare. This was a controversial stance, as no "new" text by Shakespeare had been discovered in centuries. A word-count analysis was carried out comparing the poem with Shakespeare's attributed works (the "canon"). Table 1.4 shows a small proportion of the results that involved deviance residuals:

- The analysis focused on rare words, that didn't appear often in the canon. Column "y" of the table shows 9 distinct words in the poem that had *never* appeared in the canon, "Prev" = 0; 7 that had previously appeared once each; 5 twice each; and so on, up to 5 that had appeared 80 to 99 times each.
- Column "v" gives predictions for the y values assuming Shakespearean

authorship (based on an empirical Bayes Poisson theory relating to Robbins' formula).

- "Dev" and "R" show the Poisson deviance and deviance residuals (1.64) between the counts y and predictions ν.
- a_n is the leading Bartlett correction factor in (1.65), and "RR" the partially corrected residual $R + a_n$.

Table 1.4 *Word-count deviance analysis of newly discovered poem.*

# Prev	y	ν	Dev	R	a_n	RR
0	9	6.97	0.5410	0.736	0.0631	0.799
1	7	4.21	1.5383	1.240	0.0812	1.321
2	5	3.33	0.7247	0.851	0.0913	0.943
3–4	8	5.36	1.1276	1.062	0.0720	1.134
5–9	11	10.24	0.0551	0.235	0.0521	0.287
10–19	10	13.96	1.2478	−1.117	0.0446	−1.072
20–29	21	10.77	7.5858	2.754	0.0508	2.805
30–39	16	8.87	4.6172	2.149	0.0560	2.205
40–59	18	13.77	1.1837	1.088	0.0449	1.133
60–79	8	9.99	0.4257	−0.652	0.0527	−0.600
80–99	5	7.48	0.9321	−0.965	0.0609	−0.904

The RRs should be approximately $\mathcal{N}(0, 1)$ under a hypothesis of Shakespearean authorship. There are some suspicious discrepancies, for instance $RR = 2.805$ for the 20–29 category. The sum of the Dev's is 19.98, moderately large compared to a chi-squared distribution with 11 degrees of freedom,

$$\Pr\{\chi^2_{11} > 19.98\} = 0.046.$$

Nevertheless, compared with the same analysis applied to known non-Shakespeare poems, the authors felt that Taylor's poem had at least some chance of being genuine. It remains controversial, and is not usually included in the canon.

1.9 The Saddlepoint Approximation

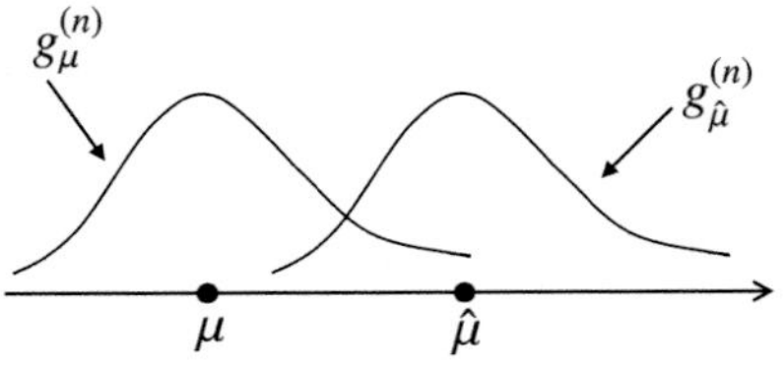

Suppose we observe a random sample of size n from some member of an exponential family $\mathcal{G}$,

$$y_1, \dots, y_n \overset{\text{iid}}{\sim} g_\mu(\cdot)$$

(now indexed by expectation parameter μ), and wish to approximate the density under $g_\mu^{(n)}$ of the sufficient statistic $\hat{\mu} = \bar{y}$ for a value of $\hat{\mu}$ perhaps far removed from μ. Let $g_\mu^{(n)}(\hat{\mu})$ denote this density.

The normal approximation

$$g_\mu^{(n)}(\hat{\mu}) \doteq \sqrt{\frac{n}{2\pi V_\mu}} e^{-\frac{1}{2}\frac{n}{V_\mu}(\hat{\mu}-\mu)^2} \tag{1.66}$$

is likely to be inaccurate if $\hat{\mu}$ is, say, several standard errors away from μ. Hoeffding's formula (1.62) provides a much better result, called the *saddlepoint approximation*. We write

$$g_\mu^{(n)}(\hat{\mu}) = g_{\hat{\mu}}^{(n)}(\hat{\mu}) e^{-nD(\hat{\mu},\mu)/2}. \tag{1.67}$$

For $\mu = \hat{\mu}$, $\bar{y}$ is approximately $\mathcal{N}(\hat{\mu}, \widehat{V}/n)$, where $\widehat{V} = \ddot{\psi}(\hat{\mu})$ is the variance of a single y under $g_{\hat{\mu}}$, giving $g_{\hat{\mu}}^{(n)}(\hat{\mu}) \doteq [n/(2\pi\widehat{V})]^{1/2}$ and, substituting in (1.67), the saddlepoint approximation

$$g_\mu^{(n)}(\hat{\mu}) \doteq \sqrt{\frac{n}{2\pi\widehat{V}}} e^{-nD(\hat{\mu},\mu)/2}. \tag{1.68}$$

Because (1.68) only involves applying the central limit theorem at the *center* of the $g_{\hat{\mu}}^{(n)}(\cdot)$ distribution, just where it is most accurate, the error in (1.68) is a factor of only $1 + O(n^{-1})$, compared to $1 + O(n^{-1/2})$ for (1.66). There is an enormous literature of extensions and improvements to the saddlepoint approximation, a good review article being Reid (1988).

Let

$$L_{\mathbf{y}}(\mu) = g_\mu^{(n)}(\mathbf{y})$$

be the likelihood function having observed data $\mathbf{y}$, expressed in terms of the expectation parameter μ. Hoeffding's formula (1.62), (1.63), gives

$$e^{-nD(\hat{\mu},\mu)/2} = \frac{L_{\mathbf{y}}(\mu)}{L_{\mathbf{y}}(\hat{\mu})},$$

so the saddlepoint approximation can be expressed as

$$g_\mu^{(n)}(\hat{\mu}) \doteq \sqrt{\frac{n}{2\pi\widehat{V}}} \frac{L_{\mathbf{y}}(\mu)}{L_{\mathbf{y}}(\hat{\mu})},$$

which provides an expression for the density of $\bar{y} = \hat{\mu}$ in terms of the likelihood function.

Since $d\hat{\mu}/d\hat{\eta} = \widehat{V}$, there is an equivalent expression for the density of $\hat{\eta}$, say $g_\eta^{(n)}(\hat{\eta})$ (abusing notation somewhat),

$$g_\eta^{(n)}(\hat{\eta}) \doteq \sqrt{\frac{n\widehat{V}}{2\pi}} \frac{L_y(\eta)}{L_y(\hat{\eta})}. \tag{1.69}$$

Barndorff-Nielsen (1980) showed approximation (1.69) holding in a variety of situations, including *curved exponential families* (Part 4), and it is sometimes known as his "magic formula".

The Lugananni–Rice Formula

The saddlepoint formula can be integrated to give an approximation to $\alpha(\mu)$, the *attained significance level* or "p-value" of parameter value μ having observed $\bar{y} = \hat{\mu}$:

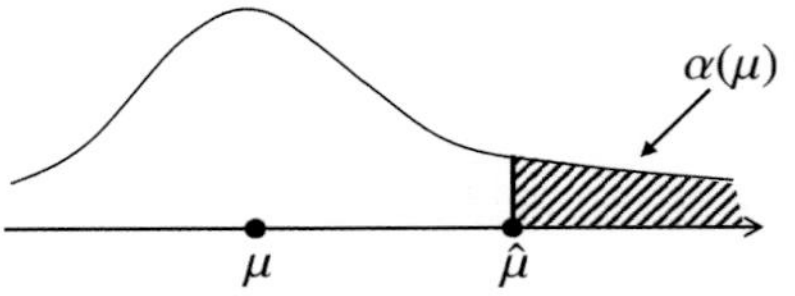

$$\alpha(\mu) = \int_{\hat{\mu}}^{\infty} g_\mu^{(n)}(t)m(dt).$$

Numerical integration is required to compute $\alpha(\mu)$ from the saddlepoint formula itself, but the *Lugananni–Rice formula* provides a highly accurate closed-form approximation:

$$\alpha(\mu) \doteq 1 - \Phi(R) - \varphi(R)\left(\frac{1}{R} - \frac{1}{Q}\right) + O\left(n^{-3/2}\right),$$

where Φ and φ are the standard normal CDF and density,

$$R = \text{sign}(\hat{\mu} - \mu)\sqrt{nD(\hat{\mu}, \mu)}$$

the deviance residual, and

$$Q = \sqrt{n\widehat{V}} \cdot (\hat{\eta} - \eta)$$

the crude form of the Pearson residual based on the canonical parameter η, not on μ. (Remember that $\widehat{\text{sd}}(\hat{\eta}) \doteq (n\widehat{V})^{-1/2}$, so $Q = (\hat{\eta} - \eta)/\widehat{\text{sd}}(\hat{\eta})$.) Reid (1988) is also an excellent reference here, giving versions of the Lugananni–Rice formula that apply not only to exponential family situations but also to general distributions of $\bar{y}$. See also Section 6 of Daniels (1983).

Homework 1.35 Suppose we observe $y \sim \lambda G_N$, G_N gamma df $= N$, with $N = 10$ and $\lambda = 1$. Use the Lugananni–Rice formula to calculate $\alpha(\mu)$ for

$y = \hat{\mu} = 15, 20, 25, 30$, and compare the results with the exact values. (You can use any expression above for R.)

Homework 1.36 Another version of the Lugananni–Rice formula is

$$1 - \alpha(\mu) \doteq \Phi(R'),$$

where

$$R' = R + \frac{1}{R} \log \frac{Q}{R}.$$

How does this relate to the first form?

Large Deviations and the Chernoff Bound

In a generic "large deviations" problem, we observe an i.i.d. sample

$$y_1, \dots, y_n \overset{\text{iid}}{\sim} g_0(\cdot)$$

from a *known* density g_0 having mean and standard deviation $y_i \sim (\mu_0, \sigma_0)$. We wish to compute

$$\alpha_n(\mu_1) = \Pr_{g_0}\{\bar{y} \geq \mu_1\}$$

for some fixed value $\mu_1 > \mu_0$. As $n \to \infty$, the number of standard errors $\sqrt{n}(\mu_1 - \mu_0)/\sigma_0$ gets big, rendering the central limit theorem useless.

Homework 1.37 ("Chernoff bound") Let $g_\eta(y) = e^{\eta y - \psi(\eta)} g_0(y)$ ("the exponential family through g_0").

(a) For any $\lambda > 0$ show that $\alpha_n(\mu_1) = \Pr_{g_0}\{\bar{y} \geq \mu_1\}$ satisfies

$$\alpha_n(\mu_1) \leq \beta_n(\mu_1) \equiv \int_{\mathcal{Y}} e^{n\lambda(\bar{y} - \mu_1)} g_0^{(n)}(\bar{y})\, d\bar{y}.$$

(b) Show that $\beta_n(\mu_1)$ is minimized at λ equal the value $\hat{\eta}$ such that

$$\dot{\psi}(\hat{\eta}) = \mu_1.$$

(c) Finally, verify Chernoff's large deviation bound

$$\Pr_{g_0}\{\bar{y} \geq \mu_1\} \leq e^{-nD(\mu_1, 0)/2}, \tag{1.70}$$

where $D(\mu_1, 0)$ is the deviance between $g_{\hat{\eta}}(y)$ and $g_0(y)$.

Notice that for fixed μ_1, neither $\hat{\eta}$ nor $D(\mu_1, 0)$ depends on n, so $\alpha_n(\mu_1) \to 0$ exponentially fast, which is typical for large deviation results.

1.10 Transformation Theory

REFERENCE DiCiccio (1984), "On parameter transformations and interval estimation", *Biometrika* 477–485.

REFERENCE Efron (1982), "Transformation theory: How normal is a family of distributions?", *Ann. Stat.* 323–339.

REFERENCE Hougaard (1982), "Parametrizations of nonlinear models", *JR SS-B* 244–252.

Power transformations are used to make exponential families more like the standard normal translation family $Y \sim \mathcal{N}(\mu, 1)$. For example, $Y \sim \text{Poi}(\mu)$ has variance $V_\mu = \mu$ depending on the expectation μ, while the transformation

$$Z = H(Y) = 2\sqrt{Y}$$

yields, approximately, $\text{Var}(Z) = 1$ for all μ. In a regression situation with Poisson responses $y_1, \dots, y_n$, we might first change to $z_i = 2{y_i}^{1/2}$ and then employ standard linear model methods. (That's *not* how we will proceed in Part 3, where generalized linear model techniques are discussed. The introduction of GLMs reduced, but did not eliminate, interest in transformation theory.)

Table 1.5, credited to unpublished work by R. Wedderburn, encompasses a considerable number of special transformations for one-parameter exponential families. Let ζ be a transformation of μ,

$$\zeta = H(\mu) \quad \text{and} \quad \hat{\zeta} = H(\hat{\mu}),$$

where $\hat{\mu}$ is the MLE of μ based on observing a single $y \sim g_\mu(\cdot)$. If we make the derivative $H'(\mu)$ satisfy

$$H'(\mu) = V_\mu^{\delta-1}, \tag{1.71}$$

then various choices of δ result in $\hat{\zeta} = H(\hat{\mu})$ satisfying the properties shown in Table 1.5, as explained next.

Table 1.5 *Wedderburn's transformations* (1.71) *and their results.*

δ	0	1/3	1/2	2/3	1
Result	Canonical parameter η	Normal likelihood	Stabilized variance	Normal density	Expectation parameter μ

The choice $\delta = 0$ has

$$\frac{d\zeta}{d\mu} = H'(\mu) = \frac{1}{\sqrt{V_\mu}}.$$

But $d\eta/d\mu = 1/V_\mu$, so in this case $\zeta = \eta$. At the other end of the scale, $\delta = 1$ has $H'(\mu) = 1$, that is, $\zeta = \mu$.

The stabilized variance result, $\delta = 1/2$, follows from the delta method:

$$\hat{\zeta} = H(\hat{\mu}), \qquad \text{with } H'(\mu) = \frac{1}{\sqrt{V_\mu}},$$

implies that

$$\text{sd}_\mu(\hat{\zeta}) \doteq \frac{\text{sd}_\mu(\hat{\mu})}{\sqrt{V_\mu}} = 1.$$

For the Poisson family, with $V_\mu = \mu$,

$$H'(\mu) = \frac{1}{\sqrt{\mu}}$$

gives

$$H(\mu) = 2\sqrt{\mu} + \text{any constant},$$

as above. For $Y \sim \text{Poi}(\mu)$, the usual approximation for expectation and variance is

$$2\sqrt{Y} \dot{\sim} (2\sqrt{\mu}, 1). \tag{1.72}$$

Homework 1.38 Numerically calculate how well (1.72) works for $\mu = 5, 8, 12, 18, 25$.

Small adjustments to the $\delta = 1/2$ formula are known to improve variance stabilization. For the binomial case

$$p \sim \text{Bi}(N, \pi)/N,$$

Anscombe's transformation

$$\hat{\zeta} = 2\sqrt{N}\sin^{-1}\left(\sqrt{\frac{Np + 3/8}{N + 3/4}}\right) \tag{1.73}$$

does a good job of making $\text{Var}_\pi(\hat{\zeta}) \doteq 1$ for n say 15 or more.

Homework 1.39 Ignoring correction terms $3/8$ and $3/4$, show that (1.71) with $\delta = 1/2$ gives (1.73).

Normal density, $\delta = 2/3$, means that $\hat{\zeta} = H(\hat{\mu})$ is approximately $\mathcal{N}(H(\mu), 1)$. (This is *not* the same as $\delta = 1/2$ in Table 1.5, where the emphasis is on constant variance rather than normality.) Its rationale is based on asymptotic expansions. Working in a repeated sampling framework, $y_1, \dots, y_n \overset{\text{iid}}{\sim} g_\mu(\cdot)$, where $\hat{\mu} = \bar{y}$, define

$$z_n = \frac{\bar{y} - \mu}{\sqrt{V/n}} \qquad (V = V_\mu),$$

so that z_n has expectation 0, variance 1, and skewness $\gamma \cdot n^{-1/2}$, where γ is the skewness for a single y_i, as in Homework 1.7. A two-term *Cornish–Fisher expansion* suggests that the distribution of z_n can be normalized by the transformation

$$Z = z_n - \frac{\gamma}{6\sqrt{n}}(z_n^2 - 1), \tag{1.74}$$

which makes the skewness of Z approximately 0.

We want to show that $\hat{\zeta} = H(\hat{\mu})$, with $\delta = 2/3$ in (1.71), asymptotically agrees with (1.74), at least through two terms. The Taylor expansion

$$\begin{aligned}\hat{\zeta} &\doteq H(\mu) + H'(\mu)(\hat{\mu} - \mu) + H''(\mu)\frac{(\hat{\mu} - \mu)^2}{2} \\ &= H(\mu) + H'(\mu)\sqrt{\frac{V}{n}} z_n + H''(\mu)\frac{V}{2n} z_n^2\end{aligned}$$

gives $\hat{\zeta}$ approximate expectation and standard deviation

$$\begin{aligned}E_\mu\left\{\hat{\zeta}\right\} &= H(\mu) + H''(\mu)\frac{V}{2}, \\ \text{sd}_\mu\left\{\hat{\zeta}\right\} &= H'(\mu)\sqrt{\frac{V}{n}}\end{aligned}$$

(assuming $H'(\mu) > 0$). Therefore

$$\frac{\hat{\zeta} - E_\mu\left\{\hat{\zeta}\right\}}{\text{sd}_\mu\left\{\hat{\zeta}\right\}} \doteq z_n + \frac{H''(\mu)}{2H'(\mu)}\sqrt{\frac{V}{n}}(z_n^2 - 1). \tag{1.75}$$

Expansion (1.75) agrees with (1.74) if

$$\frac{H''(\mu)}{H'(\mu)}\sqrt{V} = -\frac{\gamma}{3},$$

or equivalently if

$$\frac{d \log H'(\mu)}{d\mu} = -\frac{\gamma}{3\sqrt{V}}.$$

However, $\gamma = V' \cdot V^{-1/2}$ is an exponential family (1.19), so in order to get agreement we need

$$\frac{d \log H'(\mu)}{d\mu} = -\frac{V'}{3V} = \frac{d}{d\mu} \log V^{-1/3}$$

or

$$H'(\mu) = cV^{-1/3},$$

i.e, $\delta = 2/3$, as in Table 1.5.

Normal likelihood, $\delta = 1/3$, means that the transformation $\hat{\zeta} = H(\hat{\mu})$ results in

$$\left.\frac{\partial^3 l_\mu(y)}{\partial \zeta^3}\right|_{\hat{\zeta}} = 0, \qquad (1.76)$$

where $l_\mu(y) = \log g_\mu(y)$. This makes the log likelihood look parabolic near its maximum at $\zeta = \hat{\zeta}$. Efron (1982) gives an argument connecting $H'(\mu) = V_\mu^{-2/3}$ and (1.76).

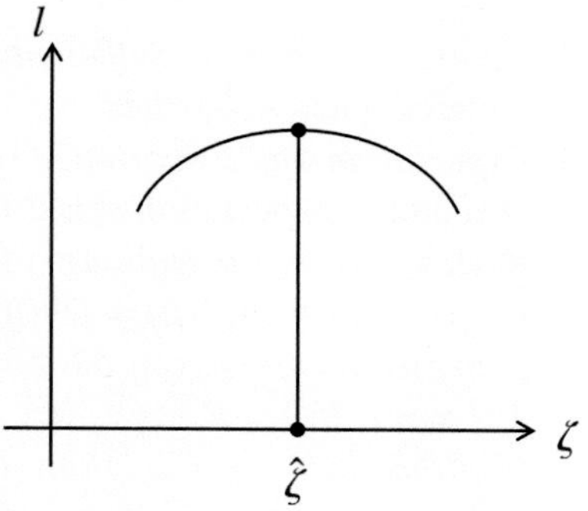

For the Poisson case $Y \sim \text{Poi}(\mu)$, the three choices $\delta = 1/3, 1/2$, or $2/3$ correspond to the transformations $3/2 Y^{2/3}, 2Y^{1/2}$, or $3Y^{1/3}$. All three have been referred to in the literature as "the" Poisson transformation.

Homework 1.40 We observe independent χ^2 variables

$$\hat{\sigma}_i^2 \sim \frac{\sigma_i^2 \chi_{\nu_i}^2}{\nu_i},$$

the ν_i being known degrees of freedom, and wish to regress $\hat{\sigma}_i^2$ versus some known covariates. Two frequently suggested transformations are $\log(\hat{\sigma}_i^2)$ and $(\hat{\sigma}_i^2)^{1/3}$, the latter being the "Wilson–Hilferty" transformation. Discuss the two transformations in terms of Table 1.5.

2

Multiparameter Exponential Families

2.1 Natural Parameters, Sufficient Statistics, CGF

One-parameter families are too restrictive for most real data analysis problems. It is easy, however, to extend exponential families to multiparametric situations. One of the charms of exponential families is that multiparameter theory is quite simply related to its univariate little brother. Expectations, variances, repeated sampling, likelihoods, score functions, Cramér–Rao lower bounds, deviances, and Hoeffding's formula require only small changes in notation from what we saw in Part 1.

This is less true for maximum likelihood estimation, where multiple dimensions make for a more informative picture of the MLE. It is definitely not true for specific applications. The examples in Section 2.8 of multiparameter families such as the Dirichlet and multivariate normal show some of the implications of multidimensional geometry. The most common multiparameter exponential family, the multinomial (Section 2.9), is not a simple extension of univariate ideas. We defer until Part 3, generalized linear models, the most successful modern addition to exponential family applications.

A *p-parameter exponential family* is a collection of probability densities

$$\mathcal{G} = \left\{ g_\eta(y), \eta \in A, y \in \mathcal{Y} \right\} \qquad (A \text{ and } \mathcal{Y} \in \mathcal{R}^p)$$

of the form

$$g_\eta(y) = e^{\eta^\top y - \psi(\eta)} g_0(y) m(dy). \tag{2.1}$$

- η is the $p \times 1$ *natural*, or *canonical*, parameter vector.
- y is the $p \times 1$ vector of *sufficient statistics*, range space $y \in \mathcal{Y} \subset \mathcal{R}^p$.
- $g_0(y)$ is the *carrying density*, defined with respect to some *carrying measure* $m(dy)$ on $\mathcal{Y}$.
- A is the *natural parameter space*: all η having $\int_{\mathcal{Y}} e^{\eta^\top y} g_0(y) m(dy) < \infty$.
- $\psi(\eta)$ is the *normalizing function* or *cumulant generating function*.

Except for the transpose sign above η, all of this is the same as (1.2) in Part 1.

For any point η_0 in A we can express $\mathcal{G}$ as

$$g_\eta(y) = e^{(\eta-\eta_0)^\top y - [\psi(\eta) - \psi(\eta_0)]} g_{\eta_0}(y).$$

$\mathcal{G}$ consists of exponential tilts of g_{η_0}. The log tilting functions are linear in the p sufficient statistics $y = (y(1), \ldots, y(p))^\top$, offering richer possibilities than the single-directional tilts of Part 1.

In most applications, y is a function of a data set $\boldsymbol{x}$, usually much more complicated than a p-vector, as in Homework 2.1. Then y earns the name "sufficient vector". The mapping $y = t(\boldsymbol{x})$ from the full data to the sufficient vector is crucial to the statistical analysis. It says which parts of the problem are important, and which can be ignored.

Homework 2.1 Show that $x_1, \ldots, x_n \overset{\text{iid}}{\sim} \mathcal{N}(\lambda, \Gamma)$ can be written in form (2.1) with $y = (\bar{x}, \bar{x^2})$.

2.2 Expectation and Covariance

The *expectation vector* $\mu = E_\eta\{y\}$ is the p-vector given by

$$\mu = E_\eta\{y\} = \underset{p\times 1}{\dot{\psi}}(\eta) = \begin{pmatrix} \vdots \\ \partial\psi(\eta)/\partial\eta_i \\ \vdots \end{pmatrix}, \tag{2.2}$$

while the $p \times p$ *covariance matrix* V equals the second derivative matrix of ψ,

$$V = \mathrm{Cov}_\eta\{y\} = \underset{p\times p}{\ddot{\psi}}(\eta) = \begin{pmatrix} \vdots \\ \partial^2\psi(\eta)/\partial^2\eta_i\partial\eta_j \\ \vdots \end{pmatrix}. \tag{2.3}$$

Both μ and V are functions of η, usually suppressed in our notation. (They can just as well be thought of as functions of μ.)

Homework 2.2 Verify the expressions for μ and V.

The Relationship of μ and η

In the one-parameter case of Part 1, η and μ are 1:1 functions of each other, with μ smoothly increasing in η according to the differential relationship $d\mu = V\,d\eta$, (1.14)–(1.15). Things are basically the same in multiparameter exponential families, but with a more interesting interpretation of what "increasing" means.

From $\mu = \dot{\psi}(\eta)$ and $V = \ddot{\psi}(\eta)$ we see that the $p \times p$ derivative matrix[1] $d\mu/d\eta = (\partial\mu_j/\partial\eta_i)$ equals V. As in the one-parameter case, we can express the differential relationship as

$$d\mu = V\,d\eta. \tag{2.4}$$

This looks linear until you remember that V depends on η. Going in the other direction, $d\eta/d\mu = V^{-1}$, or

$$d\eta = V^{-1}\,d\mu. \tag{2.5}$$

Here we are assuming that the carrying density $g_0(y)$ in (2.1) is of full rank, in the sense that it is not entirely supported on any lower-dimensional subspace of $\mathcal{R}^p$, implying that V will be positive definite for all η in A.

[1] We follow the matrix notational convention that i indexes rows and j indexes columns, but that isn't important here since V is symmetric.

Homework 2.3 Verify the preceding statement.

The set A of all vectors η for which $\int_{\mathcal{Y}} e^{\eta^\top y} g_0(y)$ is finite maps into B, the set of all μ vectors,

$$B = \left\{\mu = E_\eta\{y\}, \eta \in A\right\}. \tag{2.6}$$

The one-to-one mappings $\eta \to \mu$ and $\mu \to \eta$ are schematically illustrated in Figure 2.1, making use of the differential relationships (2.4) and (2.5). Notice that

$$d\eta^\top\, d\mu = d\eta^\top V\, d\eta > 0,$$

since V is positive definite, so the angle between $d\eta$ and $d\mu$ is less than 90°. In this sense, μ is an increasing function of η, and vice versa. We will be seeing many elaborations of Figure 2.1 as we investigate maximum likelihood estimation of η and μ.

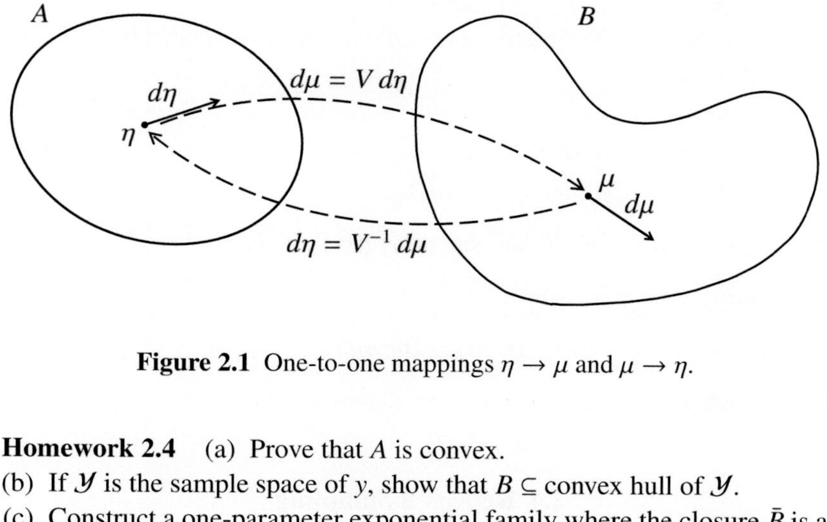

Figure 2.1 One-to-one mappings $\eta \to \mu$ and $\mu \to \eta$.

Homework 2.4 (a) Prove that A is convex.
(b) If $\mathcal{Y}$ is the sample space of y, show that $B \subseteq$ convex hull of $\mathcal{Y}$.
(c) Construct a one-parameter exponential family where the closure $\bar{B}$ is a proper subset of $\mathcal{Y}$.

REFERENCE Efron (1978), "Geometry of exponential families", *Ann. Stat.*, Section 2, provides an example of non-convex B.

2.3 Review of Transformations

This review applies to general 1:1 transformations, though we'll be interested in (η, μ) as previously defined. Suppose η and μ are vectors in $\mathcal{R}^p$, smoothly related to each other,

$$\eta \overset{1:1}{\longleftrightarrow} \mu,$$

and that $h(\eta) = H(\mu)$ is some smooth real-valued function. Let D be the $p \times p$ derivative matrix (Hessian)

$$D = \begin{pmatrix} & \vdots & \\ \cdots & \partial\eta_j/\partial\mu_i & \cdots \\ & \vdots & \end{pmatrix}, \qquad i \downarrow \quad \text{and} \quad j \rightarrow .$$

($D = V^{-1}$ in (2.5).) Letting $\cdot$ indicate derivatives with respect to η, and $'$ indicate derivatives with respect to μ, we have

$$H'(\mu) = D\dot{h}(\eta), \tag{2.7}$$

where $\dot{h}(\eta) = (\cdots \partial h/\partial\eta_j \cdots)^\top$ and $H'(\mu) = (\cdots \partial H/\partial\mu_i \cdots)^\top$.

The second derivative matrices of $h(\eta)$ and $H(\mu)$ are related by

$$\begin{array}{ccccc} H''(\mu) & = & D\ddot{h}(\eta)D^\top & + & D_2\dot{h}(\eta). \\ \uparrow & & \uparrow & & \\ \dfrac{\partial^2 H}{\partial\mu_i\partial\mu_j} & & \dfrac{\partial^2 h}{\partial\eta_i\partial\eta_j} & & \end{array} \tag{2.8}$$

Here D_2 is the $p \times p \times p$ three-way array $(\partial^2\eta_k/\partial\mu_i\partial\mu_j)$.

Important: At a point where $\dot{h}(\eta) = 0$,

$$H''(\mu) = D\ddot{h}(\eta)D^\top. \tag{2.9}$$

Homework 2.5 Prove the important statement (2.9).

2.4 Repeated Sampling

A key to what makes exponential families so useful both in theory and application is the simple way definition (2.1) accommodates repeated sampling. Suppose we observe repeated samples from (2.1),

$$\boldsymbol{y} = (y_1, \ldots, y_n) \overset{\text{iid}}{\sim} g_\eta(y).$$

Let $\bar{y}$ denote the average of the vectors y_i,

$$\bar{y} = \sum_{i=1}^{n} y_i/n \qquad \left(\text{component-wise, } \bar{y}_j = \sum_{i=1}^{n} y_{ij}\right).$$

(It will not surprise the reader that $\bar{y}$ is the MLE of μ; see Section 2.6.) Then

$$g_\eta^{(n)}(\mathbf{y}) = e^{n[\eta^\top \bar{y} - \psi(\eta)]} g_0^{(n)}(\mathbf{y}), \tag{2.10}$$

as in (1.22), with $g_0^{(n)}(\mathbf{y}) = \prod_{i=1}^n g_0(y_i)$. This is a p-dimensional exponential family (2.1), with:

- natural parameter $\eta^{(n)} = n\eta$;
- sufficient vector $y^{(n)} = \bar{y}$;
- expectation vector $\mu^{(n)} = \mu$;
- variance matrix $V^{(n)} = V/n \quad$ (since $\mathrm{Cov}(\bar{y}) = \mathrm{Cov}(y)/n$);
- CGF $\psi^{(n)}(\eta^{(n)}) = n\psi(\eta^{(n)}/n)$;
- sample space = product space $\mathcal{Y}_1 \otimes \mathcal{Y}_2 \otimes \cdots \otimes \mathcal{Y}_n$.

Since $\bar{y}$ is sufficient, we can consider it on its own sample space, say $\mathcal{Y}^{(n)}$, with densities

$$g_\eta^{(n)}(\bar{y}) = e^{n[\eta^\top \bar{y} - \psi(\eta)]} g_0^{(n)}(\bar{y}), \tag{2.11}$$

where $g_0^{(n)}(\bar{y})$ is the density of $\bar{y}$ for $\eta = 0$. (Notice that $g_\eta^{(n)}(\mathbf{y})$ is not the same density as $g_\eta^{(n)}(\bar{y})$, and similarly for $g_0^{(n)}(\mathbf{y})$ and $g_0^{(n)}(\bar{y})$; we'll depend on context to distinguish which is which.)

From $\eta^{(n)} = n\eta$ and $\mu^{(n)} = \mu$ we see that

$$A^{(n)} = nA \quad \text{and} \quad B^{(n)} = B.$$

In what follows, we will parameterize family (2.11) with η rather than $\eta^{(n)} = n\eta$. Then we can use Figure 2.1 relating A and B *exactly as drawn.* This unchanging geometry greatly simplifies asymptotic computations for exponential families. As n grows larger, the MLE $\bar{y} = \hat{\mu}$ moves closer to μ, allowing us to use differential relationships (2.4)–(2.5) to quite simply calculate the asymptotic accuracy of the MLE (Section 2.6).

Homework 2.6 Is $d\mu^{(n)} = V^{(n)} d\eta^{(n)}$ the same as $d\mu = V d\eta$?

2.5 Likelihoods, Score Functions, Cramér–Rao Lower Bounds

Maximum likelihood estimation is a classical inference methodology that continues to thrive in the Big Data era. As the name suggests, MLE theory focuses on properties of the likelihood function. These take a particularly simple form in exponential families. The multiparameter situation is much the same as the one-parameter development of Section 1.4, with just enough notational difference to justify restatement.

Working in the repeated sampling framework of Section 2.4, expression (2.10) shows that the log likelihood function for $\mathbf{y} = (y_1, \dots, y_n)$ is

$$l_\eta(\mathbf{y}) = n\left[\eta^\top \bar{y} - \psi(\eta)\right]. \tag{2.12}$$

The *score function* is defined to be the component-wise derivative with respect to η,

$$\dot{l}_\eta(\mathbf{y}) = \frac{\partial l_\eta(\mathbf{y})}{\partial \eta_j} = n(\bar{y} - \mu) \tag{2.13}$$

(remembering that $\dot{\psi}(\eta) = \mu$), and the second derivative matrix is

$$\ddot{l}_\eta(\mathbf{y}) = -nV \tag{2.14}$$

(since $\dot{\mu} = \ddot{\psi} = V$). The *Fisher information* for η in $\mathbf{y}$ is the outer product[2]

$$i_\eta^{(n)} = E_\eta\left\{\dot{l}_\eta(\mathbf{y})\dot{l}_\eta(\mathbf{y})^\top\right\} = nV, \tag{2.15}$$

Since $E_\eta\{\dot{l}_\eta(\mathbf{y})\} = 0$, (2.15) says that nV is the covariance matrix of $\dot{l}_\eta(\mathbf{y})$. It can also be shown that

$$i_\eta^{(n)} = E_\eta\left\{-\ddot{l}_\eta(\mathbf{y})\right\}, \tag{2.16}$$

a general result not restricted to exponential families.

We can also consider the score function with respect to μ,

$$\frac{\partial l_\eta(\mathbf{y})}{\partial \mu} = \frac{\partial \eta}{\partial \mu}\frac{\partial l_\eta(\mathbf{y})}{\partial \eta} = V^{-1}\dot{l}_\eta(\mathbf{y}) = nV^{-1}(\bar{y} - \mu);$$

the Fisher information for μ, denoted $i_\eta^{(n)}(\mu)$ ("at η, for μ, sample size n") is

$$i_\eta^{(n)}(\mu) = E\left\{\frac{\partial l_\eta(\mathbf{y})}{\partial \mu}\frac{\partial l_\eta(\mathbf{y})^\top}{\partial \mu}\right\} = n^2 V^{-1}\operatorname{Cov}(\bar{y})V^{-1} = nV^{-1}. \tag{2.17}$$

Cramér–Rao Lower Bound (CRLB)

Between 1922 and 1934, Fisher developed the central elements of statistical estimation theory: sufficiency, efficiency, information, ancillarity, and the asymptotic superiority of the MLE. A surprise development – part of the surprise being that Fisher didn't do it – was an information lower bound on the variance of an unbiased estimator: the *Cramér–Rao lower bound*. It appeared right after World War II in separate papers by Harald Cramér, C. R. Rao, Maurice Frechet, and George Darmois (the war having disrupted scholarly communications).

[2] The outer product $ab^\top$ of p-vector a and q-vector b is the $p \times q$ matrix $(a_i b_j)$.

The multiparameter CRLB is a straightforward generalization of the univariate bound (1.28). Suppose $\zeta = t(\eta)$ is a smoothly defined q-dimensional function of the p-dimensional parameter η, and $\dot{t}$ its $p \times q$ derivative matrix

$$\dot{t} = \frac{\partial t_j}{\partial \eta_i}.$$

If $\bar{\zeta}$ is an unbiased estimator of ζ, the CRLB for ζ is

$$\operatorname{Cov}\{\bar{\zeta}\} \geq \dot{t}^\top \left(i_\eta^{(n)^{-1}}\right) \dot{t} = \frac{\dot{t}^\top V_\eta^{-1} \dot{t}}{n} \tag{2.18}$$

("$\geq$" meaning that $\operatorname{Cov}\{\bar{\zeta}\} = \dot{t}^\top V_\eta^{-1} \dot{t}/n$ is a non-negative definite matrix).

If ζ is μ, the expectation parameter, the derivative matrix $\dot{t}$ is $d\mu/d\eta = V$, so (2.18) gives

$$\operatorname{Cov}(\bar{\mu}) \geq \frac{V}{n}.$$

But $\operatorname{Cov}(\bar{y}) = V/n$, so the MLE $\hat{\mu} = \bar{y}$ attains the CRLB. This only happens for μ, or linear transformations of μ. So for example, the MLE $\hat{\eta}$ does not attain

$$\operatorname{CRLB}(\eta) = \frac{V^{-1}}{n}.$$

Homework 2.7 Let ζ be a scalar function of η or μ, say $\zeta = t(\eta) = s(\mu)$, with gradient vector $\dot{t}(\eta) = (\cdots \partial t/\partial \eta_i \cdots)$, and likewise $s'(\mu) = (\cdots \partial s/\partial \mu_j \cdots)$. Having observed $\boldsymbol{y} = (y_1, \ldots, y_n)$, show that the lower bound on the variance of an unbiased estimate of ζ is

$$\operatorname{CRLB}(\zeta) = \frac{\dot{t}(\eta)^\top V^{-1} \dot{t}(\eta)}{n} = \frac{s'(\mu)^\top V s'(\mu)}{n}. \tag{2.19}$$

In applications, $\hat{\eta}$ is substituted for η, or $\hat{\mu}$ for μ in (2.19), to get an approximate variance for the MLE $\hat{\zeta} = t(\hat{\eta}) = s(\hat{\mu})$. Even though ζ is generally not unbiased for ζ, the variance approximation – which is equivalent to using the delta method – is usually a reasonable guide.

2.6 Maximum Likelihood Estimation

In the one-parameter families of Section 1.4, we saw that the MLE of μ was directly available as $\hat{\mu} = \bar{y}$, while the MLE of η required the nonlinear mapping $\hat{\eta} = \dot{\psi}^{-1}(\hat{\mu})$. Things are formally the same in multiparameter exponential families, but higher dimensions make the mapping picture more interesting and useful.

Working in the repeated sampling context of Section 2.4, $\boldsymbol{y} = (y_1, \dots, y_n) \overset{\text{iid}}{\sim} g_\eta(y)$, the score function $\dot{l}_\eta(\boldsymbol{y}) = n(\bar{y} - \mu)$ (2.13) shows that the maximum likelihood estimate $\hat{\eta}$ for η must satisfy

$$\mu_{\hat{\eta}} = \bar{y}. \tag{2.20}$$

That is, $\hat{\eta}$ is the value of η that makes the theoretical expectation μ of the average equal to the observed value $\bar{y}$. Moreover, the matrix of second derivatives $\ddot{l}_\eta(\boldsymbol{y}) = -nV$ (2.14) shows that the log likelihood $l_\eta(\boldsymbol{y})$ is a *concave* function of η (since V is positive definite for all η) so that there can be no local maxima lurking elsewhere.

From $\partial/\partial\mu l_\eta(\boldsymbol{y}) = nV^{-1}(\bar{y} - \mu)$, we see again that the MLE of μ is $\hat{\mu} = \bar{y}$, this also following from the fact that MLEs map correctly,

$$\hat{\mu} = \mu_{\hat{\eta}} = \bar{y}.$$

Figure 2.2 schematically illustrates the relationship between $\hat{\eta}$ and $\hat{\mu}$. Here the expectation space B and the sample space $\mathcal{Y}$ have been superimposed, which makes intuitive sense since points μ in B are the expectations of points y sampled from $g_\mu(\cdot)$. The little dots represent the individual observations y_i going into $\hat{\mu} = \bar{y} = \sum y_i/n$.

The mapping from $\hat{\mu}$ to $\hat{\eta}$ inverts $\hat{\mu} = \dot{\psi}(\hat{\eta})$ (2.2),

$$\hat{\eta} = \dot{\psi}^{-1}(\hat{\mu}); \tag{2.21}$$

(2.21) is almost always nonlinear, and sometimes – as for the generalized linear models of Part 3 – not representable in closed form. As we'll see, the local differential expression from (2.5),

$$d\hat{\eta} = \widehat{V}^{-1}\, d\hat{\mu} \qquad \left(\widehat{V} = V_{\hat{\eta}}\right), \tag{2.22}$$

is convenient for computational purposes.

A crucial point, and one that greatly simplifies maximum likelihood calculations in exponential families, is that nothing in Figure 2.2 depends on the sample size n. A, B, $\mathcal{Y}$, and the mapping $\hat{\eta} = \dot{\psi}^{-1}(\hat{\mu})$ stay the same. What does change is how well $\hat{\mu} = \bar{y}$ estimates μ. As n goes to infinity, the central limit theorem says that

$$\hat{\mu} = \bar{y} \mathrel{\dot\sim} \mathcal{N}_p(\mu, V/n), \tag{2.23}$$

V equaling V_μ, the covariance of y at the true expectation μ; normality is approximate in (2.23), but the expectation and covariance are exact; $\bar{y}$ approaches μ at rate $n^{-1/2}$ according to (2.23), making the local linear approximation $d\eta = V^{-1}\, d\mu$ increasingly accurate. Table 2.1 summarizes some

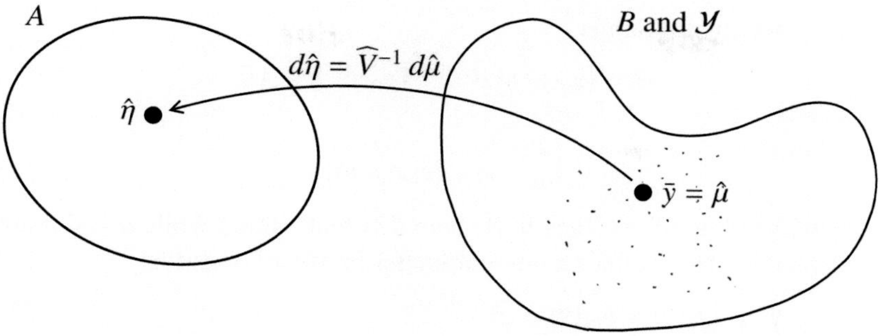

Figure 2.2 Relationship between $\hat{\eta}$ and $\hat{\mu}$.

basic properties of maximum likelihood estimation in multiparameter exponential families.

Table 2.1 *Summary of estimation properties.*

	η	μ
Score $\dot{l}$	$n(\bar{y} - \mu)$	$nV^{-1}(\bar{y} - \mu)$
Information $i^{(n)}$	nV	nV^{-1}
CRLB	V^{-1}/n	V/n
MLE	$\hat{\eta} = \dot{\psi}^{-1}(\bar{y})$	$\hat{\mu} = \bar{y}$

Homework 2.8 Use Figure 2.2 to derive an approximation for $\text{Cov}(\hat{\eta})$. In what sense is it not as good as approximation (2.23)? How does it relate to the CRLB for $\hat{\eta}$?

Homework 2.9 Show that the $p \times p$ second derivative matrix of the log likelihood with respect to μ, evaluated at $\hat{\mu}$, is

$$\left. \left(\frac{\partial^2 l_\eta(\mathbf{y})}{\partial \mu_i \partial \mu_j} \right) \right|_{\hat{\mu}} = -n\widehat{V}^{-1}.$$

One-parameter Subfamilies

A puzzling multiparameter exponential family problem can sometimes be clarified by considering one-parameter subfamilies. For now we will consider just a single observation y rather than repeated samples. Beginning with a p-parameter family

$$\mathcal{G} = \left\{g_\eta(y) = e^{\eta^\top y - \psi(\eta)} g_0(y), \eta \in A\right\},$$

let

$$\{\eta_\theta = a + b\theta, \theta \in \Theta\} \tag{2.24}$$

be a straight line of η values in A; here θ is real-valued while a and b are fixed vectors. This defines a one-parameter family of densities

$$\mathcal{F} = \left\{f_\theta(y) = g_{\eta_\theta}(y) = e^{(a+b\theta)^\top y - \psi(a+b\theta)} g_0(y), \theta \in \Theta\right\}; \tag{2.25}$$

(2.25) is a one-parameter exponential family with:

- natural parameter θ;
- sufficient statistic $x = b^\top y$;
- CGF $\phi(\theta) = \psi(a + b\theta)$; (2.26)
- carrier $f_0(y) = e^{a^\top y} g_0(y)$;
- natural parameter space Θ comprising θ values having η_θ within A.

To simplify the notation we write μ_θ for μ_{η_θ}, V_θ for V_{η_θ}, etc. As θ increases, η_θ moves through A in a straight line parallel to b, but μ_θ usually traces out a curve, the differential relationship being

$$d\mu_\theta = V_\theta \, d\eta_\theta. \tag{2.27}$$

Since $d\eta_\theta = bd\theta$ in (2.24), (2.27) becomes

$$d\mu_\theta = V_\theta \, bd\theta.$$

The dependence of V_θ on θ causes the curvature of $\{\mu_\theta, \theta \in \Theta\}$.

A portion of $\mathcal{F}$ is indicated by the straight line seen inside A in Figure 2.3. (The full set of η vectors $\{\eta_\theta, \theta \in \Theta\}$ would extend to the boundaries of A.) The corresponding portion of the curve $\{\mu_\theta, \theta \in \Theta\}$ is shown within B.

We can apply the moment relationships of Section 1.2 to the one-parameter family $f_\theta(y)$ (2.25) by differentiating $\phi(\theta) = \psi(a + b\theta)$ with respect to θ, giving the cumulants of the sufficient statistic x:

$$\begin{aligned} E_\theta\{x\} &= \frac{d\phi}{d\theta} = b^\top \dot{\psi}(\eta_\theta) = b^\top \mu_\theta, \\ \mathrm{Var}_\theta\{x\} &= \frac{d^2\phi}{d\theta^2} = b^\top \ddot{\psi}(\eta_\theta) b = b^\top V_\theta b, \\ E_\theta\{x - E_\theta(x)\}^3 &= \frac{d^3\phi}{d\theta^3} = \sum_i \sum_j \sum_k \dddot{\psi}_{ijk}(\eta_\theta) b_i b_j b_k. \end{aligned} \tag{2.28}$$

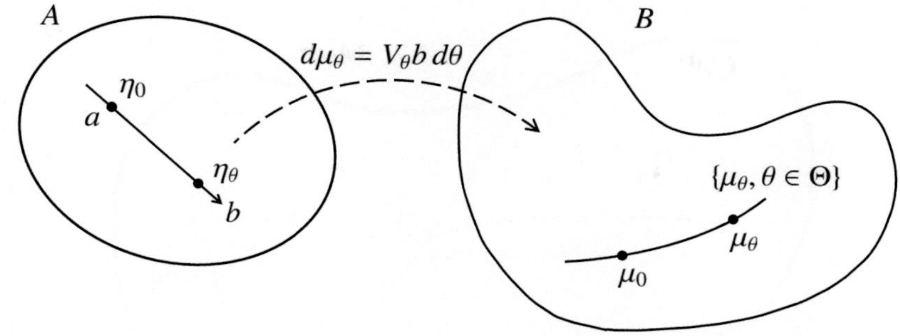

Figure 2.3 Differential relationship (2.27); portion of $\mathcal{F}$ (2.25) within A, corresponding portion of curve $\{\mu_\theta, \theta \in \Theta\}$ within B.

Here

$$\dddot{\psi}_{ijk}(\eta_\theta) = \left.\frac{\partial^3 \psi(\eta)}{\partial \eta_i \partial \eta_j \partial \eta_k}\right|_{\eta=\eta_\theta}.$$

Homework 2.10 (a) Show that the last line of (2.28) implies that

$$\dddot{\psi}_{ijk}(\eta) = E_\eta \left\{(y_i - \mu_i)(y_j - \mu_j)(y_k - \mu_k)\right\}, \tag{2.29}$$

that is, $\dddot{\psi}(\eta)$ is the $p \times p \times p$ array of third central mixed moments for the components of y.

(b) What is $\ddddot{\psi}_{ijkl}$?

Relationship (2.29) is an example of one-parameter subfamilies being used to derive properties of the p-parameter family $\mathcal{G}$.

Suppose we observe an i.i.d. sample $\boldsymbol{y} = (y_1, \ldots, y_n)$ from $g_{\eta_\theta}(y)$ in $\mathcal{F}$ (2.25). Again this reduces to a one-parameter exponential family $f_\theta(\boldsymbol{y})$, as in (2.26) but now with

$$x = b^\top \bar{y} \quad \text{and} \quad \phi(\theta) = n\psi(a + b\theta). \tag{2.30}$$

The score function is

$$\dot{l}_\theta(\boldsymbol{y}) = \frac{\partial \log f_\theta(\boldsymbol{y})}{\partial \theta} = nb^\top(\bar{y} - \mu_\theta). \tag{2.31}$$

Figure 2.4 illustrates maximum likelihood estimation of θ. If $\hat{\theta}$ is the MLE, then setting $\dot{l}_{\hat{\theta}}(\bar{y}) = 0$ says that $\bar{y}$ must lie on the flat surface

$$\overset{\perp}{\mathcal{L}}_{\hat{\theta}} = \{\bar{y} : b^\top (\bar{y} - \mu_{\hat{\theta}}) = 0\}, \tag{2.32}$$

i.e., the $(p-1)$-dimensional hyperplane passing through $\mu_{\hat{\theta}}$ orthogonally

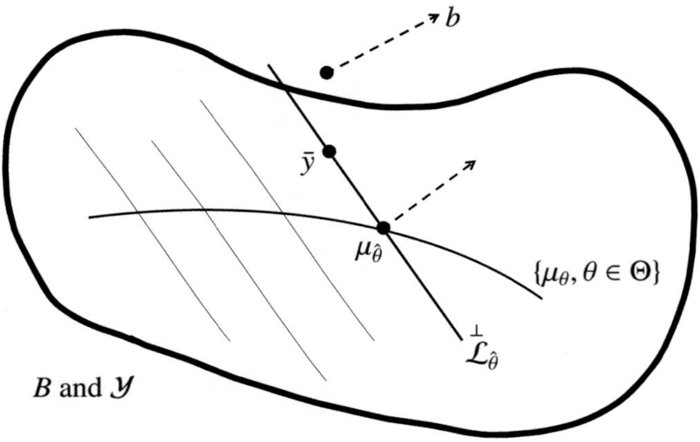

Figure 2.4 Maximum likelihood estimation of θ.

to the vector b in (2.24). Maximum likelihood estimation of θ amounts to projecting $\bar{y}$ onto the curve $\{\mu_\theta, \theta \in \Theta\}$, orthogonally to b. If the curve were a straight line we could find $\hat{\theta}$ by ordinary linear projection calculations, but it usually isn't, making iterative search calculations necessary.

The hyperplanes $\overset{\perp}{\mathcal{L}}_{\hat{\theta}}$ of constant $\hat{\theta}$ value are parallel to each other in Figure 2.4. *Curved exponential families*, in Part 4, have b changing with θ, making the projection directions variable.

Homework 2.11 (a) Verify statements (2.30)–(2.32).
(b) How would you estimate the standard deviation of $\hat{\theta}$?

Stein's Least Favorable Family (LFF)

Stein (1956)'s "least favorable family" is a particular one-parameter subfamily that can be used to reduce multidimensional estimation and testing problems to more tractable single-parameter form. The idea is not restricted to exponential families but it is especially useful in them, as we will see in Part 5 with its development of bootstrap confidence intervals.

Suppose that in the p-parameter family $\mathcal{G}$ we wish to do estimation or testing for a real-valued parameter ζ, which can be evaluated as a function of either η or μ,

$$\zeta = s(\eta) = t(\mu). \tag{2.33}$$

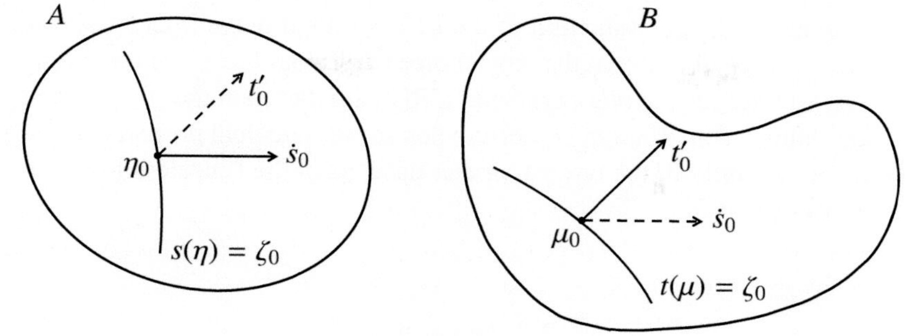

Figure 2.5 Set of η vectors $\{s(\eta) = \zeta_0\}$ forming a curved $(p-1)$-dimensional surface in A, passing through η_0; likewise the set of μ vectors $\{t(\mu) = \zeta_0\}$ forming a curved surface in B.

Let the true value of η be η_0, and $\mu_0 = \dot{\psi}(\eta_0)$ the true value of μ. Then the true value of ζ is $\zeta_0 = s(\eta_0) = t(\mu_0)$.

The set of η vectors such that $\{s(\eta) = \zeta_0\}$ is a curved $(p-1)$-dimensional surface in A, passing through η_0. Likewise $\{t(\mu) = \zeta_0\}$ defines a curved surface in B, both surfaces shown in Figure 2.5. Let $\dot{s}_0$ be the gradient vector of $s(\eta)$ at η_0, and t'_0 be the gradient of $t(\mu)$ at μ_0,

$$\dot{s}_0 = \left(\frac{\partial s(\eta)}{\partial \eta_j}\right)\bigg\|_{\eta_0}, \qquad t'_0 = \left(\frac{\partial t(\mu)}{\partial \mu_j}\right)\bigg\|_{\mu_0}; \tag{2.34}$$

$\dot{s}_0$ is orthogonal to the surface $\{s(\eta) = \zeta_0\}$ at η_0, and t'_0 orthogonal to $\{t(\mu) = \zeta_0\}$ at μ_0, as in Figure 2.5.

Stein's least favorable family (LFF) is defined to be the one-parameter subfamily (2.24) with $a = \eta_0$ and $b = t'_0$,

$$\eta_\theta = \eta_0 + t'_0\theta \tag{2.35}$$

(*not* $\eta_0 + \dot{s}_0\theta$), for θ in an open interval containing 0. The name "least favorable" is justified by the following property:

Homework 2.12 (a) Show that the one-parameter CRLB for estimating ζ in the LFF, evaluated at $\theta = 0$, is the same as the p-parameter CRLB for estimating ζ in $\mathcal{G}$, evaluated at $\eta = \eta_0$.

(b) Show that both of these equal $t_0'^{\top} V_0 t'_0$, where V_0 is the variance matrix evaluated at η_0 or μ_0.

In other words, the reduction to the LFF does not make it any easier to estimate ζ. It can be shown that any choice other than $b = t_0'$ for the family $\eta_\theta = \eta_0 + b\theta$ makes the one-parameter CRLB smaller than the p-parameter CRLB. Stein's construction is useful when some statistical property is easily calculated only in the one-parameter case, as in the "acceleration" calculations of Part 5.

2.7 Deviance

Deviance and Hoeffding's formula are at their most useful in multiparameter exponential families, where deviance is the natural generalization of Euclidean distance. The one-parameter development in Section 1.8 holds almost unchanged here, as briefly summarized next.

Just as in (1.60), the deviance $D(\eta_1, \eta_2)$ between g_{η_1} and g_{η_2} in family $\mathcal{G}$ (2.1) is

$$\begin{aligned} D(\eta_1, \eta_2) &= 2E_{\eta_1}\left\{\log \frac{g_{\eta_1}(y)}{g_{\eta_2}(y)}\right\} \\ &= 2\left[(\eta_1 - \eta_2)^\top \mu_1 - \psi(\eta_1) + \psi(\eta_2)\right], \end{aligned} \tag{2.36}$$

which we also can denote as $D(\mu_1, \mu_2)$ or just $D(1, 2)$. It is non-negative and greater than zero if $\eta_1 \neq \eta_2$. Usually $D(\eta_1, \eta_2) \neq D(\eta_2, \eta_1)$. For a repeated sample $y_i \overset{\text{iid}}{\sim} g_\eta(\cdot)$, the observed data $\boldsymbol{y} = (y_1, \dots, y_n)$ has deviance

$$D^{(n)}(\eta_1, \eta_2) = nD(\eta_1, \eta_2). \tag{2.37}$$

Hoeffding's Formula

Indexing the family by μ rather than η, and remembering that the MLE $\hat{\mu} = \bar{y}$, we have Hoeffding's formula

$$\begin{aligned} \frac{g_\mu^{(n)}(\boldsymbol{y})}{g_{\hat{\mu}}^{(n)}(\boldsymbol{y})} &= e^{-nD(\hat{\mu},\mu)/2} \\ &= e^{-nD(\bar{y},\mu)/2}, \end{aligned} \tag{2.38}$$

the proof being the same as for Lemma 1.3 of Section 1.8. Since $D(\hat{\mu}, \mu) > 0$ unless $\mu = \hat{\mu}$, this shows that the log likelihood of μ, having observed $\boldsymbol{y}$, falls off as $nD(\hat{\mu}, \mu)$.

Given $\hat{\mu}$, the isopleths of equal values of $D(\hat{\mu}, \mu)$, as a function of μ, are the curves of equal likelihood $g_{\mu}^{(n)}(\mathbf{y})$, as indicated in the sketch at right. Since $D(\hat{\mu}, \mu)$ doesn't depend on n, the same diagram applies to all values of n. However, the value of the likelihood assigned to any one curve declines to zero as

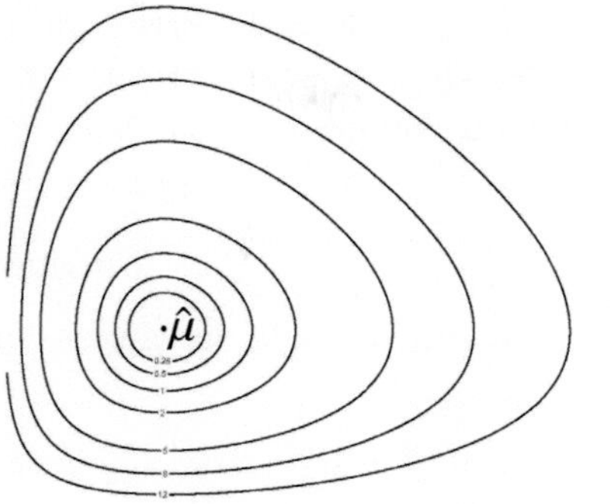

$$g_{\mu}^{(n)}(\mathbf{y}) = e^{-nD(\hat{\mu},\mu)/2}.$$

A Taylor series expansion of the deviance begins

$$D(\mu_1, \mu_2) = (\mu_2 - \mu_1)^\top V_{\mu_1}^{-1}(\mu_2 - \mu_1) + o(\|\mu_2 - \mu_1\|^2), \tag{2.39}$$

the notation indicating an error term going to zero as $\mu_2 \to \mu_1$ at a rate faster than $\|\mu_2 - \mu_1\|^2$. Hoeffding's formula and (2.39) says that for large values of n the likelihood takes on a normal form,

$$g_{\mu}^{(n)}(\mathbf{y}) \doteq \text{constant} \cdot e^{-n/2(\mu-\hat{\mu})' V_{\hat{\mu}}^{-1}(\mu-\hat{\mu})}. \tag{2.40}$$

Homework 2.13 (a) Derive (2.39).

(b) Suppose that $y_i \overset{\text{ind}}{\sim} \mathcal{N}_p(\mu, V)$ for $i = 1, \dots, n$, with V a fixed and known $p \times p$ positive definite matrix. Show that (2.38) gives (2.40).

Homework 2.14 Show that

$$(\eta_2 - \eta_1)^\top(\mu_2 - \mu_1) = \frac{D(1,2) + D(2,1)}{2}.$$

Since the right-hand side is positive, this proves that the relationship between η and μ is "globally monotonic": the angle between $(\mu_2 - \mu_1)$ and $(\eta_2 - \eta_1)$ is always less than 90°.

2.8 Examples of Multiparameter Exponential Families

There is an interesting list of named multiparameter families that show up frequently in applications. They aren't as familiar as their one-parameter cousins (binomial, Poisson, gamma...) but, when needed, can be an invaluable part of the statistician's toolbox. As in the one-parameter case, the named families are ones where the CGF $\psi(\eta)$ has a convenient closed-form expression. Modern computing power allows us to venture beyond the closed-form world; Part 3 considers generalized linear models, including logistic regression, the most widely used computer-based exponential family technique.

Univariate Normal

We observe a single one-dimensional normal variate $X \sim \mathcal{N}(\lambda, \Gamma)$, with density

$$g(x) = \frac{1}{\sqrt{2\pi\Gamma}} e^{-\frac{1}{2\Gamma}(x-\lambda)^2} = \frac{1}{\sqrt{2\pi\Gamma}} e^{\left\{-\frac{x^2}{2\Gamma} + \frac{\lambda}{\Gamma}x - \frac{\lambda^2}{2\Gamma}\right\}}.$$

In exponential family form (2.1),

$$g_\eta(x) = e^{\eta_1 y_1 + \eta_2 y_2 - \psi(\eta)} g_0(x),$$

with:

- $\eta = \begin{pmatrix}\eta_1\\ \eta_2\end{pmatrix} = \begin{pmatrix}\lambda/\Gamma\\ -1/2\Gamma\end{pmatrix}$;
- $\mu = \begin{pmatrix}\lambda\\ \lambda^2 + \Gamma\end{pmatrix}$;
- $y = \begin{pmatrix}y_1\\ y_2\end{pmatrix} = \begin{pmatrix}x\\ x^2\end{pmatrix}$; (2.41)
- $\psi = \frac{1}{2}\left(\frac{\lambda^2}{\Gamma} + \log\Gamma\right)$;
- $g_0(x) = 2\pi^{-1/2}$ with respect to uniform measure on $(-\infty, \infty)$.

Homework 2.15 Use $\dot{\psi}$ and $\ddot{\psi}$ to derive μ and V.

It seems like we have lost ground: our original univariate observation x is now represented two-dimensionally by $y = (x, x^2)$. However, if we have an i.i.d. sample $x_1, \ldots, x_n \overset{\text{iid}}{\sim} \mathcal{N}(\lambda, \Gamma)$, then

$$\bar{y} = \begin{pmatrix}\sum \frac{x_i}{n}\\ \sum \frac{x_i^2}{n}\end{pmatrix} \tag{2.42}$$

is still a two-dimensional sufficient statistic (though not the more usual form $(\bar{x}, \hat{\sigma}^2)$). Figure 2.6 shows diagrams of A and of B and $\mathcal{Y}$.

Homework 2.16 An i.i.d. sample $x_1, \ldots, x_n \sim \mathcal{N}(\lambda, \Gamma)$ gives $y_1, \ldots, y_n$, with $y_i = (x_i, x_i^2)$. Draw a diagram of B and $\mathcal{Y}$ indicating the points y_i and the sufficient vector $\bar{y}$.

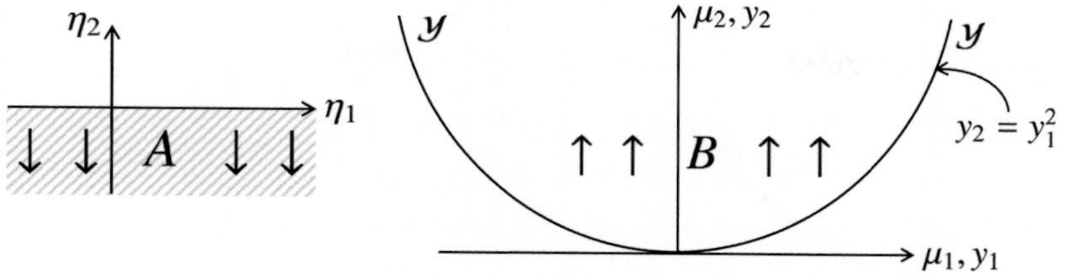

Figure 2.6 Diagrams of A, B, and $\mathcal{Y}$.

Beta

A Beta random variable $X \sim \text{Be}(\alpha_1, \alpha_2)$ is univariate with density on $[0, 1]$,

$$\frac{x^{\alpha_1 - 1}(1 - x)^{\alpha_2 - 1}}{\text{be}(\alpha_1, \alpha_2)}, \tag{2.43}$$

where

$$\text{be}(\alpha_1, \alpha_2) = \frac{\Gamma(\alpha_1)\Gamma(\alpha_2)}{\Gamma(\alpha_1 + \alpha_2)}, \quad \text{and} \quad \alpha_1, \alpha_2 \text{ positive.}$$

The density is defined with respect to Lebesgue (i.e., uniform) measure on $[0, 1]$. This is written in exponential family form as

$$g_\alpha(x) = e^{\alpha^\top y - \psi(\alpha)} g_0(x) \begin{cases} \alpha = (\alpha_1, \alpha_2) \\ y = [\log x, \log(1 - x)]^\top \text{ and } \mathcal{Y} = \mathcal{R}^2 \\ \psi = \log\left[\text{be}(\alpha_1, \alpha_2)\right] \\ g_0(x) = [x(1 - x)]^{-1}. \end{cases}$$

Graphs of some beta distributions are shown in Figure 2.7.

Homework 2.17 (a) Verify by integration of (2.43) that x has

$$\text{mean } \frac{\alpha_1}{\alpha_1 + \alpha_2} \quad \text{and} \quad \text{variance } \frac{\alpha_1 \alpha_2}{(\alpha_1 + \alpha_2)^2(\alpha_1 + \alpha_2 + 1)}.$$

(b) Find an expression for $E\{\log x\}$ in terms of the digamma function, digamma $(z) = \Gamma'(z)/\Gamma(z)$. A useful fact: if $X_1 \sim G_{\nu_1}$ and $X_2 \sim G_{\nu_2}$ then

$$\frac{X_1}{X_1 + X_2} \sim \text{Be}(\nu_1, \nu_2).$$

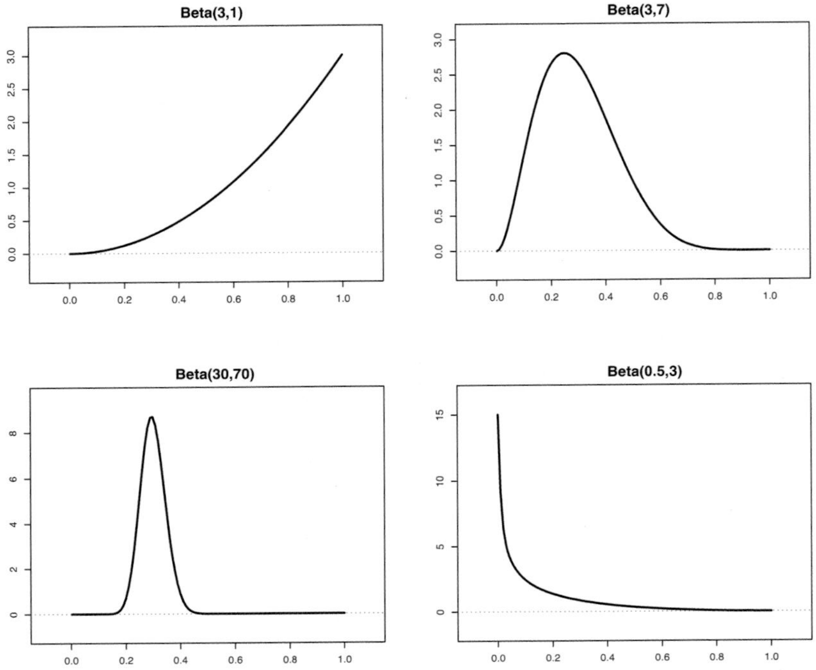

Figure 2.7 Beta distributions (2.43) as a function of x.

Beta is the conjugate prior distribution to the binomial (Section 1.6). Suppose

$$\theta \sim \text{Be}(\alpha_1, \alpha_2) \quad \text{and} \quad x \mid \theta \sim \text{Bi}(n, \theta).$$

Then the posterior density of θ given x is an updated Beta density,

$$\begin{aligned} \pi(\theta \mid x) = c\pi(\theta)g_\theta(x) &= c\theta^x(1-\theta)^{n-x}\theta^{\alpha_1-1}(1-\theta)^{\alpha_2-1} \\ &= c\theta^{x+\alpha_1-1}(1-\theta)^{n-x+\alpha_2-1} \\ &\sim \text{Be}(x+\alpha_1, n-x+\alpha_2), \end{aligned} \tag{2.44}$$

with posterior expectation

$$E\{\theta \mid x\} = \frac{x+\alpha_1}{n+\alpha_1+\alpha_2},$$

which is equivalent to having α_1 prior observations of 1 and α_2 observations of 0.

Example $(\alpha_1, \alpha_2) = (2, 2)$, $n = 10$, $x = 8$. Then $\hat{\theta} = x/n = 0.80$ but

$$E\{\theta \mid x\} = \frac{10}{14} = 0.71.$$

Dirichlet

The binomial distribution relates to dichotomies, where there are only two possibilities, say "East" or "West". If we need to deal with more than two categories – East, West, South, Midwest – then we are in the realm of the *multinomial*, discussed in Section 2.9. The Dirichlet distribution is the big brother of the beta, playing the role of conjugate prior for multinomials.

The sample space for a Dirichlet distribution is the *p-dimensional simplex* $\mathcal{S}_p$ of vectors having non-negative coordinates adding to 1,

$$\mathcal{S}_p = \left\{ x = (x_1, \dots, x_p) : x_i \geq 0 \text{ and } \sum_1^p x_i = 1 \right\}; \quad (2.45)$$

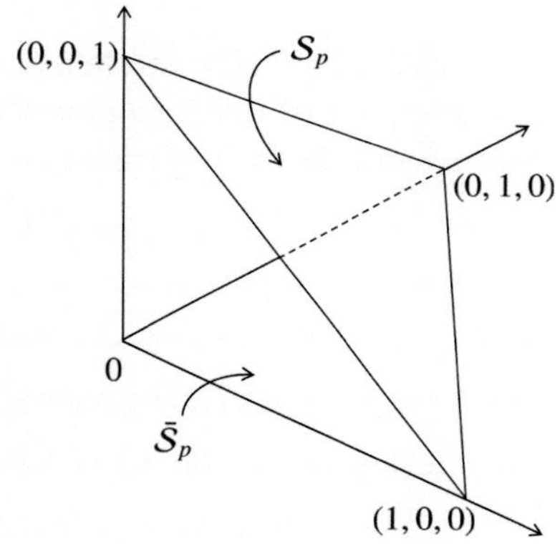

$\mathcal{S}_p$ is a $(p-1)$-dimensional flat surface set at an angle in p-dimensional Euclidean space. We can eliminate the superfluous coordinate by setting $x_p = 0$, which amounts to projecting $\mathcal{S}_p$ down to

$$\overline{\mathcal{S}}_p = \left\{ x = (x_1, \dots, x_p) : x_i \geq 0, x_p = 0, \sum_1^{p-1} x_i \leq 1 \right\},$$

as indicated in the diagram above.

Let $\alpha = (\alpha_1, \dots, \alpha_p)$ be a parameter vector having all positive coordinates. A Dirichlet random variable $X \sim \text{Di}(\alpha_1, \dots, \alpha_p)$ has density

$$g_\alpha(x) = \frac{1}{\sqrt{p}} \frac{\prod_{i=1}^p x_i^{\alpha_i - 1}}{\text{di}(\alpha)} \qquad \left(\text{di}(\alpha) = \frac{\prod_{i=1}^p \Gamma(\alpha_i)}{\Gamma\left(\sum_1^p \alpha_i\right)} \right), \quad (2.46)$$

for $x \in \mathcal{S}_p$. Rather than (2.46), the Dirichlet density is usually represented with $x \in \overline{\mathcal{S}}_p$,

$$g_\alpha(x) = x_1^{\alpha_1 - 1} x_2^{\alpha_2 - 1} \cdots x_{p-1}^{\alpha_{p-1} - 1}, \frac{\left(1 - \sum_1^{p-1} x_i\right)^{\alpha_p - 1}}{\text{di}(\alpha)}. \quad (2.47)$$

(The $p^{-1/2}$ factor in (2.46) has disappeared because $\overline{\mathcal{S}}_p$ has $p^{1/2}$ less volume than $\mathcal{S}_p$; see the following.) Now $g_\alpha(x)$ is a density for $(x_1, \ldots, x_{p-1})$, with the redundant coordinate x_p set to $1 - \sum_1^{p-1} x_i$. Notice that the Beta density (2.43) is (2.47) for the case $p = 2$.

Suppose C is a subset of $\mathcal{S}_p$. We can represent its probability as an integral of $g_\alpha(x)$ in (2.46),

$$\Pr_\alpha\{C\} = \int_C g_\alpha(x)\, dx,$$

but if we want to use ordinary calculus methods to evaluate it, we need to project C down into $\overline{\mathcal{S}}_p$, say to $\overline{C}$, and compute

$$\Pr_\alpha\left\{\overline{C}\right\} = \int_{\overline{C}} g_\alpha(x)\, dx$$

using density (2.47). The fact that (2.47) integrates to 1 if $\overline{C} = \overline{\mathcal{S}}_p$ goes back to Dirichlet, one of Europe's premier mathematicians in the mid-1800s.[3]

We can write both (2.46) and (2.47) as a p-parameter exponential family

$$g_\alpha(x) = e^{\alpha^\top y - \psi(\alpha)} g_0(x) m(dx),$$

where:

- $\alpha = (\alpha_1, \ldots, \alpha_p)$ is the natural parameter;
- $y = \left(\log x_1, \ldots, \log x_p\right)$ is the sufficient vector;
- $\psi(\alpha) = \log(\mathrm{Di}(\alpha))$ is the CGF; (2.48)
- $g_0(x) = \left(\prod_1^p x_i\right)^{-1}$;
- $m(dx) = 1$ is the uniform measure on $\overline{\mathcal{S}}_p$ or uniform$p^{-1/2}$ on $\mathcal{S}_p$.

Homework 2.18 What are μ and V for the Dirichlet? Compare with Homework 2.17. What is rank(V)?

Here is a short list of useful facts concerning Dirichlet calculations:[4]

1. The $(p-1)$-dimensional Euclidean volume of $\mathcal{S}_p$ is $p^{1/2}(\Gamma(p))^{-1}$, while that of $\overline{\mathcal{S}}_p$ is $\Gamma(p)^{-1}$.

[3] Gauss said of Dirichlet's comparatively slim CV, "You can't measure jewels on a grocery scale."

[4] See Kotz et al. (2000) and also the excellent Wikipedia entry on Dirichlet distributions.

2. Choosing $\alpha = (1, \ldots, 1)$ gives $X \sim \text{Di}(\alpha)$ a uniform distribution, over $\mathcal{S}_p$ with (2.46) or $\overline{\mathcal{S}}_p$ with (2.47). That is, for $\overline{C}$ in $\overline{\mathcal{S}}_p$, $\Pr_\alpha\{\overline{C}\}$ is the $(p-1)$-dimensional Euclidean volume of $\overline{C}$ divided by the volume of $\overline{\mathcal{S}}_p$; likewise, for C in $\mathcal{S}_p$, $\Pr_\alpha\{C\}$ is the $(p-1)$-dimensional volume of C divided by the volume of $\mathcal{S}_p$.
3. Suppose $Z_1, \ldots, Z_p$ are independent Gamma variates,

$$Z_i \overset{\text{ind}}{\sim} \lambda G_{\alpha_i}, \qquad i = 1, \ldots, p, \tag{2.49}$$

as in Section 1.5. Let $S = \sum_1^p Z_i$. Then

$$X = (Z_1, \ldots, Z_p)/S \sim \text{Di}(\alpha_1, \ldots, \alpha_p). \tag{2.50}$$

Moreover, $S \sim \lambda G_{\alpha_1+\alpha_2+\cdots+\alpha_p}$ is independent of X.
4. If $X \sim \text{Di}(\alpha_1, \ldots, \alpha_p)$ then

$$Y = \sum_1^k X_i \sim \text{Be}(\alpha_1 + \alpha_2 + \cdots + \alpha_k, \alpha_{k+1} + \alpha_{k+2} + \cdots + \alpha_p). \tag{2.51}$$

Homework 2.19 Verify (2.49)–(2.50) for the case $p = 2$.

Homework 2.20 Use item 3 above to verify item 4.

Just as the beta distribution is conjugate to the binomial (2.44), the Dirichlet is conjugate to the multinomial; see Section 2.9. This has favored the Dirichlet's use as a prior distribution in situations involving categorical responses, and played a prominent role in the theory and practice of *Bayesian nonparametrics*.

Ordinary Least Squares

In normal-theory OLS, the most widely used linear regression model, we observe an n-vector $\boldsymbol{x}$ which is assumed to be generated from a p-dimensional linear model with $p < n$,

$$\boldsymbol{x} = \boldsymbol{M}\alpha + \boldsymbol{e}, \qquad \text{with } \boldsymbol{e} \sim \mathcal{N}_n(\boldsymbol{0}, \sigma^2, \boldsymbol{I}), \tag{2.52}$$

that is, $e_i \overset{\text{iid}}{\sim} \mathcal{N}(0, \sigma^2)$ for $i = 1, \ldots, n$. Here $\boldsymbol{M}$ is a known $n \times p$ matrix of rank p, α an unknown p-dimensional parameter vector, and σ^2 an unknown positive scalar. The MLE of α is $\hat{\alpha} = (\boldsymbol{M}^\top\boldsymbol{M})^{-1}\boldsymbol{M}^\top\boldsymbol{x}$, while $\hat{\sigma}^2 = \|\boldsymbol{x} - \boldsymbol{M}\hat{\alpha}\|^2/n$ is the MLE (not the unbiased estimate) of σ^2.

Homework 2.21 Show that (2.52) is a $(p+1)$-parameter exponential family with

$$\eta = \begin{pmatrix} \alpha/\sigma^2 \\ -1/2\sigma^2 \end{pmatrix} \quad \text{and} \quad y = \begin{pmatrix} \boldsymbol{M}^\top \boldsymbol{x} \\ \|\boldsymbol{x}\|^2 \end{pmatrix}.$$

Hint: Let $\boldsymbol{r} = \boldsymbol{\Gamma}^\top \boldsymbol{x}$, where $\boldsymbol{\Gamma}$ is an $n \times (n-p)$ orthonormal matrix spanning the $(n-p)$-dimensional space that is orthogonal to the column space of $\boldsymbol{M}$, in which case

$$\boldsymbol{r} \sim \mathcal{N}_{n-p}(\boldsymbol{0}, \sigma^2 I),$$
$$\text{independent of} \quad \hat{\alpha} \sim \mathcal{N}_p\left(\alpha, \sigma^2(\boldsymbol{M}^\top \boldsymbol{M})^{-1}\right).$$

Multivariate Normal

This is another big brother case, little brother being the univariate normal (2.41). We observe n independent observations from a d-dimensional normal distribution with expectation vector λ and $d \times d$ covariance matrix Γ,

$$x_1, \ldots, x_n \overset{\text{iid}}{\sim} \mathcal{N}_d(\lambda, \Gamma). \tag{2.53}$$

(This is the generic starting point for classical multivariate analysis.) Stipulating the exponential family quantities y, η, μ, and ψ requires some special notation:

- For H a $d \times d$ symmetric matrix, let $h = H^{(v)}$ be the $d(d+1)/2$ vector that strings out the on-or-above diagonal elements,

  $$h = (H_{11}, H_{12}, \ldots, H_{1d}, H_{22}, H_{23}, \ldots, H_{2d}, H_{31}, \ldots, H_{dd})^\top,$$

  and let $h^{(m)}$ be the inverse mapping from h back to H.
- Also let $\text{diag}(H)$ = matrix with on-diagonal elements those of H and off-diagonal elements 0.
- Straightforward but tedious (and easily bungled) calculations show that the density of $\boldsymbol{x} = (x_1, \ldots, x_n)$ forms a $[p = d(d+3)/2]$-dimensional exponential family

  $$g_\eta(\boldsymbol{x}) = e^{n[\eta^\top y - \psi(\eta)]} g_0(\boldsymbol{x}),$$

  described as follows:

  - $y^\top = (y1^\top, y2^\top)$, where $y1 = \bar{x}$ and $y2 = \left(\sum_{i=1}^{n} \frac{x_i x_i^\top}{n} \right)^{(v)}$, (2.54) dimensions d and $d(d + 1)/2$, respectively;

- $\eta^\top = (\eta 1^\top, \eta 2^\top)$,
  where $\eta 1 = n\Gamma^{-1}\lambda$ and $\eta 2 = n\left(\text{diag}(\Gamma^{-1})/2 - \Gamma^{-1}\right)^{(v)}$;
- $\mu^\top = (\mu 1^\top, \mu 2^\top)$, where $\mu 1 = \lambda$ and $\mu 2 = (\lambda\lambda^\top + \Gamma)^{(v)}$;
- $\psi = \frac{1}{2}\left(\lambda^\top\Gamma^{-1}\lambda + \log\Gamma\right)$.

• We also have:

- $\lambda = \mu 1 = \frac{1}{n}\Gamma \cdot \eta 1$;
- $\Gamma = \mu 2^{(m)} - \mu 1 \mu 1^\top = -n\left(\text{diag}(\eta 2^{(m)}) + \eta 2^{(m)}\right)^{-1}$.

Reference DiCiccio and Efron (1992), "More accurate confidence intervals in exponential families", *Biometrika*, Section 3.

Homework 2.22 Calculate the deviance $D[\mathcal{N}_p(\lambda_1, \Gamma_1), \mathcal{N}_p(\lambda_2, \Gamma_2)]$.

Graph Models

Exponential families on graphs have become a booming topic in network theory. We have a graph with n nodes, each with its own known value θ_i. Let

$$x_{ij} = \begin{cases} 1 & \text{if an edge between } \theta_i, \theta_j \\ 0 & \text{if no edge between } \theta_i, \theta_j. \end{cases}$$

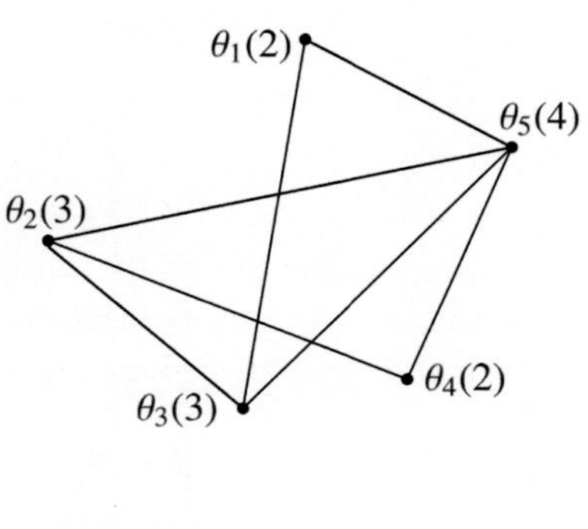

The *degree model* assumes the x_{ij} are independent Bernoulli variates, with probability $x_{ij} = 1$,

$$\pi_{ij} = \frac{e^{\theta_i + \theta_j}}{1 + e^{\theta_i + \theta_j}}.$$

Homework 2.23 (a) Show that the degree model is an n-parameter exponential family with sufficient vector $\boldsymbol{y}$,

$$y_i = \#\{\text{edges entering node } i\} \qquad (\text{"degree } i\text{"}).$$

(b) Describe η, μ, and V.

If all θ_is are the same, the degree model reduces to a one-parameter exponential family, density $e^{\eta E-\psi(\eta)}$ with E the total number of edges. There are more complicated models, for instance

$$e^{\eta_1 E+\eta_2 T-\psi(\eta_1,\eta_2)},$$

where T is the total number of triangles. Very large graphs – the kinds of greatest interest these days – make it difficult to compute the normalizing factor $\psi(\eta_1, \eta_2)$, or to directly sample from the family–giving rise to Markov chain Monte Carlo (MCMC) techniques.

Suppose now that the nodes of the graph represent teams, perhaps the $n = 32$ professional football teams in the National Football League. At any given time in the season we would like to rate the teams' relative strengths. Let P_{ij} be the probability that Team i defeats Team j. The *Bradley–Terry model* assumes that each team has a strength parameter θ_i, and that

$$P_{ij} = \frac{e^{\theta_i}}{e^{\theta_i} + e^{\theta_j}}. \tag{2.55}$$

If w_{ij} is the number of times Team i has defeated Team j ($w_{ij} = 0$, 1, or 2 for the NFL), it is assumed that

$$w_{ij} \sim \text{Bi}(n_{ij}, P_{ij}),$$

where n_{ij} is the number of times they played. Maximum likelihood is used to estimate the parameters θ_i, hence the probabilities P_{ij}. See Section 3.5.

Homework 2.24 Write the Bradley–Terry model in exponential family form.

Truncated Data

Suppose $y \sim g_\eta(y) = e^{\eta^\top y-\psi(\eta)} g_0(y)$, but we only get to observe y if it falls into a given subset $\mathcal{Y}_0$ of the sample space $\mathcal{Y}$. (This is the usual case in astronomy, where only sufficiently bright objects can be observed.) The conditional density of y given that it is observed is then

$$g_\eta(y \mid \mathcal{Y}_0) = e^{\eta^\top y-\psi(\eta)} \frac{g_0(y)}{G_\eta(\mathcal{Y}_0)},$$

where

$$G_\eta(\mathcal{Y}_0) = \int_{\mathcal{Y}_0} g_\eta(y) m(dy).$$

But this is still an exponential family,

$$g_\eta(y \mid \mathcal{Y}_0) = e^{\eta^\top y-\psi(\eta)-\log G_\eta(\mathcal{Y}_0)} g_0(y). \tag{2.56}$$

Homework 2.25 In what is sometimes called the "decapitated Poisson distribution", $x \sim \text{Poi}(\mu)$ but we only observe x if it is > 0. (a) Describe $g_\mu(x)$ in exponential family form. (b) Differentiate its CGF to get μ and V.

Conditional Families

REFERENCE Lehmann and Romano (2005), *Testing Statistical Hypotheses*, 3rd edition, Section 2.7.

In a multiparameter exponential family $g_\eta(y)$, it is sometimes advantageous to condition inferences on part of the y vector. Historically, this was a main tactic in developing the theory of most powerful tests (including the t-test). Here we will show how the same tactic can simplify analysis in a variety of special situations.

We have a p-parameter exponential family, where now the carrying density $g_0(y)m(dy)$ is represented in Stieltjes form "$dG_0(y)$",

$$\mathcal{G} = \left\{g_\eta(y) = e^{\eta^\top y - \psi(\eta)} dG_0(y), \eta \in A\right\}.$$

We partition η and y into two parts, of dimensions p_1 and p_2,

$$\eta = (\eta_1, \eta_2) \quad \text{and} \quad y = (y_1, y_2).$$

Lemma 2.1 *The conditional distributions of y_1 given y_2 form a p_1-parameter exponential family, with densities*

$$g_{\eta_1}(y_1 \mid y_2) = e^{\eta_1^\top y_1 - \psi(\eta_1 | y_2)} dG_0(y_1 \mid y_2), \tag{2.57}$$

natural parameter vector η_1, sufficient statistic y_1, and CGF $\psi(\eta_1 \mid y_2)$ that depends on y_2 but not *on η_2. Here $dG_0(y_1 \mid y_2)$ represents the conditional distribution of y_1 given y_2 when $\eta_1 = 0$.*

Less usefully, the *marginal* distributions of y_2 form a p_2-parameter exponential family that depends on η_1 but not on y_1,

$$g^{Y_2}_{\eta_1,\eta_2}(y_2) = e^{\eta_2^\top y_2 - \psi_{\eta_1}(\eta_2)}\, dG_{\eta_1,0}(y_2).$$

Proof We will consider, for simplicity, the case where $dG_0(y) = g_0(y)\,dy$, that is, where the family is defined with respect to Lebesque measure. Integrating out y_1, the marginal density of y_2 is

$$\begin{aligned} g^{Y_2}_\eta(y_2) &= \int_{\mathcal{Y}_1} e^{\eta_1^\top y_1 + \eta_2^\top y_2 - \psi(\eta)} g_0(y_1 \mid y_2) g^{Y_2}_0(y_2) \\ &= e^{\eta_2^\top y_2 - \psi(\eta)} \left(\int_{\mathcal{Y}_1} e^{\eta_1^\top y_1} g_0(y_1 \mid y_2)\, dy_1 \right) g^{Y_2}_0(y_2). \end{aligned}$$

With $\psi(\eta_1 \mid y_2)$ the log of the term in square brackets above, this becomes

$$g_\eta(y_1 \mid y_2) = \frac{g_\eta(y_1, y_2)}{g_\eta^{Y_2}(y_2)} = \frac{e^{\eta_1^\top y_1 + \eta_2^\top y_2 - \psi(\eta)} g_0(y_1)}{e^{\eta_2^\top y_2 - \psi(\eta) + \psi(\eta_1 \mid y_2)} g_0^{Y_2}(y_2)}$$
$$= e^{\eta_1^\top y_1 - \psi(\eta_1 \mid y_2)} g_0(y_1 \mid y_2),$$

This is (2.57). A general result we are using here is the fact that

$$dG_0(y_1 \mid y_2) = \frac{g_0(y_1)}{g_0^{Y_2}(y_2)} m(dy_1 \mid y_2),$$

where $m(dy)$ is the original carrying measure for the family $\mathcal{G}$ (hidden in the notation of (2.57)). ■

Familiar applications of Lemma 2.1 involve transformations of the original family $\mathcal{G}$: for M a $p \times p$ nonsingular matrix, let

$$\tilde{\eta} = (M^{-1})^\top \eta \quad \text{and} \quad \tilde{y} = My. \tag{2.58}$$

Since $\tilde{\eta}^\top \tilde{y} = \tilde{\eta}^\top y$, we see that the transformed densities $\tilde{g}_{\tilde{\eta}}(\tilde{y})$ also form an exponential family

$$\tilde{\mathcal{G}} = \left\{ \tilde{g}_{\tilde{\eta}}(\tilde{y}) = e^{\tilde{\eta}^\top \tilde{y} - \psi(M\tilde{\eta})} d\tilde{G}_0(\tilde{y}), \tilde{\eta} \in M^{-1}A \right\},$$

to which we can apply Lemma 2.1. What follows are three useful examples.

Example 1 (Fisher's exact test for a 2×2 table) As in the discussion of 2×2 tables in Section 1.5, counts for a four-category multinomial vector,

$$X = (X_1, X_2, X_3, X_4),$$

are observed for N cases, with row totals r_1 and r_2, and column totals c_1 and c_2. In the ulcer data example, $X = (9, 12, 7, 17)$, $N = 45$, $r_1 = 21$, and $c_1 = 16$. The (unknown) true probabilities for the four categories are $\pi = (\pi_1, \pi_2, \pi_3, \pi_4)$; we wish to test the null hypothesis of independence (1.39),

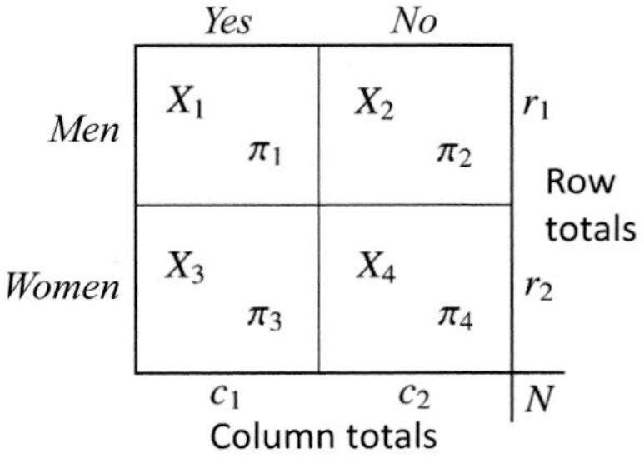

$$H_0 : \log\left(\frac{\pi_1/\pi_2}{\pi_3/\pi_4}\right) = 0.$$

Section 2.9 shows that X has an exponential family distribution with sufficient vector $y = X$ and natural parameter η which can be taken to be

$$\eta = (\eta_1, \eta_2, \eta_3, \eta_4) = (\log \pi_1, \log \pi_2, \log \pi_3, \log \pi_4). \tag{2.59}$$

Let the matrix M in transformations (2.58) be

$$M = \frac{1}{4}\begin{pmatrix} 1 & -1 & -1 & 1 \\ 1 & -1 & 1 & -1 \\ 1 & 1 & -1 & -1 \\ 1 & 1 & 1 & 1 \end{pmatrix}, \tag{2.60}$$

in which case

$$M^{-1} = 4M^{\top}.$$

The transformed natural parameter $\tilde{\eta} = M^{-1}\eta$ has first coordinate

$$\tilde{\eta}_1 = \log\left(\frac{\pi_1/\pi_2}{\pi_3/\pi_4}\right) = \theta,$$

the log odds parameter (1.39).

Applying Lemma 2.1 (2.57) to the $(\tilde{\eta}, \tilde{y})$ family, with $y_1 = \tilde{y}_1$ and $y_2 = (\tilde{y}_2, \tilde{y}_3, \tilde{y}_4)$, we see that y_1 has a one-parameter exponential family distribution (conditional on y_2) with natural parameter $\tilde{\eta}_1$; the other parameters $(\tilde{\eta}_2, \tilde{\eta}_3, \tilde{\eta}_4)$ play no role in the distribution of $\tilde{\eta}_1$.

Some algebra shows that

$$\begin{aligned} \tilde{y}_1 &= X_1 - \frac{r_1}{2} - \frac{c_1}{2} + \frac{N}{4}, \\ (\tilde{y}_2, \tilde{y}_3, \tilde{y}_4) &= \left(\frac{c_1}{2} - \frac{N}{4}, \frac{r_1}{2} - \frac{N}{4}, \frac{N}{4}\right). \end{aligned} \tag{2.61}$$

This implies that conditioning on $(\tilde{y}_2, \tilde{y}_3, \tilde{y}_4)$ is the same as conditioning on the margins of the 2×2 table, and that the conditional distribution of y_1 follows a one-parameter exponential family with natural parameter θ.

Homework 2.26 (a) Verify (2.61).
(b) Why does the derivation justify the *tilted hypergeometric* argument of Section 1.5?

Example 2 (The Wishart statistic) Given a sample from a multivariate normal distribution (2.53), $x_1, \ldots, x_n \overset{\text{iid}}{\sim} \mathcal{N}_d(\lambda, \Gamma)$, let

$$y1 = \bar{x} \quad \text{and} \quad y2 = \sum \frac{x_i x_i^{\top}}{n}$$

be the sufficient statistics (2.54).[5] These have dimensions d and $d(d+1)/2$. Then Lemma 2.1 says that $y2 \mid y1$ follows a $p = [d(d+1)/2]$-dimensional exponential family with natural parameter vector

$$\eta 2 = n\left(\frac{1}{2}\,\text{diag}(\Gamma^{-1}) - \Gamma^{-1}\right)^{(v)}.$$

But the Wishart statistic

$$W = \sum_1^n \frac{(x_i - \bar{x})(x_i - \bar{x})^\top}{n} = \sum_1^n \frac{x_i x_i^\top}{n} - \bar{x}\bar{x}^\top \tag{2.62}$$

is, given $\bar{x}$, a function of $y2 = \sum x_i x_i^\top / n$, which shows that W follows a $[d(d+1)/2]$-parameter exponential family. Since $\eta 2$ depends only on Γ and not on λ, the same is true for the distribution of W.

Note In this case W turns out to be independent of $y1 = \bar{x}$; W is downwardly biased as an estimate of Γ,

$$E\{W\} = \frac{n-p}{n}\Gamma.$$

Example 3 (Poisson/multinomial connection: the "Poisson trick")
Suppose that $\boldsymbol{s} = (s_1, s_2, \ldots, s_L)$ is a vector of L independent Poisson variates,

$$s_l \overset{\text{ind}}{\sim} \text{Poi}(\mu_l), \tag{2.63}$$

$l = 1, \ldots, L$. Then the conditional distribution of $\boldsymbol{s} = (s_1, s_2, \ldots, s_L)$, given that $\sum s_l = n$, is multinomial:

$$\boldsymbol{s} \mid n \sim \text{Mult}_L(n, \boldsymbol{\pi}); \tag{2.64}$$

the notation indicates an L-category multinomial, n draws, and true probabilities $\boldsymbol{\pi} = (\pi_1, \ldots, \pi_L)$, with

$$\pi_l = \frac{\mu_l}{\sum_1^L \mu_i}.$$

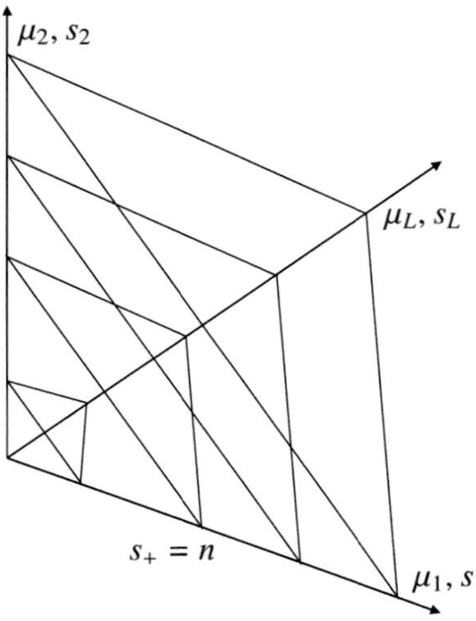

For the 2×2 table of Example 1, $L = 4$, $n = 45$, and $\boldsymbol{s} = (9, 12, 7, 17)$.

Homework 2.27 Directly verify (2.64).

[5] The notation (y_1, y_2) in (2.54) requires the reverse of that in Lemma 2.1.

Another way to say this reverses the order.

$$\begin{aligned} &\text{If} \quad \boldsymbol{s} \mid n \sim \text{Mult}_L(n, \boldsymbol{\pi}) \quad \text{and} \quad n \sim \text{Poi}(\mu_+), \\ &\text{then} \quad s_l \overset{\text{ind}}{\sim} \text{Poi}(\mu_+ \cdot \pi_l), \qquad l = 1, \dots, L. \end{aligned} \tag{2.65}$$

If you are having trouble with a multinomial calculation, usually because of multinomial correlation among the s_l, then thinking of sample size n as Poisson instead of fixed makes the components independent. This "Poisson trick" is often used for quick multinomial approximations.

Homework 2.28 For the 2×2 table of Example 1 or of Section 1.5, use the Poisson trick to find the delta-method approximation of the standard error for the log odds ratio, $\log(\pi_1\pi_4/\pi_2\pi_3)$.

The vector of proportions in the L categories, $\boldsymbol{p} = \boldsymbol{s}/n$, is an unbiased estimator of the true probability vector $\boldsymbol{\pi}$. Its distribution

$$\boldsymbol{p} \sim \text{Mult}_i(n, \boldsymbol{\pi})/n$$

takes values in the simplex $\mathcal{S}_L$ (2.45). Writing $\boldsymbol{s} \sim \text{Poi}(\boldsymbol{\mu})$ for (2.63), the Poisson/multinomial connection is

$$\boldsymbol{s} \sim \text{Poi}(\boldsymbol{\mu}) \Longrightarrow \boldsymbol{p} = \boldsymbol{s}/n \sim \text{Mult}_L(n, \boldsymbol{\pi})/n, \tag{2.66}$$

independent of s_+. Section 2.9 has more interesting applications of the Poisson trick.

Homework 2.29 (a) How is (2.66) analogous to the Gamma/Dirichlet relation (2.49)–(2.50)? (b) Is there a "Dirichlet trick"?

2.9 The Multinomial as an Exponential Family

Traditional multivariate analysis focused on the multivariate normal distribution. A growing interest in nonparametric methods has increased attention to the multinomial distribution, which comprises all possible probability distributions supported on a discrete and finite set of points. The multinomial is a multiparameter exponential family, but one that requires some care in its description.

We assume that each of n subjects has been put independently into one of L categories. In the 2×2 ulcer data table of Section 1.5, $n = 45$ (called N there), $L = 4$, with categories *Treatment-Success*, *Treatment-Failure*, *Control-Success*, and *Control-Failure*.

It is convenient to code the L categories with L-dimensional indicator vectors,

$$e_l = (0, \dots, 0, 1, 0, \dots, 0)^\top \qquad \text{(1 in the } l\text{th place)},$$

indicating category l. The data can be written as $\boldsymbol{y} = (y_1, \dots, y_n)$, where $y_i = e_{l_i}$, l_i the ith subject's category.

The multinomial probability model says that the category of each subject is chosen independently with probability π_l for category l,

$$\pi_l = \Pr\{y_i = e_l, l = 1, \dots, L\}. \tag{2.67}$$

The vector of probabilities $\boldsymbol{\pi} = (\pi_1, \dots, \pi_L)^\top$ takes its values in the simplex $\mathcal{S}_L$,

$$\mathcal{S}_L = \left\{\boldsymbol{\pi} : \pi_L \geq 0, \sum_{l=1}^{L} \pi_l = 1\right\};$$

see (2.45). $\mathcal{S}_L$ represents *all* possible probability distributions on L categories. As in the binomial case (1.31), we need to avoid zero probabilities, leaving

$$\mathcal{S}_L^+ = \left\{\boldsymbol{\pi} : \pi_L > 0, \sum_{l=1}^{L} \pi_l = 1\right\} \tag{2.68}$$

as the parameter space for π.

Define

$$\begin{aligned} \eta_l &= \log \pi_l, \\ s_l &= \#\,\{y_i = e_l\} \qquad \text{(count for category } l\text{)}, \end{aligned} \tag{2.69}$$

with corresponding vectors $\eta = (\eta_1, \dots, \eta_L)^\top$ and $s = (s_1, \dots, s_L)^\top$. The probability of the observed data $\boldsymbol{y} = (y_1, \dots, y_n)$ is

$$g_\pi(\boldsymbol{y}) = \prod_{i=1}^{n} \pi_{l_i} = \prod_{l=1}^{L} \pi_l^{s_l} = e^{\eta^\top s}. \tag{2.70}$$

This looks to be in exponential family form but isn't since $\eta = (\eta_1, \dots, \eta_L)^\top$ is constrained to lie in a *nonlinear* subset of $\mathcal{R}^L$,

$$\sum_{l=1}^{L} e^{\eta_l} = \sum_{l=1}^{L} \pi_l = 1.$$

To circumvent this difficulty we let η be *any* vector in $\mathcal{R}^L$, and define

$$\pi_l = \frac{e^{\eta_l}}{\sum_{j=1}^{L} e^{\eta_j}}, \tag{2.71}$$

for $l = 1, \ldots, L$, so

$$\log \pi_l = \eta_l - \log \sum_{l=1}^{L} e^{\eta_l}.$$

Now the multinomial density can be written in genuine exponential family form,

$$g_\eta(\mathbf{y}) = e^{\eta^\top s - n\psi(\eta)}, \qquad \text{with} \quad \psi(\eta) = \log \sum_{l=1}^{L} e^{\eta_l}. \tag{2.72}$$

More familiarly, the multinomial is expressed in terms of the vector p of observed proportions in the L categories,

$$p = (p_1, \ldots, p_L)^\top = (s_1/n, \ldots, s_L/n)^\top,$$

as

$$g_\eta(p) = e^{n[\eta^\top p - \psi(\eta)]} g_0(p). \tag{2.73}$$

Here $g_0(p)$ is the multinomial coefficient defined with respect to counting measure on the lattice of points $(s_1, \ldots, s_L)^\top/n$,

$$g_0(p) = \binom{n}{np_1, \ldots, np_L} = \frac{n!}{\prod_{l=1}^{L} s_l!},$$

with the s_l being non-negative integers summing to n. This distribution will be denoted

$$p \sim \text{Mult}_L(n, \pi)/n. \tag{2.74}$$

Each y_i can take on L possible values, giving $\mathbf{y}$ a discrete sample space with L^n points for $g_\eta(\mathbf{y})$ in (2.72); the carrier $g_0(y)$ equals 1 for all L^n points, excusing its omission from the formula. Expression (2.73) has collapsed $g_\eta(\mathbf{y})$ down to the density $g_\eta(p)$ for the sufficient statistic p (with p playing the role of y in the general notation (2.1)). Instead of L^n, the sample space B for p has

$$B_n = \frac{(2n-1)!}{n!(n-1)!}$$

points of support[6] in (2.73), forming a discrete lattice in the simplex $\mathcal{S}_L$.

In formulation (2.68)–(2.69), the natural parameter η has space A equaling all of $\mathcal{R}^L$. The curved surface A of Figure 2.8 represents those parameter

[6] For the ulcer data 2×2 table of Section 1.5, $L^n = 1.24 \cdot 10^{27}$ while $B_n = 5.19 \cdot 10^{25}$; B_n is the formula for the number of distinct bootstrap samples of size n.

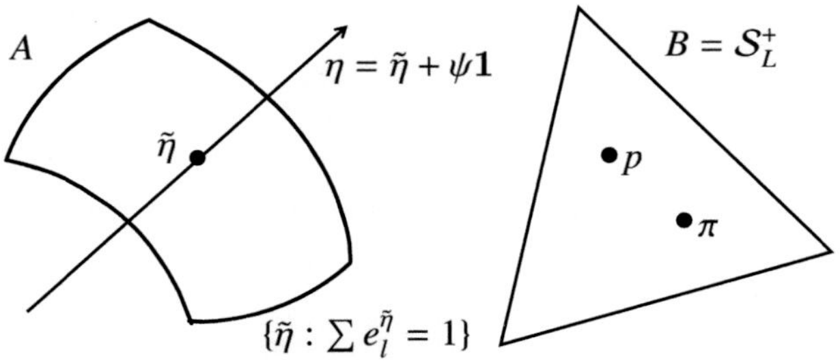

Figure 2.8 Natural parameter η has space A equaling all of $\mathcal{R}^L$; each point π in $B = \mathcal{S}_L^+$ corresponds to a line of points in $\mathcal{R}^p$.

vectors $\tilde\eta$ in $\mathcal{R}^L$ satisfying

$$\sum_{l=1}^{L} e^{\tilde\eta_l} = 1.$$

For a given $\tilde\eta$, every vector

$$\eta = \tilde\eta + \psi\mathbf{1}, \tag{2.75}$$

ψ any constant and $\mathbf{1} = (1, \dots, 1)^\top$, maps into the same probability vector π in $\mathcal{S}_L^+$ according to definition (2.71). Each point π in $\mathcal{S}_L^+$ corresponds to a line of points (2.75) in $\mathcal{R}^p$.

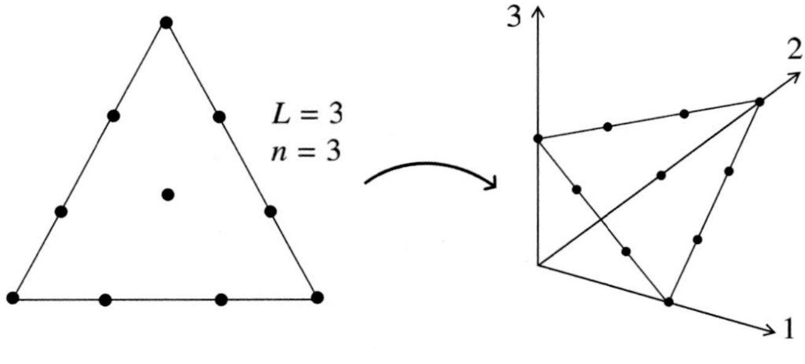

The vector of observed proportions $p = (s_1, \dots, s_L)/n$ takes its values in $\mathcal{S}_L$, now with zero coordinates allowed. The diagram above illustrates the case $L = 3$ and $n = 3$, where there are 10 possible lattice points, nine on the boundary of $\mathcal{S}_3$.

Homework 2.30 Re-express $g_\eta(p)$ so that $\eta = 0$ corresponds to $p \sim \text{Mult}(n, \pi^0)/n$, where $\pi^0 = (1, \dots, 1)^\top/L$ (the centerpoint of simplex $\mathcal{S}_L$).

Homework 2.31 Differentiate $\psi(\eta)$ to show that:

- p has expectation vector π (playing the role of μ in our original notation);
- p has covariance matrix $V_p = V^{(1)}/n$ (where $V^{(1)} = D_\pi - \pi\pi^\top$ and D_π is the diagonal matrix with elements π_l) so that

$$V_{ij}^{(1)} = \begin{cases} \pi_i(1-\pi_i)/n & \text{if } i = j \\ (-\pi_i\pi_j)/n & \text{otherwise.} \end{cases}$$

Homework 2.32 We saw at (2.44) that beta is the conjugate distribution for the binomial. Show that Dirichlet is conjugate to the multinomial.

As n gets large, $p \sim \text{Mult}_L(n, \pi)/n$ becomes approximately normal,

$$p \dot{\sim} \mathcal{N}_L(\pi, V),$$

but its L-dimensional distribution is confined to the $(L-1)$-dimensional flat set $\mathcal{S}_L$. As a result, the Hessian matrix $d\pi/d\eta = V^{(1)}$ is singular,

$$V^{(1)}\mathbf{1} = (D_\pi - \pi\pi^\top)\mathbf{1} = \boldsymbol{\pi} - \boldsymbol{\pi} = \mathbf{0}.$$

For most theoretical calculations we can take the inverse matrix $d\eta/d\pi$ to be the "pseudo-inverse"

$$V^{(1)-1} = \text{diag}(1/\pi),$$

the diagonal matrix with entries π_l^{-1}. The multinomial can be written directly in $(L-1)$-dimensional form by taking

$$\pi_L = 1 - \sum_1^{L-1} \pi_l,$$

as with the Dirichlet (2.47), but the asymmetry can cause awkward difficulties.

Homework 2.33 In what way is the Poisson trick related to $V^{(1)-1}$ above?

The vector $\boldsymbol{\eta} = \log \pi$ (plus a constant) in (2.71) can have any value in $\mathcal{R}^L$. Restricting it to $\boldsymbol{\eta} = X\beta$, where X is an $L \times p$ known matrix and β is an unknown p-dimensional parameter vector, produces a p-parameter exponential subfamily of the full multinomial family (an extension of the construction of one-parameter subfamilies in Section 2.6). Multinomial subfamilies will play a role in the multicategory logistic regression model of Section 3.2, Lindsey's method of Section 3.4, and the proportional hazards model of Section 3.8.

Example A statistically minded tavern owner is keeping track of patrons' drink preferences by their age groups. The top portion of Table 2.2 shows results from the first week: $n = 200$ patrons have been cross-classified by beverage choice (beer, wine, spirits, soda) and age group (less than 30 years old, 30–39, 40–49, 50–59, and 60 or older). For example, 21 patrons that week between the ages of 30 and 39 ordered beer. This is commonly called a *4-by-5 contingency table*, and can be thought of as an $L = 20$ category multinomial sample.

The owner has a classic statistical question in mind: do the age groups differ in their drink preferences? In formal terms, the question is a test of the null hypothesis of *independence*,

H_0 : the row and column choices are independent of each other.

(For 2×2 tables, H_0 is the null hypothesis $\theta = 0$ at (1.39).) Hoping for an answer, our tavern owner computes the 4×5 *independence table* shown as the bottom portion of Table 2.2, estimating what would be seen if H_0 were true. For instance, the observed table has 83 total counts in the first row and 20 counts in the first column, giving $(83 \times 20) \div 200 = 8.3$ as the top left entry in the independence table.

Homework 2.34 What is the justification for this calculation?

The observed table on the top does not perfectly match the independence table on the bottom. Are the differences so large that we can reject H_0? In 1900, Karl Pearson proposed the *chi-squared* test of H_0 for contingency tables, effectively introducing multinomial calculations into mathematical statistics.

Extended to $I \times J$ tables, Pearson's test is usually stated as a sum of scaled differences between the observed table O and "expected" table E (the independence table),

$$C^2 = \sum_{i=1}^{I} \sum_{j=1}^{J} \frac{(O_{ij} - E_{ij})^2}{E_{ij}}. \tag{2.76}$$

The independence hypothesis H_0 is tested by seeing if C^2 is too large compared to a standard χ^2 distribution with m degrees of freedom (Section 1.5),

$$m = (I - 1)(J - 1); \tag{2.77}$$

$C^2 = 11.3$ for the tavern owner's data, $m = 12$ degrees of freedom, giving p-value

$$\Pr\{\chi^2_{12} \geq 11.3\} = 0.497,$$

Table 2.2 *Top: 4-by-5 contingency table of drink preferences by age group from survey of 200 bar patrons. Bottom: Best-fit table assuming independence of age and preferences.*

4×5 table	<30	30–39	40–49	50–59	60+
Beer	7	21	29	17	9
Wine	5	11	21	11	3
Spirits	3	5	11	8	7
Soda	5	4	15	3	5

Best-fit table	<30	30–39	40–49	50–59	60+
Beer	8.3	17.01	31.5	16.18	9.96
Wine	5.1	10.45	19.4	9.95	6.12
Spirits	3.4	6.97	12.9	6.63	4.08
Soda	3.2	6.56	12.2	6.24	3.84

which is not at all significant. So H_0 is not rejected.

Homework 2.35 Carry out the chi-squared test for the ulcer data of Section 1.5. How do the results compare with Figure 1.4?

The Poisson trick is valuable here: it is easier to think of the entries in the $I \times J$ table as $L = IJ$ independent Poisson observations than as an L-category multinomial. Section 3.3 shows the connection of the chi-squared test with a Poisson regression test for the independence hypothesis.

2.10 The Rotation Data

Figure 2.9 displays the *rotation data*, the rotational speeds of 179 A-type stars in a nearby cluster. To the eye, the histogram appears to be at least bimodal. Are two modes enough? A specific form of that question is explored in what follows, giving us a chance to use some of the multivariate exponential family techniques of Part 2, along with a peek ahead to the generalized linear models of Part 3.

Let $f(x)$ be the density of the rotational speeds, assumed to be independent,

$$x_i \stackrel{\text{ind}}{\sim} f(x), \qquad \text{for } i = 1, \dots, n = 179. \tag{2.78}$$

The bimodal hypothesis H_0 takes here the specific form that $f(x)$ is a mixture of two unimodal components,

$$f(x) = w \frac{\phi(x/c_1)}{c_1} + (1 - w) \frac{\phi(x/c_2)}{c_2}, \tag{2.79}$$

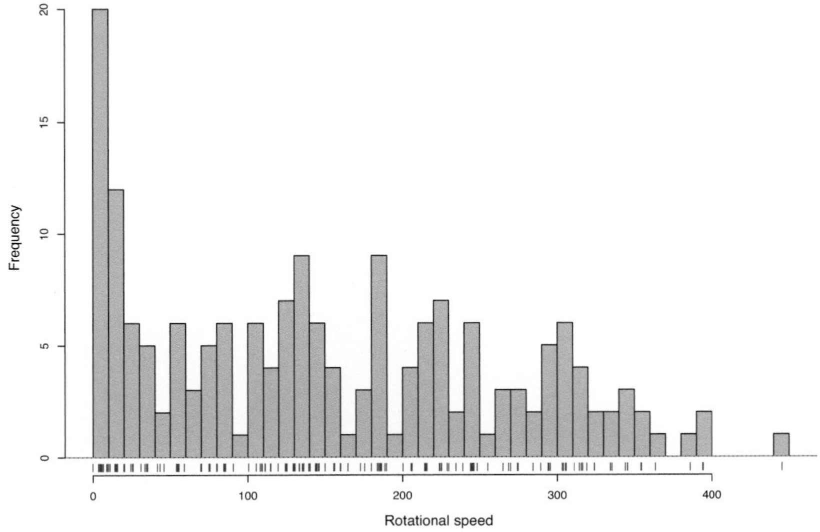

Figure 2.9 Rotational speeds for 179 A-type stars; hash marks represent unbinned data.

where w is the probability of the first component and $(1-w)$ that of the second; $\phi(\cdot)$ is a "base function" derived from physical considerations. Further, (2.79) supposes that $f(x)$ is a mixture of two scaled versions of $\phi(x)$, with scale factors c_1 and c_2, respectively.

There are, in fact, two competing candidates for $\phi(x)$:

$$\phi_1(x) = 2xe^{-x^2}, \qquad \sim (\chi_2^2/2)^{1/2} \text{ (square root of a } G_1\text{);} \tag{2.80a}$$

$$\phi_2(x) = 4x^2e^{-x^2}\pi^{-1/2}, \qquad \sim (\chi_3^2)^{1/2} \text{ (a "Maxwell" density).} \tag{2.80b}$$

Figure 2.10 graphs ϕ_1 and ϕ_2, with ϕ_2 scaled to have the same mode as ϕ_1. The left tails differ, only ϕ_2 having an inflection point. Testing the bimodality hypothesis H_0 comes down to seeing if model (2.79) adequately fits the data in Figure 2.9, using either candidate for $\phi(x)$.

For reasons discussed in Section 3.5, it is convenient to *bin* the data. As in Figure 2.9, we take the bin partitions to be $0, 10, \ldots, 450$, and set y_l to be the count in bin l, for $l = 1, \ldots, L = 45$. The first column of Table 2.3 lists the bin counts, $y_1 = 16, \ldots, y_{45} = 1$. Letting π_l be the probability of bin l,

$$\pi_l = \int_{\text{bin } l} f(x)\, dx,$$

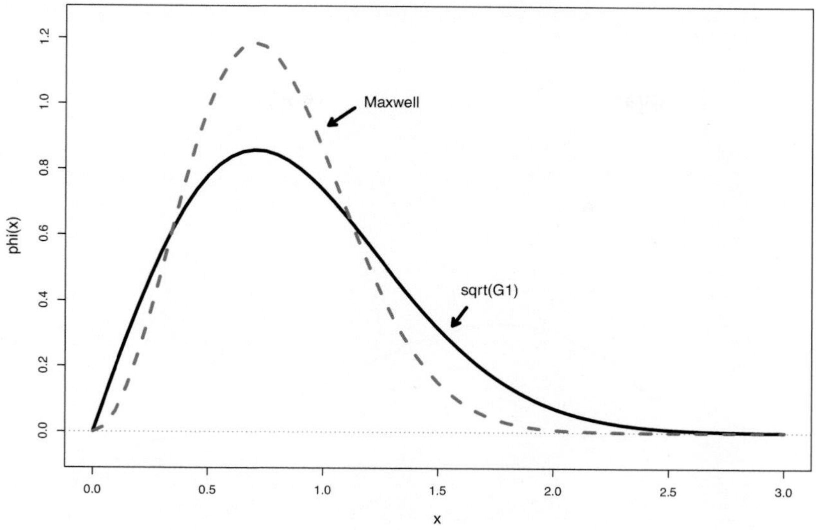

Figure 2.10 Two candidate base functions for rotation data.

$\pi = (\pi_1, \ldots, \pi_L)$, gives the multinomial model

$$y \sim \text{Mult}_L(n, \pi) \qquad (L = 45, n = 179). \tag{2.81}$$

A further convenience employs the Poisson trick: rather than (2.81), we assume that the y_l values are independent Poissons,

$$y_l \overset{\text{ind}}{\sim} \text{Poi}(\mu_l), \qquad \text{with } \mu_l = n\pi_l. \tag{2.82}$$

Any choice of (w, c_1, c_2) in (2.79) produces estimates of π_l and μ_l for $l = 1, \ldots, L$, and then the Poisson likelihood

$$L(w, c_1, c_2) = \prod_{l=1}^{L} e^{-\mu_l} \frac{\mu_l^{y_l}}{y_l}. \tag{2.83}$$

The R nonlinear maximizer `nlminb` found MLE values $(\hat{w}, \hat{c}_1, \hat{c}_2)$, maximizing $L(\hat{w}, \hat{c}_1, \hat{c}_2)$, separately for ϕ equal ϕ_1 or ϕ_2 in (2.80). The solid curve in Figure 2.11 tracks $\hat{\mu}_l$ for ϕ_1; the dashed curve tracks $\hat{\mu}_l$ for ϕ_2.

The Poisson trick works especially well in this type of calculation. It turns out in Section 3.3 that these are also the maximum likelihood estimates from the multinomial model (2.81).

The second and third columns of Table 2.3 give the fitted values $\hat{\mu}_{l1}$ using

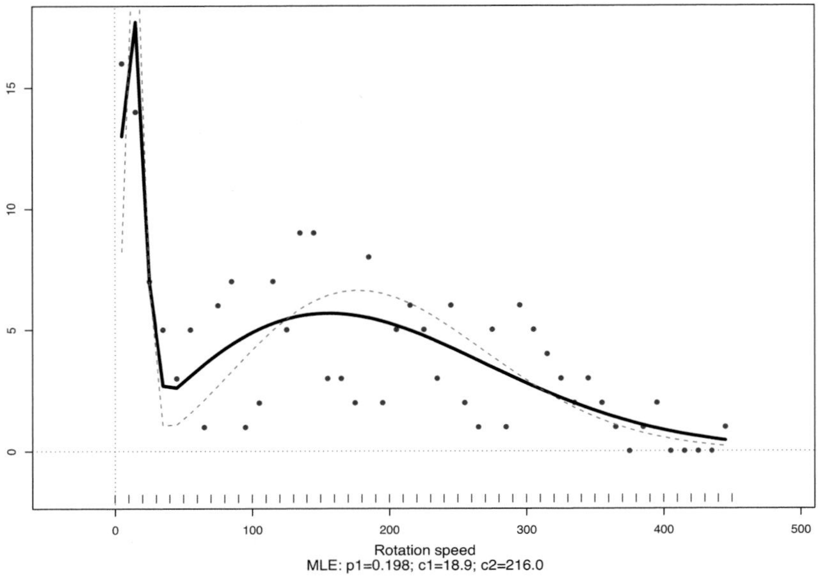

Figure 2.11 Rotation data; binned counts (dots); func_1 fit (solid black), func_2 (dashed red). Total deviation: func_1 52.8, χ^2_{42} p-value 0.123; func_2 78.3, p-value 0.0006.

ϕ_1, and the Poisson deviance residuals (1.64). The total deviance

$$D_+(\boldsymbol{y}, \boldsymbol{\mu}) = \sum_{l=1}^{45} \text{devres}_l^2 = 52.80 \tag{2.84}$$

equals Wilks' likelihood ratio test statistic $2\log(g_\mu(\boldsymbol{y})/g_{\hat{\mu}}(\boldsymbol{y}))$, as in (2.38), on $L - 3 = 42$ degrees of freedom,

$$\text{pvalue} = \Pr\{\chi^2_{42} \geq 52.80\} = 0.123.$$

By conventional criteria, we can *accept* the null hypothesis of bimodality.

More convincingly, examining the deviance residuals reveals no particularly large values or patterns of misfit. The same cannot be said for ϕ_2, where we see a worrisome pattern of large residuals in the early bins. Its total deviance 78.30 has χ^2_{42} p-value 0.0006.

A reasonable conclusion is that the histogram in Figure 2.9 can be explained by bimodal model (2.79), using base function $\phi_1(x)$, but not with the Maxwell choice ϕ_2. Binning the data loses a small amount of informa-

Table 2.3 *Binned rotation data; sum of* $devres_1^2$ = 52.80*, of* $devres_2^2$ = 78.30.

bin	y_l	$\hat{\mu}_{l1}$	devres_1	$\hat{\mu}_{l2}$	devres_2
[1,]	16	13.01	0.80	8.24	2.39
[2,]	14	17.72	−0.92	23.46	−2.11
[3,]	7	6.94	0.02	6.79	0.08
[4,]	5	2.71	1.25	1.07	2.75
[5,]	3	2.63	0.23	1.10	1.49
[6,]	5	3.11	0.99	1.58	2.17
[7,]	1	3.58	−1.62	2.12	−0.86
[8,]	6	4.01	0.92	2.70	1.73
[9,]	7	4.40	1.14	3.29	1.77
[10,]	1	4.74	−2.09	3.89	−1.75
[11,]	2	5.03	−1.54	4.45	−1.31
[12,]	7	5.26	0.72	4.98	0.85
[13,]	5	5.44	−0.19	5.45	−0.20
[14,]	9	5.57	1.33	5.85	1.21
[15,]	9	5.65	1.30	6.17	1.06
[16,]	3	5.67	−1.23	6.41	−1.51
[17,]	3	5.65	−1.23	6.56	−1.56
[18,]	2	5.59	−1.75	6.62	−2.11
[19,]	8	5.48	1.00	6.60	0.53
[20,]	2	5.34	−1.66	6.49	−2.07
[21,]	5	5.17	−0.08	6.32	−0.54
[22,]	6	4.97	0.45	6.08	−0.03
[23,]	5	4.75	0.11	5.78	−0.33
[24,]	3	4.51	−0.76	5.45	−1.15
[25,]	6	4.26	0.79	5.08	0.40
[26,]	2	4.00	−1.11	4.69	−1.41
[27,]	1	3.73	−1.68	4.30	−1.92
[28,]	5	3.47	0.77	3.89	0.54
[29,]	1	3.20	−1.44	3.50	−1.58
[30,]	6	2.94	1.56	3.12	1.45
[31,]	5	2.68	1.26	2.75	1.22
[32,]	4	2.44	0.91	2.41	0.94
[33,]	3	2.20	0.51	2.09	0.59
[34,]	2	1.98	0.01	1.80	0.15
[35,]	3	1.77	0.84	1.54	1.04
[36,]	2	1.58	0.32	1.30	0.57
[37,]	1	1.40	−0.36	1.09	−0.09
[38,]	0	1.23	−1.57	0.91	−1.35
[39,]	1	1.08	−0.08	0.75	0.27
[40,]	2	0.95	0.94	0.62	1.39
[41,]	0	0.82	−1.28	0.50	−1.00
[42,]	0	0.71	−1.19	0.41	−0.90
[43,]	0	0.61	−1.11	0.33	−0.81
[44,]	0	0.52	−1.02	0.26	−0.72
[45,]	1	0.45	0.71	0.21	1.25

tion but usually not very much. As a check, we might see if finer or coarser binnings have a noticeable effect.

Homework 2.36 Repeat the calculations for ϕ_1, now with bin limits $0, 20, 40, \ldots, 440$.

3

Generalized Linear Models

In the standard normal linear regression model the statistician observes independent normal variates

$$y_i \overset{\text{ind}}{\sim} \mathcal{N}(\mu, \sigma^2), \qquad \text{for } i = 1, \dots, N, \tag{3.1}$$

where each μ_i depends on an observed p-dimensional covariate vector x_i

and an unknown p-dimensional parameter vector β,

$$\mu_i = x_i^\top \beta, \qquad i = 1, \ldots, N. \tag{3.2}$$

The least squares algorithm for estimating β goes back to Gauss and Legendre in the early 1800s.

The normality assumption in (3.1) can be dubious, obviously so if the y_i are counts from a discrete process. Generalized linear models (GLMs) began their development in the 1960s as a methodology for extending normal linear regression (3.1)–(3.2) to situations where the responses are binomial, Poisson, gamma, or from any other one-parameter exponential family. Generalized linear models have blossomed as the great success story of exponential family applications. In particular, logistic regression, where the responses are binomial rather than normal, has become a star in the machine learning world as well as a workhorse in traditional settings.

3.1 Exponential Family Regression Models

As a generalization of the normal regression model (3.1)–(3.2) we assume that the observations $y_1, \ldots, y_N$ have come from a given one-parameter exponential family

$$\mathcal{G} = \left\{ g_\eta(y) = e^{\eta y - \psi(\eta)} g_0(y),\ \eta \in A,\ y \in \mathcal{Y} \right\},$$

but with each y_i having its own natural parameter η_i,

$$y_i \stackrel{\text{ind}}{\sim} g_{\eta_i}(y_i) = e^{\eta_i y_i - \psi(\eta_i)} g_0(y_i), \qquad \text{for } i = 1, \ldots, N. \tag{3.3}$$

A *generalized linear model* (GLM) expresses the η_i as linear functions of an unknown p-dimensional parameter vector β and a known p-dimensional covariate vector x_i,

$$\eta_i = x_i^\top \beta, \qquad \text{for } i = 1, \ldots, N. \tag{3.4}$$

If $\mathcal{G}$ is the normal family $y \sim \mathcal{N}(\mu, \sigma^2)$ (σ^2 known) then η equals μ/σ^2 (1.30), making normal linear regression (3.1)–(3.2) a special case of (3.3)–(3.4). The "general" in GLM gives statisticians a principled framework for carrying out linear regression with non-normal responses.

The density for $\boldsymbol{y} = (y_1, \ldots, y_N)^\top$, obtained by multiplying factors (3.3)–(3.4), depends on the parameter vector $\boldsymbol{\beta}$ and the $N \times p$ matrix of covariate vectors $X = (x_1, \ldots, x_N)^\top$,

$$\begin{aligned} g_\beta(\boldsymbol{y}) = e^{\sum_i (\eta_i y_i - \psi(\eta_i))} \prod_i g_0(y_i) &= e^{\beta^\top X^\top \boldsymbol{y} - \sum_i \psi(x_i^\top \beta)} \prod_i g_0(y_i) \\ &= e^{\beta^\top z - \phi(\beta)} g_0(\boldsymbol{y}). \end{aligned} \tag{3.5}$$

This is a p-parameter exponential family (2.1) where:

- β is the $p \times 1$ natural parameter vector;
- $z = X^\top \boldsymbol{y}$ is the $p \times 1$ sufficient vector;
- $\phi(\beta) = \sum_{i=1}^{N} \psi(x_i^\top \beta)$ is the CGF; (3.6)
- $g_0(\boldsymbol{y}) = \prod_{i=1}^{N} g_0(y_i)$ is the carrying density.

Notation

Boldface vectors will be used for N-dimensional quantities such as $\boldsymbol{y} = (y_1, \dots, y_n)^\top$ and $\boldsymbol{\eta} = (\eta_1, \dots, \eta_N)^\top$; likewise $\boldsymbol{\mu}$ for the vector of expectations

$$\boldsymbol{\mu} = (\mu_1, \dots, \mu_N)^\top, \tag{3.7}$$

$\mu_i = \dot{\psi}(\eta_i)$ for $i = 1, \dots, N$, or more succinctly, $\boldsymbol{\mu} = \dot{\psi}(\boldsymbol{\eta})$. Similarly, $\boldsymbol{V}$ will denote the $N \times N$ *diagonal* matrix of variances, written

$$\boldsymbol{V} = \ddot{\psi}(\boldsymbol{\eta}) = \text{diag}\left(\ddot{\psi}(\eta_i)\right), \tag{3.8}$$

with $\boldsymbol{V}_\beta = \text{diag}(\ddot{\psi}(x_i^\top \beta))$ denoting the variance matrix evaluated at $\boldsymbol{\eta} = X\beta$.

Homework 3.1 Show the following.

(a) $E_\beta\{z\} = \dot{\phi}(\beta) = \underset{p\times N}{X^\top} \underset{N\times 1}{\boldsymbol{\mu}} (\beta)$.

(b) $\text{Cov}_\beta\{z\} = \ddot{\phi}(\beta) = \underset{p\times N}{X^\top} \underset{N\times N}{\boldsymbol{V}_\beta} \underset{N\times p}{X} = i_\beta$ (the Fisher information for β).

(c) $\underset{p\times N}{\dfrac{d\hat{\beta}}{d\boldsymbol{y}}} = (X^\top \boldsymbol{V}_{\hat{\beta}} X)^{-1} X^\top$.

(The matrix $M = d\hat{\beta}/d\boldsymbol{y}$ is the *influence function* of $\hat{\beta}$: a small change in y_i, say to $y_i + dy_i$, changes $\hat{\beta}$ by approximately $M_i dy_i$ where M_i is the ith column of M.)

The *score function* for a p-parameter model $f_\beta(\boldsymbol{y})$ is the gradient of $l_\beta = \log f_\beta(\boldsymbol{y})$ with respect to β, i.e., the vector with jth component $\partial \log f_\beta(\boldsymbol{y}) / \partial\beta_j$. From (3.5)–(3.6) we get

$$l_\beta(\boldsymbol{y}) = \beta' X^\top y - \sum_{i=1}^{N} \psi(x_i^\top \beta) \tag{3.9}$$

(plus a term not depending on β).

Using dot notation to indicate differentiation with respect to the components of β, the score function for GLM (3.3)–(3.4) is calculated to be

$$\dot{l}_\beta(\boldsymbol{y}) = X^\top (\boldsymbol{y} - \boldsymbol{\mu}(\beta)), \tag{3.10}$$

where $\boldsymbol{\mu}(\beta)$ is the N-vector of expectation (3.7) evaluated at $\boldsymbol{\eta}_\beta = X\beta$.

Since $\boldsymbol{\mu} = \dot{\psi}(\boldsymbol{\eta})$ and $\boldsymbol{V} = \ddot{\psi}(\boldsymbol{\eta})$, (3.8) says that

$$\frac{d\boldsymbol{\mu}}{d\boldsymbol{\eta}} = \boldsymbol{V}. \tag{3.11}$$

Differentiating (3.10) with respect to β then gives

$$-\ddot{l}_\beta(\boldsymbol{y}) = X^\top \boldsymbol{V}_\beta X. \tag{3.12}$$

This doesn't depend on $\boldsymbol{y}$ so, as in (2.16), the $p \times p$ Fisher information matrix for β is

$$i_\beta = E\left\{-\ddot{l}_\beta(\boldsymbol{y})\right\} = X^\top \boldsymbol{V}_\beta X. \tag{3.13}$$

From (3.10) the MLE equation for β is

$$X^\top (\boldsymbol{y} - \boldsymbol{\mu}(\beta)) = \boldsymbol{0}, \tag{3.14}$$

$\boldsymbol{0}$ being a vector of p zeros. Solving the nonlinear equation (3.14) for the MLE $\hat{\beta}$ is usually carried out by some form of Newton–Raphson iteration, as discussed later.[1] The $p \times p$ information matrix i_β will usually increase with N, the number of cases. As i_β grows, the general theory of maximum likelihood estimations yields the normal approximation

$$\hat{\beta} \dot{\sim} \mathcal{N}_p(\beta, i_\beta^{-1}), \tag{3.15}$$

though a formal statement depends on boundedness properties of the x_i as $N \to \infty$.

The regression model $\{\eta_i = x_i^\top \beta\}$ is a p-parameter subfamily of the N-parameter family that lets each η_i take on any value in A (that is, *not* assuming $\boldsymbol{\eta} = X\beta$),

$$g_{\boldsymbol{\eta}}(\boldsymbol{y}) = e^{\boldsymbol{\eta}^\top \boldsymbol{y} - \sum_i \psi(\eta_i)} g_0(\boldsymbol{y}). \tag{3.16}$$

If the original one-parameter family $g_\eta(y)$ has natural space $\eta \in A$ and expectation space $\mu \in B$, then $g_{\boldsymbol{\eta}}(\boldsymbol{y})$ has spaces A_N and B_N as N-fold products, $A_N = A^N$ and $B_N = B^N$, and similarly $\mathcal{Y}_N = \mathcal{Y}^N$ for the sample spaces.

[1] In the normal case (3.1)–(3.2), equation (3.14) is linear, permitting the familiar least squares solution $\hat{\beta} = (X^\top X)^{-1} X^\top \boldsymbol{y}$.

The natural parameter vectors for the GLM, $\boldsymbol{\eta} = X\beta$, lie in the p-dimensional linear subspace of A_N generated by the columns of $X = (\boldsymbol{x}_{(1)}, \ldots, \boldsymbol{x}_{(p)})$. This flat space maps into a curved p-dimensional manifold in B_N,

$$\{\boldsymbol{\mu}(\beta)\} = \left\{\boldsymbol{\mu} = \dot{\psi}(\boldsymbol{\eta}), \boldsymbol{\eta} = X\beta\right\}, \tag{3.17}$$

as pictured below. (In the normal regression situation, $y_i \overset{\text{ind}}{\sim} \mathcal{N}(x_i^\top\beta, \sigma^2)$, $\{\boldsymbol{\mu}(\beta)\}$ is flat, but this is essentially the only such case.)

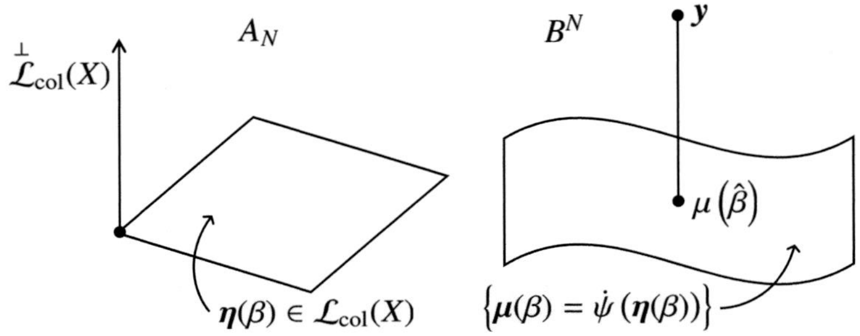

The data vector $\boldsymbol{y}$ will usually *not* lie in the curved manifold $\{\boldsymbol{\mu}(\beta)\}$. From the MLE equations $X^\top(\boldsymbol{y}-\boldsymbol{\mu}(\hat{\beta})) = \boldsymbol{0}$ we see that $\boldsymbol{y}-\boldsymbol{\mu}(\hat{\beta})$ must be orthogonal to the columns $\boldsymbol{x}_{(j)}$ of X. In other words, the maximum likelihood estimate $\hat{\boldsymbol{\mu}} = \boldsymbol{\mu}(\hat{\beta})$ is obtained by projecting $\boldsymbol{y}$ into $\{\boldsymbol{\mu}(\beta)\}$ orthogonally to the columns of X. Letting $\mathcal{L}_{\text{col}}(X)$ denote the column space of X,

$$\mathcal{L}_{\text{col}}(X) = \{\boldsymbol{\eta} : \boldsymbol{\eta} = X\beta,\ \beta \in \mathcal{R}^p\}, \tag{3.18}$$

the residual $\boldsymbol{y} - X\hat{\beta}$ must lie in the space $\overset{\perp}{\mathcal{L}}_{\text{col}}(X)$ of N-vectors orthogonal to $\mathcal{L}_{\text{col}}(X)$. Notice that the *same* space $\overset{\perp}{\mathcal{L}}_{\text{col}}(\boldsymbol{X})$ applies to maximum likelihood estimation for any choice of $\boldsymbol{y}$.

In the normal case, where $\{\boldsymbol{\mu}(\beta)\}$ is flat, $\boldsymbol{\mu}(\hat{\beta})$ is obtained from the usual OLS (ordinary least squares) equations. Iterative methods are necessary for GLMs: if β^0 is an interim guess, we update to $\beta^1 = \beta^0 + d\beta$ where

$$d\beta = (X^\top \boldsymbol{V}_{\beta^0} X)^{-1}\left(\boldsymbol{y} - \boldsymbol{\mu}(\beta^0)\right)$$

(see Homework 3.1(c)), continuing until $d\beta$ is sufficiently close to 0. Because $g_\beta(\boldsymbol{y})$ is an exponential family (3.5) there are no local maxima to worry about.

Modern computer packages such as `glm` in R find $\hat{\beta}$ quickly and painlessly. Having found it they use the asymptotic normal approximation

$$\hat{\beta} \mathrel{\dot\sim} \mathcal{N}_p\left(\beta, (X^\top \boldsymbol{V}_\beta X)^{-1}\right) \tag{3.19}$$

(from Homework 3.1(b)) to report approximate standard errors for the components of $\hat{\beta}$.

Warning The resulting approximate confidence intervals, e.g., $\hat{\beta}_j \pm 1.96 \cdot \widehat{\text{se}}_j$, may be quite inaccurate, as shown by comparison with better bootstrap intervals. See the discussion following Table 3.1 in the next section.

Two Useful Extensions

The basic GLM model (3.3)–(3.4) can be extended to cover a wider range of applications. First of all, we can add a known "offset" vector $\boldsymbol{a}$ to the definition of $\boldsymbol{\eta}$ in the GLM,

$$\boldsymbol{\eta} = \boldsymbol{a} + X\beta. \tag{3.20}$$

Everything said previously remains valid, except now

$$\mu_i(\beta) = \dot{\psi}(a_i + x_i^\top \beta) \quad \text{and} \quad V_i(\beta) = \ddot{\psi}(a_i + x_i^\top \beta). \tag{3.21}$$

Homework 3.2 Write the offset situation in form $g_\beta(\boldsymbol{y}) = e^{\beta^\top z - \phi(\beta)} g_0(\boldsymbol{y})$. Show that Homework 3.1(a) and (b) still hold true, with these changes.

As a second extension, suppose that corresponding to each case i we have n_i i.i.d. observations,

$$y_{i1}, \dots, y_{in_i} \overset{\text{iid}}{\sim} g_{\eta_i}(y) = e^{\eta_i y - \psi(\eta_i)} g_0(y), \tag{3.22}$$

with

$$\eta_i = x_i^\top \beta, \qquad i = 1, \dots, N. \tag{3.23}$$

This is the same situation as before, now of size $\sum_1^N n_i$. However, we can reduce to the sufficient statistics, as in Section 1.3,

$$\bar{y}_i = \sum_{j=1}^{n_i} y_{ij}/n_i,$$

having

$$\bar{y}_i \overset{\text{ind}}{\sim} e^{n_i[\eta_i \bar{y}_i - \psi(\eta_i)]} g_0^{(n)}(\bar{y}_i) \qquad (\eta_i = x_i^\top \beta), \tag{3.24}$$

which reduces the size of the GLM from $\sum n_i$ to N.

Homework 3.3 Let $\boldsymbol{n\bar{y}}$ be the N-vector with elements $n_i \bar{y}_i$, similarly $\boldsymbol{n\mu}$ for the vector of elements $n_i \mu_i(\beta)$, $\boldsymbol{nV}_\beta$ for $\text{diag}(n_i V_i(\beta))$ and $\boldsymbol{i}_\beta$ for the total Fisher information.

(a) Show that

$$g_\beta(\boldsymbol{y}) = e^{\beta^\top z - \phi(\beta)} g_0(\boldsymbol{y}) \begin{cases} z = X^\top(\boldsymbol{n}\bar{\boldsymbol{y}}) \\ \phi = \sum_{i=1}^N n_i \psi(x_i^\top \beta). \end{cases} \tag{3.25}$$

(b) Also show that

$$\dot{l}_\beta(\boldsymbol{y}) = X^\top(\boldsymbol{n}\bar{\boldsymbol{y}} - \boldsymbol{n}\boldsymbol{\mu}), \quad -\ddot{l}_\beta(\boldsymbol{y}) = X^\top \boldsymbol{n} V_\beta X = \boldsymbol{i}_\beta. \tag{3.26}$$

Note Standard errors for the components of $\hat{\beta}$ are usually based on the approximation $\mathrm{Cov}(\hat{\beta}) = \boldsymbol{i}_{\hat{\beta}}^{-1}$.

Homework 3.4 ("Self-grouping property") Suppose we *don't* reduce to the sufficient statistics $\bar{y}_i$, instead doing a GLM with X having $\sum n_i$ rows. Show that we get the same estimates of $\hat{\beta}$ and $\boldsymbol{i}_{\hat{\beta}}$.

Homework 3.5 Show that the solution to the MLE equation (3.14) minimizes the total deviance distance

$$\hat{\beta} = \arg\min_\beta \left\{ \sum_{i=1}^N n_i D\left(y_i, \mu_i(\beta)\right) \right\},$$

where $D(y, \mu)$ is the appropriate one-parameter deviance function, as in Homework 1.30, Section 1.8. In other words, "least deviance" is the GLM analogue of least squares for normal regression.

3.2 Logistic Regression

When the observations y_i are binomials we are in the realm of logistic regression, the most widely used of the generalized linear models. In the simplest formulation, the y_is are independent *Bernoulli* variables $\mathrm{Ber}(\pi_i)$ (that is, binomials with sample size 1),

$$y_i = \begin{cases} 1 & \text{with probability } \pi_i \\ 0 & \text{with probability } 1 - \pi_i, \end{cases} \tag{3.27}$$

where 1 or 0 code the outcomes of a dichotomy, perhaps male or female, success or failure, etc.

The binomial natural parameter η is the logistic transform

$$\eta = \log \frac{\pi}{1 - \pi} \qquad \left(\text{so } \pi = \frac{1}{1 + e^{-\eta})}\right), \tag{3.28}$$

Section 1.6. A logistic GLM is of the form

$$\eta_i = x_i^\top \beta, \qquad \text{for } i = 1, \dots, N, \tag{3.29}$$

or $\boldsymbol{\eta}(\beta) = X\beta$ (so $\pi_i = (1 + \exp\{-x_i^\top \beta\})^{-1}$). The vector of probabilities $\boldsymbol{\pi} = (\pi_1, \ldots, \pi_N)^\top$ is $\boldsymbol{\mu}$ in this case; the MLE equations (3.14) can be written in vector notation as

$$X^\top \left(\boldsymbol{y} - \frac{\mathbf{1}}{\mathbf{1} + e^{-\eta(\beta)}} \right) = \mathbf{0}. \tag{3.30}$$

The Fisher information matrix for β (3.13) is

$$i_\beta = \boldsymbol{X}^\top \operatorname{diag}\left[\pi_i(1 - \pi_i)\right] X. \tag{3.31}$$

Logistic regression includes the situation where $y_i \overset{\text{ind}}{\sim} \text{Bi}(n_i, \pi_i)$, n_i possibly greater than 1. Let $p_i = y_i/n_i$ denote the proportion of 1s,

$$p_i \overset{\text{ind}}{\sim} \text{Bi}(n_i, \pi_i)/n_i, \tag{3.32}$$

so p_i is $\bar{y}_i$ in the notation of Homework 3.3, with $\mu_i = \pi_i$. From Homework 3.3(b) we get[2]

$$\begin{aligned} \dot{l}_\beta(\boldsymbol{y}) &= X^\top(\boldsymbol{np} - \boldsymbol{n\pi}), \\ -\ddot{l}_\beta(\boldsymbol{y}) &= X^\top \operatorname{diag}\left[n_i \pi_i(\beta)\,(1 - \pi_i(\beta))\right] X. \end{aligned} \tag{3.33}$$

Table 3.1 concerns a logistic regression analysis of the *transplant data*: among 223 organ transplant patients, 21 subsequently suffered a severe viral infection while 202 did not. The investigators wished to predict infection ("$y = 1$"). There were 12 predictor variables, including age, gender, and initial diagnosis, four milestone dates in the patient's treatment regimen, and four viral load measurements taken during the first weeks after transplant. In this study, the covariate matrix X was 223×12, with response vector $\boldsymbol{y}$ comprising 202 0s and 21 1s. The R call `glm(y~X,binomial)` gave the results in Table 3.1. Only the final viral load measurement (vl4) is seen to be a significant predictor, but that doesn't mean that the others aren't informative. The total residual deviance from the MLE fit was $\sum D(y_i, \pi_i(\hat{\beta})) = 54.465$, compared to $\sum D(y_i, \bar{y}) = 139.189$ for the model that always predicts the average response $\bar{y} = 21/220$. (More on GLM tests of overall significance later.)

The logistic regression estimate $\hat{\beta}$ gave a predicted probability of infection

$$\hat{\pi}_i = \pi_i\left(\hat{\beta}\right) = (1 + e^{-x_i^\top \hat{\beta}})^{-1} \tag{3.34}$$

for each patient. The left panel of Figure 3.1 compares the predictions for the 202 patients who did not suffer an infection (solid blue histogram) with

[2] In R, logistic regression for (3.32)–(3.33) can be carried out by the command `glm(p~X, family=binomial, weight=n)`, where $\boldsymbol{p} = (p_1, \ldots, p_N)$ and $\boldsymbol{n} = (n_1, \ldots, n_N)$.

Table 3.1 *Output of logistic regression, transplant data. Null deviance 139.189 on 222 degrees of freedom; residual deviance 54.465 on 210 df. Variable vl4 was highly significant (***), that is, with p-value* < 0.005.

	Estimate	St. error	z-value	Pr(> $\|z\|$)
inter	−6.76	1.48	−4.57	0.00
age	−0.21	0.41	−0.52	0.60
gen	−0.61	0.42	−1.45	0.15
diag	0.57	0.41	1.40	0.16
donor	−0.68	0.46	−1.48	0.14
start	−0.07	0.61	−0.12	0.91
date	0.41	0.49	0.83	0.41
datf	0.12	0.62	0.19	0.85
datl	−1.26	0.66	−1.89	0.06
vl1	0.07	0.49	0.15	0.88
vl2	−0.71	0.47	−1.49	0.13
vl3	0.21	0.47	0.44	0.66
vl4	5.30	1.48	3.58	0.00 ***

the 21 who did (line histogram). There seems to be considerable predictive power: the rule "predict infection if $\hat{\pi}_i$ exceeds 0.2" makes only 5% errors of the first kind (predicting an infection that doesn't happen) with 90% power (predicting infections that do happen).

This is likely to be optimistic since the MLE rule was fit to the data it is trying to predict. A cross-validation analysis split the 223 patients into 11 groups of 20 each (three of the patients were excluded). Each group was omitted in turn and a logistic regression fit to the reduced set of 200, then predictions $\tilde{\pi}_i$ made for the omitted patients, based on the reduced MLE. This gave more realistic prediction estimates, with 8% errors of the first kind and 67% power.

Homework 3.6 Repeat this analysis removing vl4 from the list of covariates. Comment on your findings.

Standard Errors

Suppose $\zeta = h(\beta)$ is a real-valued function of β, having gradient $\dot{h}(\beta) = (\cdots \partial h(\beta)/\partial\beta_j \cdots)^\top$. Then the approximate standard error assigned to the MLE $\hat{\zeta} = h(\hat{\beta})$ is

$$\operatorname{se}\left(\hat{\zeta}\right) \doteq \left\{\dot{h}\left(\hat{\beta}\right)^\top i_{\hat{\beta}}^{-1}\dot{h}\left(\hat{\beta}\right)\right\}^{1/2}.$$

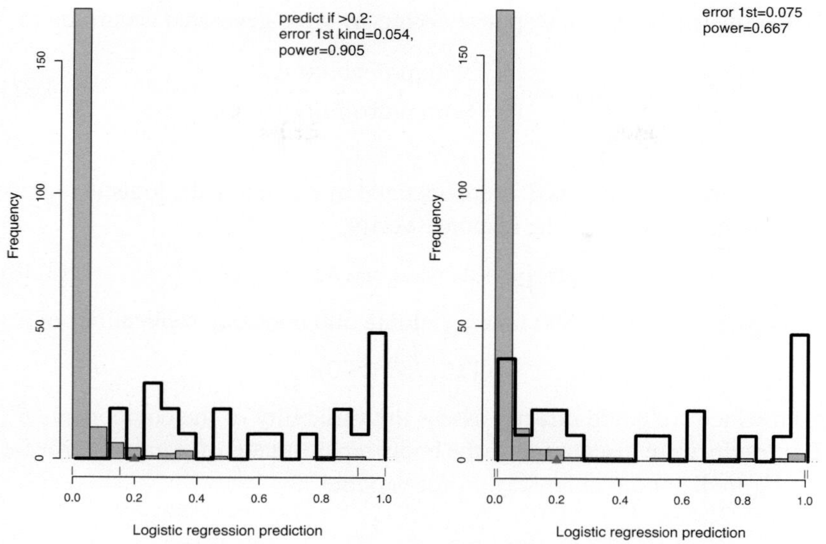

Figure 3.1 Predicted probabilities of infection, logistic regression. Line histogram represents infected, solid histogram represents not infected.

In particular, suppose we wish to estimate the probability of a 1 at a given covariate point x_0,

$$\pi_0 = \Pr\{y = 1 \mid x_0\} = (1 + e^{-x_0^\top \beta})^{-1}.$$

Then $\dot{h}(\beta) = x_0\pi_0(1 - \pi_0)$, and

$$\text{se}(\hat{\pi}_0) \doteq \hat{\pi}_0(1 - \hat{\pi}_0)\left\{x_0^\top i_{\hat{\beta}}^{-1} x_0\right\}^{1/2}, \qquad \hat{\pi}_0 = (1 + e^{-x_0^\top \hat{\beta}})^{-1}. \tag{3.35}$$

The standard error estimates in Table 3.1 are based, as usual, on the square root of the diagonal elements of $i_{\hat{\beta}}^{-1}$ (3.13). It is an unfortunate fact that these can be undependable, yielding overly optimistic assessments of accuracy. As a check, a parametric bootstrap error analysis was run on the transplant data:

1. Estimates of the probabilities π_i (3.27) were obtained using the estimate $\hat{\beta}$ from the logistic regression `glm(y~X, binomial)`,

$$\hat{\pi}_i = (1 + e^{-x_i^\top \hat{\beta}})^{-1}, \qquad i = 1, \ldots, 223. \tag{3.36}$$

2. Parametric bootstrap response vectors $\boldsymbol{y}^*$ were generated according to

$$y_i^* = \begin{cases} 1 & \text{with probability } \hat{\pi}_i \\ 0 & \text{with probability } 1 - \hat{\pi}_i, \end{cases} \tag{3.37}$$

for $i = 1, \ldots, 223$.

3. Bootstrap replications $\hat{\beta}^*$ were obtained by rerunning the logistic regression, now with $\boldsymbol{y}^*$ as the response vector,

$$\texttt{glm(}\boldsymbol{y}\texttt{*}\sim\texttt{X, binomial)} \longrightarrow \hat{\beta}^*. \tag{3.38}$$

Steps 1–3 were run 500 times, yielding 500 bootstrap replications of $\hat{\beta}$,

$$\hat{\beta}^*(1), \ldots, \hat{\beta}^*(500),$$

from which we could directly assess the variability of the components $\hat{\beta}_j^*$. The results were discouraging: the bootstrap estimates of standard error for the 12 predictor variables, say $\hat{\sigma}_j$ for variable j,

$$\hat{\sigma}_j = \left[\sum_{i=1}^{500} \frac{\left(\hat{\beta}_j^*(i) - \hat{\beta}_j^*(\cdot)\right)^2}{500} \right]^{1/2} \qquad \left(\hat{\beta}_j^*(\cdot) = \sum_{i=1}^{500} \frac{\hat{\beta}_j^*(i)}{500} \right), \tag{3.39}$$

were much bigger than those shown in column 2 of Table 3.1, some of them by substantial factors.

The line histogram in Figure 3.2 graphs the 500 bootstrap replications for the coefficient of the twelfth variable, "vl4". Even truncating the long right tail still leaves $\hat{\sigma}_{12}$ exceeding 4, compared to 1.48 in Table 3.1. In this case the `glm` estimation algorithm has turned out to be unstable as far as standard error estimation is concerned. Perhaps it is the combination of moderate sample size, low proportion of 1s, and a dozen correlated predictors at fault, but in any case some *regularization* of the MLE procedure is called for.

Penalized maximum likelihood (pMLE) regularizes maximum likelihood estimation by pulling estimates $\hat{\beta}$ toward zero: instead of maximizing the log likelihood $l_\beta(\boldsymbol{y})$ we select $\hat{\beta}$ to maximize[3]

$$l_\beta(\boldsymbol{y}) - \text{pen}(\beta), \tag{3.40}$$

where $\text{pen}(\beta)$ is a penalty term that grows larger with the size of β's coordinates. The R algorithm `glmnet` uses the *lasso* penalty

$$\text{pen}(\beta) = \lambda \frac{1}{2} \sum_{j=1}^{J} |\beta_j| \qquad (J = 12), \tag{3.41}$$

[3] See Section 4.7 for more on penalized maximum likelihood.

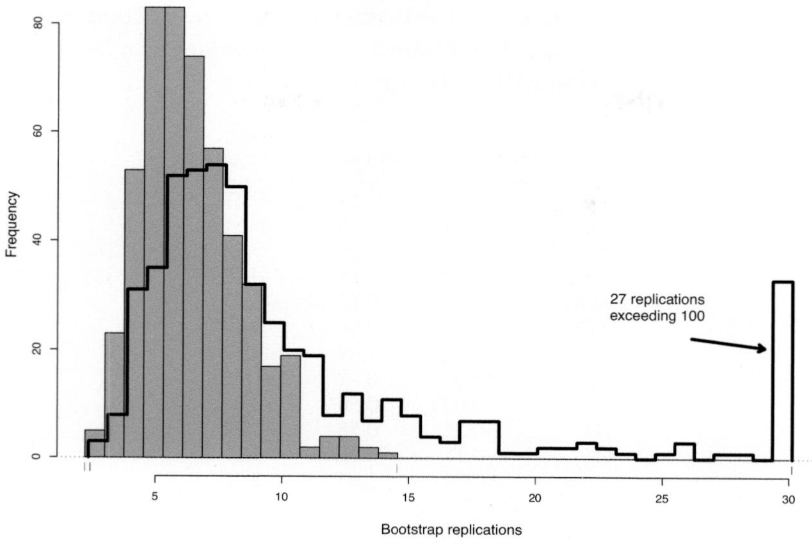

Figure 3.2 Parametric bootstrap replications of coefficient of "vl4", transplant data; `glmnet` (solid histogram), `glm` (line histogram).

for some choice of $\lambda \geq 0$. After a preliminary look, the value $\lambda = 0.001$ was selected.

The same 500 bootstrap data vectors $\boldsymbol{y}^*$ as before gave `glmnet` replications $\hat{\beta}^*_{\texttt{glmnet}}$ from (3.40)–(3.41). The solid histogram in Figure 3.2 shows that the `glmnet` bootstraps for vl4 were much more stable than the unpenalized estimates.

Table 3.2 gives the `glmnet` estimates and their bootstrap standard deviations for all 12 of the covariates in Table 3.1. The latter are still larger than the nominal standard errors in Table 3.1. The usual standard error estimates (3.21) can be thought of as one-term Taylor series approximations to parametric bootstrap standard deviations. This wasn't good enough here, where the mapping from $\boldsymbol{y}$ to $\hat{\beta}$ turned out to be highly nonlinear.

None of this says that the original unpenalized logistic regression analysis was useless. In fact, `glm` and `glmnet` gave roughly the same values for the estimated regression coefficients $\hat{\beta}_j$, with the `glmnet` values shrunk about 15% of the way back to zero, as seen in Table 3.2. Figure 3.3 shows that the estimated probabilities $\hat{\pi}_i$ (3.3) were strikingly similar. (So bootstrapping from the `glmnet` $\hat{\beta}$ vector at step (3.36) instead of the `glm` $\hat{\beta}$ pro-

Table 3.2 *Transplant data* `glmnet` *estimates and parametric bootstrap standard deviations (Figure 3.1). Column 3 shows regularized* `glmnet` *estimates shrunken about 15% of the way to zero.*

	`glmnet` Estimate	Boot sd	`glm` Estimate
age	−0.20	0.51	−0.21
gen	−0.49	0.56	−0.61
diag	0.51	0.56	0.58
donor	−0.60	0.55	−0.68
start	0.00	0.82	−0.07
date	0.35	0.56	0.40
datf	0.00	0.69	0.12
datl	−1.02	0.79	−1.26
vl1	0.02	0.56	0.07
vl2	−0.55	0.58	−0.71
vl3	0.15	0.56	0.21
vl4	4.72	1.92	5.32

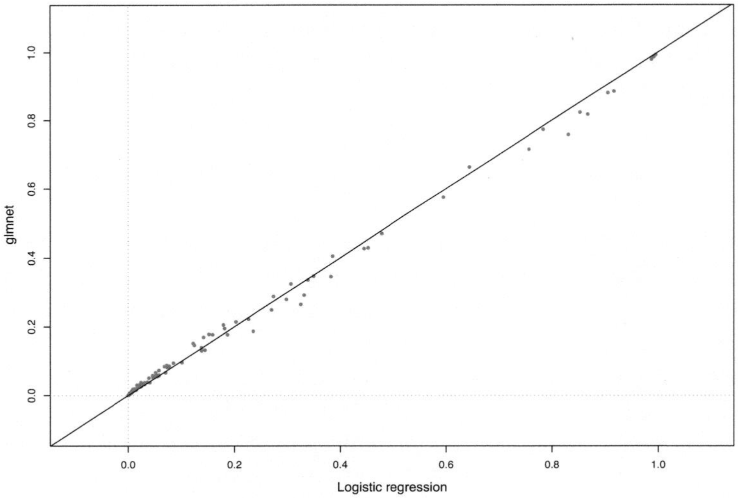

Figure 3.3 Estimates of $\Pr\{y_i = 1\}$ for 223 cases, transplant data: `glmnet` versus logistic regression.

duced almost the same results; Figure 3.1 also stayed about the same.) The main point is that a small amount of regularization can markedly decrease the variability of the $\hat{\beta}_j$ estimates, without incurring excessive bias. In any case, parametric bootstrapping can provide dependable accuracy estimates for logistic regression parameters.

In Table 3.1 the `glm` approximation $\text{Cov}(\hat{\beta}) \doteq i_{\hat{\beta}}^{-1}$ was misleading but that is certainly not always so. Table 3.3 refers to a second set of transplant data having $N = 769$ patients, 538 of whom suffered a different kind of post-operative infection. Now the bootstrap standard errors were nearly the same as the `glm` estimates, averaging only about 3% greater (except for the unexplained rogue result for "gh.symptoms3").

Table 3.3 *Second transplant study, $N = 769$ patients with cytomegalovirus, 538 with recurrence; "dr1,2,3" measures of donor/recipient match; "globuline.rx" if patient received drug treatment; 7 indicators of "graft-host" disease.*

	Estimate	glm.sterr	boot.stdev	boot.mean
treatment.year	−0.455	0.105	0.109	−0.466
donor.age	0.339	0.111	0.115	0.347
recipient.age	−0.367	0.112	0.109	−0.375
sex	0.110	0.093	0.097	0.113
race	−0.174	0.103	0.103	−0.180
dr1	−0.237	0.095	0.100	−0.245
dr2	−0.414	0.110	0.113	−0.427
dr3	−0.260	0.109	0.110	−0.266
globuline.rx	0.312	0.150	0.148	0.318
graft.host.rx1	−0.379	0.256	0.269	−0.396
graft.host.rx2	−0.862	0.213	0.228	−0.890
graft.host.rx3	−0.158	0.122	0.133	−0.159
gh.symptoms1	0.195	0.124	0.124	0.200
gh.symptoms2	0.354	0.126	0.133	0.372
gh.symptoms3	2.491	76.945	0.039	2.493
gh.symptoms4	−0.261	0.129	0.132	−0.267

Multicategory Logistic Model

The logistic regression model (3.27)–(3.29) concerns dichotomies, where each observation falls into one of two categories, e.g., whether or not the patient suffered a serious viral infection in the transplant study. "Logistic regression" also refers to multicategory response problems, where each observed case can fall into one of L distinct categories, $L \geq 2$. For instance, the drink preference data in Table 2.2, Section 2.9, has $L = 4$ categories: beer, wine, spirits, and soda.

The observed data now comprises independent multinomial count samples

$$s_i \overset{\text{ind}}{\sim} \text{Mult}_L(n_i, \pi_i), \qquad i = 1, \dots, N, \tag{3.42}$$

with s_{il} the count for category l in the ith sample ($s_{23} = 21$ in Table 2.2), and $\pi_i = (\pi_{i1}, \dots, \pi_{iL})$ an unobserved vector of true probabilities. The multicategory logistic regression model involves L p-dimensional parameter vectors ($L = 4$ for the drink preference data)

$$\gamma_1, \dots, \gamma_L \in \mathcal{R}^p \tag{3.43}$$

and N p-dimensional observed covariate vectors

$$x_1, \dots, x_N \in \mathcal{R}^p. \tag{3.44}$$

The true probabilities π_{il} are assumed to take the "softmax" form

$$\pi_{il} = \frac{e^{\gamma_l^\top x_i}}{\sum_{j=1}^L e^{\gamma_j^\top x_i}}. \tag{3.45}$$

Because the probabilities π_{il} sum to 1 for every i, we need specify only the first $L - 1$ category responses. Define, for $l = 1, \dots, L - 1$,

$$\beta_l = \gamma_l - \gamma_L \quad \text{and} \quad z_l = \sum_{i=1}^N x_i s_{il}, \tag{3.46}$$

with $\boldsymbol{\beta}$ the $p \times (L - 1)$ matrix $(\beta_1, \dots, \beta_{L-1})$. It then turns out that (3.42)–(3.45) is a $p \times (L - 1)$-dimensional exponential family,[4] with density

$$\begin{aligned} g_{\boldsymbol{\beta}}(\boldsymbol{s}) &= \exp\left\{\sum_{l=1}^{L-1} \beta_l^\top z_l - \psi(\boldsymbol{\beta})\right\} g_0(\boldsymbol{s}), \\ \psi(\boldsymbol{\beta}) &= \sum_{i=1}^N n_i \left[\log\left(1 + \sum_{l=1}^{L-1} \exp\{\beta_l^\top x_i\}\right)\right], \end{aligned} \tag{3.47}$$

$\boldsymbol{s} = (s_1, \dots, s_N)$.

Homework 3.7 (a) Verify (3.47). (b) What is $g_0(\boldsymbol{s})$? (c) Show that the $L = 2$ case reduces to our previous dichotomous logistic regression model.

[4] The `glmnet` function includes "`family=multinomial`" as a call for multicategory logistic regression models.

Conditional Logistic Regression

Suppose that the independent Bernoulli observations (3.27) are obtained in pairs (y_{i1}, y_{i2}), where y_{i1} is from a treatment of interest and y_{i2} from a control, with logits (3.28) following the model

$$\eta_{ij} = \alpha_i + \beta^\top x_{ij} \qquad (i = 1, \dots, I, j = 1, 2), \tag{3.48}$$

with the first component of x_{ij} indicating Treatment or Control. Here x_{ij} is a p-vector of covariates (including the Treatment/Control indicator), β an unknown p-dimensional parameter vector, and α_i an unknown intercept for pair i. Model (3.48) is often used in "case-control" studies, where ideally the covariates x_{i1} and x_{i2} are matched except for the Treatment/Control indicator. It could be analyzed as a standard logistic regression but the I intercept parameters, out of just $2I$ observations, destabilize estimation.

A conditioning trick gets rid of the α_i. Consider a pair (y_{i1}, y_{i2}) where $S_i = y_{i1} + y_{i2}$ equals 1. It is easy to see that

$$\begin{aligned} \Pr\{(y_{i1}, y_{i2}) = (1, 0) \mid S_i = 1\} &= \frac{e^{\beta^\top x_{i1}}}{e^{\beta^\top x_{i1}} + e^{\beta^\top x_{i2}}}, \\ \Pr\{(y_{i1}, y_{i2}) = (0, 1) \mid S_i = 1\} &= \frac{e^{\beta^\top x_{i2}}}{e^{\beta^\top x_{i1}} + e^{\beta^\top x_{i2}}}. \end{aligned} \tag{3.49}$$

Conditioning inference on those pairs where $S_i = 1$, say $i \in \mathcal{I}_1$, gives a p-parameter exponential family having natural parameter vector β and sufficient statistic $z_+ = \sum_{\mathcal{I}_1} z_i$, where

$$z_i = \begin{cases} x_{i1} & \text{if } (y_{i1}, y_{i2}) = (1, 0) \\ x_{i2} & \text{if } (y_{i1}, y_{i2}) = (0, 1). \end{cases} \tag{3.50}$$

Probit Analysis and Other Linkages

The roots of logistic regression lie in *bioassay*: to establish the toxicity of a new drug, groups of n_i mice each are exposed to an increasing sequence of doses d_i, $i = 1, \dots, K$, and the proportion p_i of deaths observed. (A customary goal is to estimate "LD50", the dose yielding 50% lethality.) The *probit model* is

$$\pi_i = \Phi(\beta_0 + \beta_1 d_i), \qquad i = 1, \dots, K,$$

where Φ is the standard normal CDF; maximum likelihood is used to solve for (β_0, β_1). Another way to say this is that each mouse has individual tolerance t for the drug, with $t \sim \mathcal{N}(-b_0/b_1, 1/b_1)$ for the population, and that dose d_i kills all mice with $t < d_i$.

Homework 3.8 Show that replacing $\Phi(x)$ above with the logistic CDF $\Lambda(x) = (1 + e^{-x})^{-1}$ reduces the bioassay problem to logistic regression.

The key idea of GLMs is to linearly model the natural parameters η_i. Since $\mu_i = \dot{\psi}(\eta_i)$, this is equivalent to linearly modeling $\dot{\psi}^{-1}(\mu_i)$. Other "links" appear in the literature. Probit analysis amounts to linearly modeling $\Phi^{-1}(\mu_i)$, sometimes called the *probit link*. But only the GLM "canonical link" allows one to make full use of exponential family theory.

Logistic regression GLMs pop up in a variety of statistical settings. An early example, mentioned in Section 2.5, is the Bradley–Terry model for paired comparisons (Bradley and Terry, 1952). Suppose we wish to rank college football teams on the basis of their losses and wins. Let y_{ij} denote the outcome if team i played team j,

$$y_{ij} = \begin{cases} 1 & \text{if team } i \text{ won} \\ 0 & \text{if team } i \text{ lost,} \end{cases}$$

with the assumption of independence among games. Each team k is assumed to have a parameter θ_k such that the probability that team i beats team j is

$$\pi_{ij} = \Pr\{y_{ij} = 1\} = \frac{e^{\theta_i}}{e^{\theta_i} + e^{\theta_j}} = \frac{1}{1 + e^{-(\theta_i - \theta_j)}}, \tag{3.51}$$

this being a logistic regression model having

$$\eta_{ij} = \log \frac{\pi_{ij}}{1 - \pi_{ij}} = \theta_i - \theta_j. \tag{3.52}$$

(One of the θ_k values is set to zero to avoid overparameterization.) The Bradley–Terry paper predated logistic regression theory, but of course it provided a correct likelihood analysis, though one more difficult to carry out then than now.

As stated before, logistic regression can claim to be the most conspicuous representation of parametric statistical theory in the world of machine learning. *Deep learning*, the spectacularly successful prediction program, uses a version of multicategory logistic regression, though with the X_i in (3.45) replaced by (very) complicated functions of the observed covariates.

3.3 Poisson Regression

The second most familar of the GLMs – and for general purposes sometimes the most useful – is Poisson regression. We observe independent

Poisson variables

$$y_i \overset{\text{ind}}{\sim} \text{Poi}(\mu_i), \qquad \text{for } i = 1, \dots, N, \tag{3.53}$$

compactly written $\boldsymbol{y} \sim \text{Poi}(\boldsymbol{\mu})$. A Poisson GLM is a linear model for the natural parameters $\eta_i = \log \mu_i$,

$$\eta_i = a_i + x_i^\top \beta, \qquad i = 1, \dots, N, \tag{3.54}$$

where β is an unknown p-dimensional parameter vector, x_i a known p-dimensional covariate vector, and a_i a known scalar "offset". (Offsets are necessary if the counts y_i are obtained under varying conditions. For example, if we are trying to analyze murder counts y_i in several cities, a_i might be the log population of city i. An offset is different from an intercept, which is an estimated parameter rather than a known constant.)

The MLE equation (3.14) is

$$X^\top(\boldsymbol{y} - e^{\boldsymbol{a}+X\hat{\beta}}) = \boldsymbol{0}, \tag{3.55}$$

the exponential notation indicating the vector with components $e^{a_i + x_i^\top \hat{\beta}}$. Since the variance V_i equals $\mu_i = e^{\eta_i}$ for Poisson variates, the asymptotic approximation (3.15) is

$$\hat{\beta} \mathrel{\dot\sim} \mathcal{N}_p\left\{\beta, \left[X^\top \operatorname{diag}(e^{a_i + x_i^\top \beta})X\right]^{-1}\right\}. \tag{3.56}$$

In practice, $\hat{\beta}$ is substituted for β on the right to obtain estimated standard errors for the coefficients $\hat{\beta}_j$–which, as in the logistic regression case, have to be interpreted cautiously.

Homework 3.9 The table here records the number of auto accidents in eight small Iowa towns during one year. The insurance company wonders whether teenage drivers are bad risks. Use Poisson regression to settle the question. (*Hint*: Offsets.)

Town	Population (1000s)	Proportion of teen drivers	# Accidents y
1	5.03	0.12	9
2	1.31	0.04	0
3	5.38	0.02	1
4	5.53	0.07	13
5	10.33	0.04	10
6	8.21	0.07	14
7	8.21	0.16	32
8	3.31	0.03	4

Table 3.4 *Counts for truncated sample of 487 galaxies, binned by magnitude 1:18 (increasingly dim) and redshift 1:15 (increasingly far).*

	1	2	3	4	5	6	7	8	9	10	11	12	13	14	15
18	1	6	6	3	1	4	6	8	8	20	10	7	16	9	4
17	3	2	3	4	0	5	7	6	6	7	5	7	6	8	5
16	3	2	3	3	3	2	9	9	6	3	5	4	5	2	1
15	1	1	4	3	4	3	2	3	8	9	4	3	4	1	1
14	1	3	2	3	3	4	5	7	6	7	3	4	0	0	1
13	3	2	4	5	3	6	4	3	2	2	5	1	0	0	0
12	2	0	2	4	5	4	2	3	3	0	1	2	0	0	1
11	4	1	1	4	7	3	3	1	2	0	1	1	0	0	0
10	1	0	0	2	2	2	1	2	0	0	0	1	2	0	0
9	1	1	0	2	2	2	0	0	0	0	1	0	0	0	0
8	1	0	0	0	1	1	0	0	0	0	1	1	0	0	0
7	0	1	0	1	1	0	0	0	0	0	0	0	0	0	0
6	0	0	3	1	1	0	0	0	0	0	0	0	0	0	0
5	0	3	1	1	0	0	0	0	0	0	0	0	0	0	0
4	0	0	1	1	1	0	0	0	0	0	0	0	0	0	0
3	0	1	0	0	0	0	0	0	0	0	0	0	0	0	0
2	0	1	0	0	0	0	0	0	0	0	0	0	0	0	0
1	0	1	0	0	0	0	0	0	0	0	0	0	0	0	0

The Galaxy Data

Table 3.4 shows counts of galaxies from a survey of a small portion of the sky: 487 galaxies have had their apparent magnitudes m and (log) redshifts r measured. Apparent brightness is a *decreasing* function of magnitude – stars of the 2nd magnitude are less bright than those of the first, etc. – while distance from Earth is an increasing function of r.

As in most astronomical studies, the galaxy data is *truncated*, with very dim galaxies lying below the threshold of detection. In this study, attention was restricted to the intervals $17.2 \leq m \leq 21.5$ and $1.22 \leq r \leq 3.32$. The range of m has been divided into 18 equal intervals, and likewise 15 equal intervals for r. Table 3.4 gives the counts y_{ij} of the 487 galaxies in the $N = 270 = 18 \times 15$ bins. The left panel of Figure 3.4 shows a perspective picture of the counts.

We can imagine Table 3.4 as the lower left corner of the much larger table we would see if the data were *not* truncated. We might then fit a bivariate normal density to the data. It seems awkward and difficult to fit part of a bivariate normal density to truncated data, but Poisson regression offers an easy solution.

We begin with the reasonable assumption that the counts are independent Poisson observations,

$$y_{ij} \overset{\text{ind}}{\sim} \text{Poi}(\mu_{ij}), \quad i = 1, \ldots, 18 \quad \text{and} \quad j = 1, \ldots, 5, \tag{3.57}$$

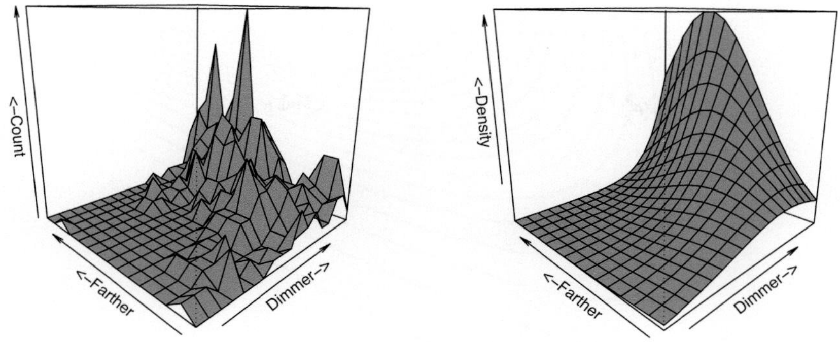

Figure 3.4 *Left*: Galaxy data, binned counts. *Right*: Poisson GLM density estimate.

particularly credible if the total count 487 was itself a Poisson variate. Let $\boldsymbol{m}$ be the 270-vector listing the m_{ij} values, say in order $(18, 17, \dots, 1)$ repeated 15 times, and similarly $\boldsymbol{r}$ for the 270 r_{ij} values. This defines the 270×15 structure matrix X,

$$X = (\boldsymbol{m}, \boldsymbol{r}, \boldsymbol{m}^2, \boldsymbol{mr}, \boldsymbol{r}^2). \tag{3.58}$$

where $\boldsymbol{m}^2$ is the 270-vector with components m_{ij}^2, etc.

Letting $\boldsymbol{y}$ denote the 270-vector of counts, the `glm` call in R

```
glm(y~X, poisson)
```

produces an estimate of the best-fit truncated normal density. We can see the estimated contours of the fitted density in Figure 3.5. The estimated density itself is shown in the right panel of Figure 3.4.

Homework 3.10 Why does this choice of X for the Poisson regression produce an estimate of a truncated bivariate normal density?

Homework 3.11 (a) Reproduce the Poisson fit. (b) Calculate the Poisson deviance residuals (1.64). Can you detect any hints of poor fit? (c) How might you supplement the model we used to improve the fit?

The Poisson variance V equals the mean μ. In some applications it may become apparent that the responses have variances exceeding their means. One solution is to replace Poisson regression with a *negative binomial* GLM, where the ratio $R = V/\mu$ can be chosen to be any desired value greater than 1. In terms of the negative binomial notation in Section 1.6, $R = 1/\theta$.

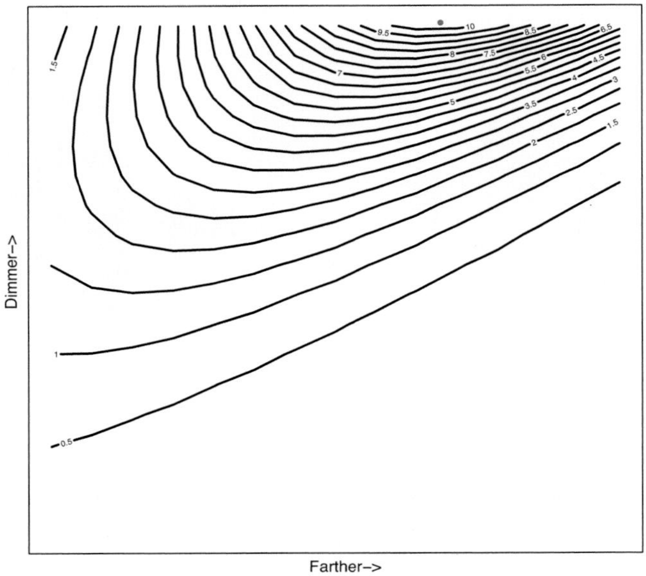

Figure 3.5 Contour curves for Poisson GLM density estimates, galaxy data; dot shows point of maximum density. Contours describe lower left corner of a bivariate normal density.

3.4 Lindsey's Method

For the *prostate data* of Figure 1.5, Section 1.7, a smooth curve $\hat{g}(z)$ was fit to the histogram of the $N = 6033$ z values, this being the crucial step for the empirical Bayes estimate (1.58). *Lindsey's method* is an ingenious algorithm that uses Poisson GLMs to short-circuit apparently intricate maximum likelihood calculations.

We wish to estimate a univariate density function $g(z)$, and begin with an assumption that the true density $g(z)$ is a member of a p-parameter exponential family,

$$g_\beta(z) = e^{\beta^\top t(z) - \psi(\beta)} g_0(z), \tag{3.59}$$

where β and $t(z)$ are in $\mathcal{R}^p$. In Figure 1.5 of Section 1.4, $t(z)$ is a fifth-degree polynomial

$$t(z) = (z, z^2, z^3, z^4, z^5). \tag{3.60}$$

(Choosing a second-degree polynomial, with $g_0(z)$ flat amounts to fitting a normal density; going up to degree 5 permits us to accommodate non-normal tail behavior.)

How can we find the MLE $\hat{\beta}$ in family (3.59)? There is no closed form for $\psi(\beta)$ or $\mu = \dot{\psi}(\beta)$ except in a few special cases such as the normal and gamma families. This is where Lindsey's method comes in. As a first step we partition the sample space $\mathcal{Z}$ (an interval of $\mathcal{R}^1$) into K subintervals $\mathcal{Z}_k$,

$$\mathcal{Z} = \bigcup_{k=1}^{K} \mathcal{Z}_k,$$

with $\mathcal{Z}_k$ having length Δ_k and centerpoint x_k. For simplicity we will take $\Delta_k = \Delta$ for all k, and $g_0(z) = 1$ in what follows.

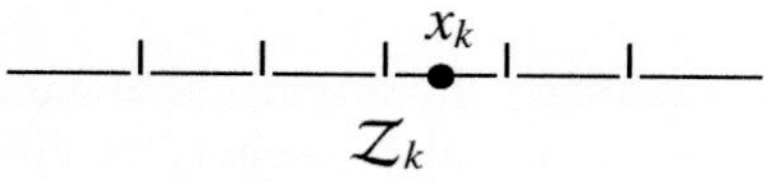

Define $\boldsymbol{\pi}(\beta) = (\pi_1(\beta), \dots, \pi_K(\beta))$ as the vector having kth component

$$\begin{aligned}\pi_k(\beta) = \Pr_\beta\{z \in \mathcal{Z}_k\} &= \int_{\mathcal{Z}_k} g_\beta(z)\, dz \\ &\doteq \Delta e^{\beta^\top t_k - \psi(\beta)} \qquad (3.61)\\ &= \frac{e^{\beta^\top t_k}}{\sum_1^K e^{\beta^\top t_j}},\end{aligned}$$

where $t_k = t(x_k)$, the last equality following from $\sum \pi_k(\beta) = 1$. Also let $\boldsymbol{y} = (y_1, \dots, y_k)$ be the vector of counts $y_k = \#\{z_i \in \mathcal{Z}_k\}$. If the z_is are independent observations from $g_\beta(z)$ then $\boldsymbol{y}$ will be a multinomial sample of size N, Section 2.9,

$$\boldsymbol{y} \sim \text{Mult}_K(N, \boldsymbol{\pi}(\beta)). \qquad (3.62)$$

For small values of Δ, the multinomial MLE will be nearly the same as the actual $\hat{\beta}$, but it doesn't seem any easier to find. Poisson regression and the Poisson trick come to the rescue.

Define

$$\mu_k(\beta_0, \beta) = e^{\beta_0 + \beta^\top t_k}, \qquad (3.63)$$

where β_0 is a free parameter, and let $\mu_+(\beta_0, \beta) = \sum_k \mu_k(\beta_0, \beta)$. Then

$$\frac{\mu_k(\beta_0, \beta)}{\mu_+(\beta_0, \beta)} = \pi_k(\beta).$$

We can now invoke the Poisson trick and use standard GLM software to find the Poisson MLE $(\hat{\beta}_0, \hat{\beta})$ in model (3.63),

$$\boldsymbol{y} \sim \text{Poi}(\boldsymbol{\mu}(\beta_0, \beta))$$

(the scale factor e^{β_0} in (3.62) allows $\mu_+(\beta_0, \beta)$ to match the sample size N). Since $\log \mu_k(\beta_0, \beta) = \beta_0 + \beta^\top t_k$, this is a Poisson GLM and is solvable directly in R.

Homework 3.12 (a) Show that $\hat{\beta}$ is the MLE in the multinomial model above. What does $e^{\hat{\beta}_0}$ equal? (b) How is Lindsey's method applied if the Δ_k are unequal or $g_0(z)$ is not constant?

The discrete family $\boldsymbol{\pi}(\beta)$, with components

$$\pi_k(\beta) = \frac{e^{\beta^\top t_k}}{\sum_{j=1}^K e^{\beta^\top t_j}},$$

approximates the original continuous family (3.59). It is a log linear subfamily of the full K-category multinomial family, as mentioned in the paragraph following Homework 2.33 in Section 2.9. Let $\boldsymbol{T}$ be the $K \times p$ matrix that has $t_k^\top$ as its kth row. Then the subfamily has $\boldsymbol{\eta} = \log(\boldsymbol{\pi})$ equaling a linear function of β, $\boldsymbol{\eta}(\beta) = \boldsymbol{T}\beta$.

3.5 Analysis of Deviance

ANOVA, the analysis of variance, was Fisher's "computer package" for the analysis of multi-factor experiments, assuming normal distributions for the response variables. The analysis of deviance extends ANOVA to the GLM situation where, as in logistic regression, the responses follow a given one-parameter exponential family $\mathcal{G}$.

We begin with observations

$$y_i \overset{\text{ind}}{\sim} g_{\mu_i}(\cdot), \qquad \text{for } i = 1, \dots, N, \tag{3.64}$$

as at (3.3) but now where it will be convenient to label the densities with μ rather than η. A GLM model

$$\underset{N\times 1}{\boldsymbol{\eta}} = \underset{N\times p}{X} \; \underset{p\times 1}{\beta} \tag{3.65}$$

is given. This is a "big" model, that is, p is large, and we wonder if a smaller submodel might give an adequate fit to the data. In the analysis of deviance, model fit is measured by *total deviance*, an analogue of ANOVA's total residual sum of squares. For a proposed estimate $\hat{\boldsymbol{\mu}} = (\hat{\mu}_1, \dots, \hat{\mu}_N)$ of the true expectation vector $\boldsymbol{\mu} = (\mu_1, \dots, \mu_N)$, the total deviance from the observed data $\boldsymbol{y} = (y_1, \dots, y_N)$ is defined to be

$$D_+(\boldsymbol{y}, \hat{\boldsymbol{\mu}}) = \sum_{i=1}^N D(y_i, \hat{\mu}_i). \tag{3.66}$$

Homework 3.13 Verify Hoeffding's formula,

$$\frac{g_y(y)}{g_{\hat{\mu}}(y)} = \prod_{i=1}^{N} \frac{g_{y_i}(y_i)}{g_{\hat{\mu}_i}(y_i)} = e^{D_+(y,\hat{\mu})/2}. \tag{3.67}$$

Suppose that β is partitioned into $\beta = (\beta^{(1)}, \beta^{(2)})$ with lengths $p^{(1)}$ and $p^{(2)}$, $p^{(1)} + p^{(2)} = p$, and likewise

$$\underset{N\times p}{X} = \left(\underset{N\times p^{(1)}}{X^{(1)}}, \underset{N\times p^{(2)}}{X^{(2)}}\right),$$

so

$$\boldsymbol{\eta} = X^{(1)}\beta^{(1)} + X^{(2)}\beta^{(2)}.$$

We wonder if the smaller model having $\beta^{(2)} = 0$ is adequate. In Table 3.1 of Section 3.2 for example, we might try eliminating vl1, vl2, and vl3, leaving $p^{(1)} = 9$ predictors for the model $\boldsymbol{\eta} = X^{(1)}\beta^{(1)}$.

The bigger model $\boldsymbol{\eta} = X\beta$ defines a p-dimensional curved manifold $\mathcal{M}$ of possible expectation vectors $\boldsymbol{\mu}$, while the smaller model gives a $p^{(1)}$-dimensional submanifold $\mathcal{M}^{(1)}$:

$$\mathcal{M} = \left\{\boldsymbol{\mu} = \dot{\psi}(X\beta)\right\} \quad \text{and} \quad \mathcal{M}^{(1)} = \left\{\boldsymbol{\mu} = \dot{\psi}\left(X^{(1)}\beta^{(1)}\right)\right\}. \tag{3.68}$$

(The notation $\dot{\psi}(X\beta)$ indicates $\dot{\psi}(\boldsymbol{\eta})$ evaluated at $\boldsymbol{\eta} = X\beta$, and likewise for $\dot{\psi}(X^{(1)}\beta^{(1)})$.) $\mathcal{M}$ and $\mathcal{M}^{(1)}$ are subsets of the full unconstrained expectation space $\boldsymbol{\mu} = \dot{\psi}(\boldsymbol{\eta})$, $\boldsymbol{\eta} \in A^N \subseteq \mathcal{R}^N$.

The figure at right shows the big model MLE $\hat{\boldsymbol{\mu}} = \dot{\psi}(X\hat{\beta})$ as the projection of $\boldsymbol{y}$ into $\mathcal{M}$ – projected orthogonally to the columns of X – as in the figure that follows (3.17) in Section 3.1. Likewise the MLE for the smaller model $\hat{\boldsymbol{\mu}}^{(1)} = \dot{\psi}(X^{(1)}\hat{\beta}^{(1)})$ is the projection of $\boldsymbol{y}$ into $\mathcal{M}^{(1)}$, orthogonal to the columns of $X^{(1)}$.

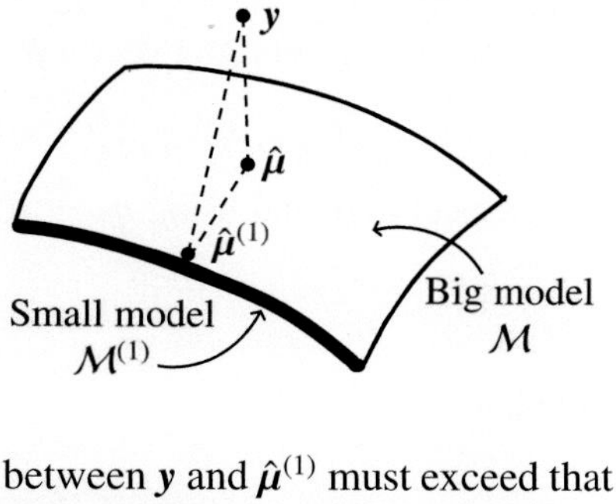

In any reasonable sense the distance between $\boldsymbol{y}$ and $\hat{\boldsymbol{\mu}}^{(1)}$ must exceed that between $\boldsymbol{y}$ and $\hat{\boldsymbol{\mu}}$. Testing the null hypothesis that the smaller model is as good as the larger amounts to testing the size of the excess distance.

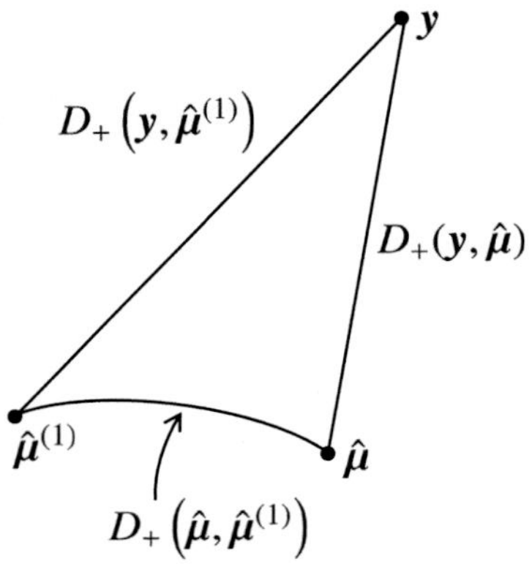

The analysis of deviance uses total deviance (3.66) as distance measure, or more precisely as distance^2. The three points $\boldsymbol{y}$, $\hat{\boldsymbol{\mu}}$, and $\hat{\boldsymbol{\mu}}^{(1)}$ form a "triangle", quotes reflecting the curvature of the side from $\hat{\boldsymbol{\mu}}^{(1)}$ to $\hat{\boldsymbol{\mu}}$. The total deviances of two of the sides are computed directly from definition (3.66). For ordinary linear regression, where $D_+(\boldsymbol{y}, \boldsymbol{\mu}) = \|\boldsymbol{y} - \boldsymbol{\mu}\|^2$, the third side would have

$$D_+(\hat{\boldsymbol{\mu}}, \hat{\boldsymbol{\mu}}^{(1)}) = D_+(\boldsymbol{y} - \hat{\boldsymbol{\mu}}^{(1)}) - D_+(\boldsymbol{y} - \hat{\boldsymbol{\mu}}), \tag{3.69}$$

according to Pythagoras. The direct definition of total deviance between $\hat{\boldsymbol{\mu}}$ and $\hat{\boldsymbol{\mu}}^{(1)}$ is, as in Section 1.8,

$$D_+(\hat{\boldsymbol{\mu}}, \hat{\boldsymbol{\mu}}^{(1)}) = 2 \sum_{i=1}^{N} E_{\hat{\mu}_i} \left\{ \log \frac{g_{\hat{\mu}_i}(y_i)}{g_{\hat{\mu}_i^{(1)}}(y_i)} \right\}. \tag{3.70}$$

The *deviance additivity theorem*, which follows next, says that the two definitions are the same – (3.69) equals (3.70) – and in fact can be stated more generally:

Theorem 3.1 (G. Simon[5]) *For any point $\boldsymbol{\mu}_\beta$ in $\mathcal{M}$ we have*

$$D_+(\boldsymbol{y}, \boldsymbol{\mu}_\beta) - D_+(\boldsymbol{y}, \hat{\boldsymbol{\mu}}) = D_+(\hat{\boldsymbol{\mu}}, \boldsymbol{\mu}_\beta).$$

(So the theorem does not require $\boldsymbol{\mu}_\beta = \hat{\boldsymbol{\mu}}^{(1)}$.)

Proof The vector $\boldsymbol{\mu}_\beta = (\cdots \mu_{\beta_i} \cdots)^\top$ equals $\dot{\psi}(\boldsymbol{\eta}_\beta)$, where $\boldsymbol{\eta}_\beta = X\beta$ for some choice of β. For any vector $\boldsymbol{\mu} = (\cdots \mu_i \cdots)^\top$, differentiating

$$D(\mu_i, \mu_{\beta i}) = 2\left[(\eta_i - \eta_{\beta i})\mu_i - \psi(\eta_i) + \psi(\eta_{\beta i})\right]$$

with respect to β, and using $\eta_{\beta i} = x_i^\top \beta$, gives

$$\frac{\partial}{\partial \beta} D(\mu_i, \mu_{\beta i}) = 2x_i(\mu_{\beta i} - \mu_i),$$

and then

$$\frac{\partial}{\partial \beta} D_+(\boldsymbol{\mu}, \boldsymbol{\mu}_\beta) = 2X^\top(\boldsymbol{\mu}_\beta - \boldsymbol{\mu}). \tag{3.71}$$

[5] See Efron (1978, Sect. 4)

Choosing $\boldsymbol{\mu} = \boldsymbol{y}$ and then $\boldsymbol{\mu} = \hat{\boldsymbol{\mu}}$ in (3.71) shows that

$$\frac{\partial}{\partial \beta}\left\{D_+(\boldsymbol{y}, \boldsymbol{\mu}_\beta) - D_+(\hat{\boldsymbol{\mu}}, \boldsymbol{\mu}_\beta)\right\} = 2\boldsymbol{X}^\top(\hat{\boldsymbol{\mu}} - \boldsymbol{y}) = \boldsymbol{0},$$

the last equality following from the maximum likelihood equation for $\hat{\boldsymbol{\mu}}$, so that $D_+(\boldsymbol{y}, \boldsymbol{\mu}_\beta) - D_+(\hat{\boldsymbol{\mu}}, \boldsymbol{\mu}_\beta)$ is a constant. The choice of $\beta = \hat{\beta}$ shows that the constant is $D_+(\boldsymbol{y}, \hat{\boldsymbol{\mu}})$, giving $D_+(\hat{\boldsymbol{\mu}}, \boldsymbol{\mu}_\beta) = D_+(\boldsymbol{y}, \boldsymbol{\mu}_\beta) - D_+(\boldsymbol{y}, \hat{\boldsymbol{\mu}})$ as desired. ■

Homework 3.14 Verify (3.71).

Testing $H_0 : \beta^{(2)} = 0$

If H_0 is true then (3.70) and Wilks' theorem say that

$$D_+(\hat{\boldsymbol{\mu}}, \hat{\boldsymbol{\mu}}^{(1)}) = 2 \log \frac{g_{\hat{\mu}}(\boldsymbol{y})}{g_{\hat{\mu}^{(1)}}(\boldsymbol{y})} \dot{\sim} \chi^2_{p^{(2)}}, \tag{3.72}$$

so we reject H_0 if $D_+(\hat{\boldsymbol{\mu}}, \hat{\boldsymbol{\mu}}^{(1)})$ exceeds $\chi^{2(\alpha)}_{p^{(2)}}$ for, say, $\alpha = 0.95$. (More accurate approximations are available, as in the corollary in Section 1.8.) Since

$$D_+(\hat{\boldsymbol{\mu}}, \hat{\boldsymbol{\mu}}^{(1)}) = \sum_{i=1}^{N} D(\hat{\mu}_i, \hat{\mu}_i^{(1)}),$$

if $D_+(\hat{\boldsymbol{\mu}}, \hat{\boldsymbol{\mu}}^{(1)})$ is significantly too large we can examine the individual components to see if any one observation is causing a bad fit to the smaller model.

For the drinking preference example in Table 2.2, Section 2.9, we can take the big model estimate $\hat{\boldsymbol{\mu}}$ to be the observed 4×5 table of observed counts on the left, and the smaller model estimate $\hat{\boldsymbol{\mu}}^{(1)}$ the 4×5 independence table on the right. The owner wished to test the null hypothesis of independence between age and drink preference. We need to compute $D_+(\hat{\boldsymbol{\mu}}, \hat{\boldsymbol{\mu}}^{(1)})$ in order to evaluate the test statistic (3.72). The Poisson trick makes this easy: we assume that N, the total number surveyed, was itself Poisson, in which case the entries y_{ij} are independently Poisson (μ_{ij}).

Direct calculation of (3.72) gives Poisson deviance

$$D_+(\hat{\boldsymbol{\mu}}, \hat{\boldsymbol{\mu}}^{(1)}) = 11.7; \tag{3.73}$$

the approximate p-value

$$\Pr\{\chi^2_{12} \geq 11.7\} = 0.477$$

is almost the same as the value 0.497 from Pearson's χ^2 test. There is no reason to reject the null hypothesis of independence. None of the Poisson deviance residuals $R_{ij} = \text{sign}(\hat{\mu}_{ij} - \hat{\mu}_{ij}^{(1)}) D(\hat{\mu}_{ij}, \hat{\mu}_{ij}^{(1)})^{1/2}$ exceeded 1.5 in absolute value, and there was no pattern of unusual values visible in the table.

We may have more than two models to consider. Suppose now β and X are partitioned into J parts,

$$\underset{p\times 1}{\beta} = \left(\beta^{(1)}_{p^{(1)}}, \ldots, \beta^{(J)}_{p^{(J)}}\right) \quad \text{and} \quad \underset{N\times p}{X} = \left(\underset{N\times p^{(1)}}{X^{(1)}}, \ldots, \underset{N\times p^{(J)}}{X^{(J)}}\right). \tag{3.74}$$

Let $\hat{\beta}(j)$ be the MLE for β assuming that $\beta^{(j+1)} = \beta^{(j+2)} = \cdots = \beta^{(J)} = 0$. An analysis of deviance table is obtained by differencing the successive maximized log likelihoods $l_{\hat{\beta}(j)}(\boldsymbol{y}) = \sum_{i=1}^{N} \log g_{\mu_i(\hat{\beta}(j))}(y_i)$ as diagrammed in Table 3.5. There are two common testing strategies for the jth coefficient vector $\beta^{(j)}$:

$$\text{Compare } D_+(\hat{\boldsymbol{\mu}}^{(j+1)}, \hat{\boldsymbol{\mu}}^{(j)}) \text{ with } \chi^2_{p^{(j+1)}}. \tag{3.75a}$$

$$\text{Compare } D_+(\hat{\boldsymbol{\mu}}^{(J)}, \hat{\boldsymbol{\mu}}^{(j)}) \text{ with } \chi^2_{p(j)}, \tag{3.75b}$$

where $p(j) = \sum_{k=j+1}^{J} p^{(k)}$. Rejection in (3.75a) implies that the jth factor is significant, while acceptance in (3.75b) implies that the model ignoring all factors past the jth one is adequate. (It is common to adjoin a "zero-th column" of all ones to X, in which case $\hat{\beta}(0)$ is taken to be the value making $\hat{\boldsymbol{\mu}}(0)$ a vector with all entries $\bar{y}$, this being the smallest model in Table 3.5.)

Table 3.5 *Analysis of deviance table.*

MLE		Twice max log like		Difference	Compare with
$\hat{\beta}(0) = \bar{y}$	$\longrightarrow$	$2l_{\hat{\beta}(0)}$			
			$\searrow$		
			$\nearrow$	$D_+(\hat{\mu}(1), \hat{\mu}(0))$	$\chi^2_{p(1)}$
$\hat{\beta}(1)$	$\longrightarrow$	$2l_{\hat{\beta}(1)}$			
			$\searrow$		
			$\nearrow$	$D_+(\hat{\mu}(2), \hat{\mu}(1))$	$\chi^2_{p(2)}$
$\hat{\beta}(2)$	$\longrightarrow$	$2l_{\hat{\beta}(2)}$			
			$\searrow$		
$\vdots$		$\vdots$		$\vdots$	$\vdots$
			$\nearrow$	$D_+(\hat{\mu}(J), \hat{\mu}(J-1))$	$\chi^2_{p(J)}$
$\hat{\beta}(J) = \hat{\beta}$	$\longrightarrow$	$2l_{\hat{\beta}(J)}$			

Empirical Bayes analysis of the prostate cancer data in Figure 1.5, Section 1.7, depended on an estimate $\hat{g}(z)$ fit to the histogram of the 6033 z-values. The fitting method depended on Poisson regression, Lindsey's method, and the analysis of deviance:

- The data was binned as in Figure 1.5, the bins each of width 0.2 and with centers x_i equally spaced,

$$\boldsymbol{x} = (-4.4, -4.2, \ldots, 5.0, 5.2).$$

- The counts y_i of z-values in bin i were calculated,

$$\boldsymbol{y} = (y_1, \ldots, y_{49}).$$

- A Poisson regression model with $\boldsymbol{\mu}^{(j)}$ being a polynomial of degree j in the x values was fit for for $j = 2, \ldots, 8$,

```
glm(y~poly(x,j), poisson),
```

  giving estimates $\hat{\boldsymbol{\mu}}^{(j)}$, for $j = 2, \ldots, 8$.
- Residual total deviances $D_+(\boldsymbol{y}, \hat{\boldsymbol{\mu}}^{(j)})$ were calculated for $j = 2, \ldots, 8$. These are shown in Table 3.6.

Table 3.6 *Residual deviance and AIC for prostate data fits* `glm(y~poly(x,df), poisson)`.

df	2	3	4	5	6	7	8
Dev	139	137	65.3	64.3	63.8	63.8	59.6
AIC	143	143	73.3	74.3	75.8	77.8	75.6

Because the models are successively bigger, the residual deviance $D_+(\boldsymbol{y}, \hat{\boldsymbol{\mu}}^{(j)})$ must decrease with increasing j. It cannot be that bigger models are always better, they just appear so. Akaike's information criterion (AIC) suggests a penalty for increased model size,

$$\text{AIC}^{(j)} = D_+^{(j)} + 2j,$$

a nearly unbiased estimate of the true expected log likelihood for model j; see Efron (1986b), Remark R. We see that $j = 4$ minimizes AIC for the prostate data.

Homework 3.15 (a) Construct the deviance table and give the significance levels for the chi-square tests. (b) Construct the analogous table using natural splines instead of polynomials,

```
glm(y~ns(x,j), poisson).
```

Table 3.7 *Survival times in days from two arms of NCOG study; + signs indicate censored observations, known only to exceed the given number.*

Arm A: chemotherapy				Arm B: chemotherapy+radiation			
7	133	185+	440	37	159	519	1771+
34	139	218	523	84	169+	528+	1776
42	140	225	523+	92	173	547+	1897+
63	140	241	583	94	179	613+	2023+
64	146	248	594	110	194	633	2146+
74+	149	273	1101	112	195	725	2297+
83	154	277	1116+	119	209	759+	
84	157	279+	1146	127	249	817	
91	160	297	1226+	130	281	1092+	
108	160	319+	1349+	133	319	1245+	
112	165	405	1412+	140	339	1331+	
129	173	417	1417	146	432	1557	
133	176	420		155	469	1642+	

3.6 Survival Analysis

The roots of survival analysis go back hundreds of years to insurance and actuarial science. Imported into statistics, it flourished in the post-war period, becoming particularly prominent in the 1980s when the "War on Cancer and HIV" studies made it crucial. Exponential families played only a minor role in this development but, as this and the next section show, they have something incisive to say both about the theory and its applications. We begin with an example and a brief review of survival methods.

A randomized clinical trial conducted by the Northern California Oncology Group (NCOG) compared two treatments for head and neck cancer: chemotherapy (Arm A of the trial, $n = 51$ patients) and chemotherapy plus radiation (Arm B, $n = 45$ patients). The results are reported in Table 3.7 in terms of the survival time in number of days past treatment. The numbers followed by + indicate patients still alive on their final day of observation. For example, the sixth patient in Arm A was alive on day 74 after his treatment, and then "lost to follow-up"; we only know that his survival time *exceeded* 74 days.[6] This is an example of *censored data*, an endemic prob-

[6] He received his treatment 74 days before the experiment ended and was alive on its last day; most of the +s in Table 3.7 are of this type.

lem in clinical trials. A powerful methodology for the analysis of censored data was developed between 1955 and 1975. Here we will discuss only a bit of the theory; a more expansive survey appears in Chapter 9 of Efron and Hastie (2016).

Hazard Rates

Survival analysis theory requires stating probability distributions in terms of hazard rates rather then densities. Suppose X is a positive discrete random variable, with probability density

$$f_i = \Pr\{X = i\}, \qquad \text{for } i = 1, 2, \ldots, \tag{3.76}$$

and *survival function* (the probability of exceeding time i)

$$S_i = \Pr\{X \geq i\} = \sum_{j \geq i} f_j. \tag{3.77}$$

Then h_i, the *hazard rate* at time i, is

$$h_i = f_i / S_i = \Pr\{X = i \mid X \geq i\}. \tag{3.78}$$

In words, h_i is the probability of dying at time i after having survived up until time i. Some algebra shows that

$$S_i = \prod_{j=1}^{i-1} (1 - h_j). \tag{3.79}$$

Homework 3.16 Prove (3.79) and give an intuitive explanation.

Life Tables

Table 3.8 presents the Arm A data in *life table* form. Now the time unit is months rather than days. Three statistics are given for each month:

- n_i = number of patients under observation at the beginning of month i;
- y_i = number of patients observed to die during month i;
- l_i = number of patients lost to follow-up at the end of month i.

So, for instance, $n_{10} = 19$ patients were under observation ("at risk") at the beginning of their 10th month after treatment, $y_{10} = 2$ died, $l_{10} = 1$ was lost to follow-up, leaving $n_{11} = 16$ at risk for month 11. Patients can be lost to follow-up for various reasons – moving away, dropping out of the study,

Table 3.8 *Arm A of NCOG study, binned by month: n = # at risk at beginning of month, y = # deaths, l = lost to follow-up,* $\hat{h}$ *= hazard rate y/n;* $\widehat{S}$ *= life table survival estimate, i.e., estimated probability of surviving past month 1, 2, etc. Note that time is measured separately for each patient as months since initial treatment.*

Month	n	y	l	$\hat{h}$	$\widehat{S}$	Month	n	y	l	$\hat{h}$	$\widehat{S}$
1	51	1	0	0.020	0.980	25	7	0	0	0.000	0.184
2	50	2	0	0.040	0.941	26	7	0	0	0.000	0.184
3	48	5	1	0.104	0.843	27	7	0	0	0.000	0.184
4	42	2	0	0.048	0.803	28	7	0	0	0.000	0.184
5	40	8	0	0.200	0.642	29	7	0	0	0.000	0.184
6	32	7	0	0.219	0.502	30	7	0	0	0.000	0.184
7	25	0	1	0.000	0.502	31	7	0	0	0.000	0.184
8	24	3	0	0.125	0.439	32	7	0	0	0.000	0.184
9	21	2	0	0.095	0.397	33	7	0	0	0.000	0.184
10	19	2	1	0.105	0.355	34	7	0	0	0.000	0.184
11	16	0	1	0.000	0.355	35	7	0	0	0.000	0.184
12	15	0	0	0.000	0.355	36	7	0	0	0.000	0.184
13	15	0	0	0.000	0.355	37	7	1	1	0.143	0.158
14	15	3	0	0.200	0.284	38	5	1	0	0.200	0.126
15	12	1	0	0.083	0.261	39	4	0	0	0.000	0.126
16	11	0	0	0.000	0.261	40	4	0	0	0.000	0.126
17	11	0	0	0.000	0.261	41	4	0	1	0.000	0.126
18	11	1	1	0.091	0.237	42	3	0	0	0.000	0.126
19	9	0	0	0.000	0.237	43	3	0	0	0.000	0.126
20	9	2	0	0.222	0.184	44	3	0	0	0.000	0.126
21	7	0	0	0.000	0.184	45	3	0	1	0.000	0.126
22	7	0	0	0.000	0.184	46	2	0	0	0.000	0.126
23	7	0	0	0.000	0.184	47	2	1	1	0.500	0.063
24	7	0	0	0.000	0.184						

etc. – but most often because they entered the study late and were still alive when it closed.[7]

The key assumption of survival analysis is that, given n_i, the number of deaths y_i is binomial with probability of death the hazard rate h_i,

$$y_i \mid n_i \overset{\text{ind}}{\sim} \text{Bi}(n_i, h_i). \tag{3.80}$$

Time in a survival analysis – the "month" column in Table 3.8 – is measured from each patient's entry into the study. Model (3.80) makes a powerful assumption: everything that happened before time i is summarized by

7 The computational convention is that losses occur at the *end* of their period, after deaths.

n_i, at least as far as the distribution of y_i is concerned, and is independent of all preceding events.

Homework 3.17 Suppose patients can sense if the end is near, and drop out of the study just before they die. How would this affect model (3.80)?

The unbiased hazard rate estimate based on (3.80) is

$$\hat{h}_i = y_i/n_i; \tag{3.81}$$

(3.79) then gives the survival estimate

$$\widehat{S}_i = \prod_{j=1}^{i-1} \left(1 - \hat{h}_j\right) \tag{3.82}$$

(so $\widehat{S}_1 = 0.980$ is the estimated probability of *not* dying in the first month following treatment).

Figure 3.6 compares the estimated survival curves for the two arms of the NCOG study. The more aggressive treatment seems better: the one-year survival rate estimate for Arm B is about 50%, compared with 35% for Arm A.

Estimated survival curves are customarily called *Kaplan–Meier curves* in the literature. (Formally speaking, the name applies to estimates (3.82) where the time unit, months in our example, is decreased to zero.) Suppose the observed death times are

$$t_{(1)} < t_{(2)} < t_{(3)} < \cdots < t_{(m)} \tag{3.83}$$

(assuming no ties). Then the Kaplan–Meier curve $\widehat{S}(t)$ is flat between death times, with downward jumps at the observed $t_{(i)}$ values.

Homework 3.18 What is the downward jump at $t_{(i)}$?

The binomial model (3.80) leads, after considerable work, to *Greenwood's formula*, an approximate standard error for $\widehat{S}_i$,

$$\text{sd}\left\{\widehat{S}_i\right\} \doteq \widehat{S}_i \left(\sum_{j\le i} \frac{y_j}{n_j(n_j - y_j)}\right)^{1/2}.$$

The vertical bars in Figure 3.6 indicate $\pm 1.96\,\text{sd}_i$, approximate 95% confidence limits for S_i. There is overlap between the bars for the two curves; at no one time point can we say that Arm B is significantly better than Arm A (though more sophisticated two-sample tests do in fact show B's superiority).

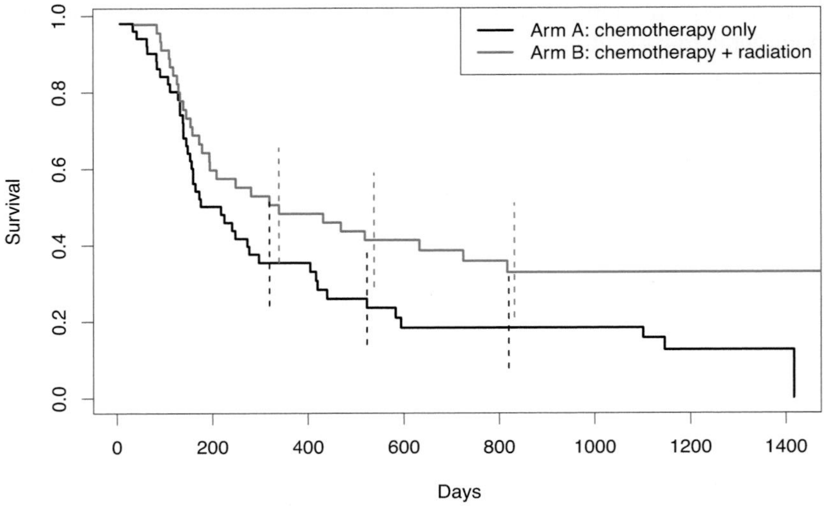

Figure 3.6 NCOG estimated survival curves; lower is Arm A (chemotherapy only); upper is Arm B (chemotherapy+radiation). Vertical lines indicate approximate 95% confidence intervals.

Parametric Survival Analysis

Life table survival curves are nonparametric in the sense that the true hazard rates h_i are not assumed to follow any particular model. A parametric approach can greatly improve the estimation accuracy of the curves. In particular, we can use a logistic GLM, letting η_i be the logistic transform of h_i,

$$\eta_i = \log \frac{h_i}{1 - h_i}, \tag{3.84}$$

and assuming that $\boldsymbol{\eta} = (\eta_1, \ldots, \eta_N)^\top$ satisfies

$$\boldsymbol{\eta} = X\beta, \tag{3.85}$$

as in Section 3.1 and Section 3.2.

Consider the Arm A data of Table 3.8, which provided $N = 47$ binomial observations $y_i \overset{\text{ind}}{\sim} \text{Bi}(n_i, h_i)$ (3.81). For the analysis in Figure 3.7, we took X in (3.85) to be the 47×4 matrix having ith row

$$x_i = \left[1, i, (i - 11)_-^2, (i - 11)_-^3,\right]^\top, \tag{3.86}$$

where $(i - 11)_-$ equals $i - 11$ for $i \leq 11$ and 0 for $i > 11$. Then $\boldsymbol{\eta} = X\beta$

describes a cubic-linear spline with the knot at 11. This choice allows for more detailed modeling of the early months, when there is the most data and the greatest variation in response, as well as allowing stable estimation in the low-data right tail.

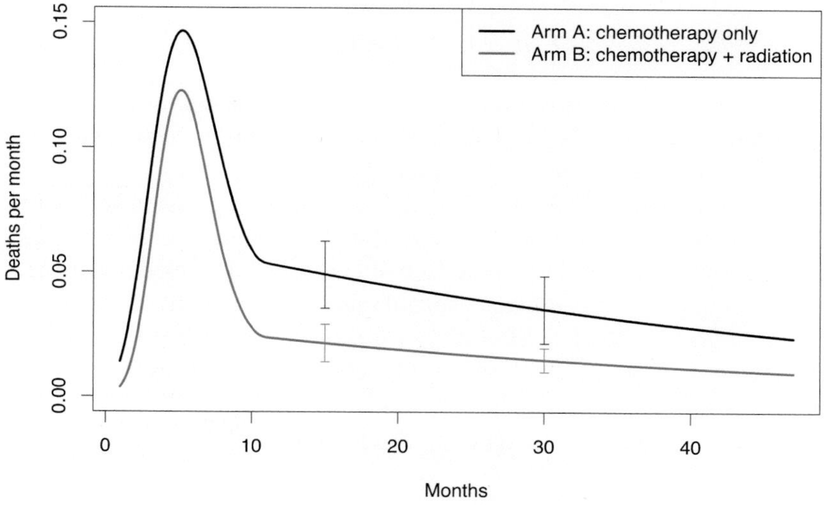

Figure 3.7 Parametric hazard rate estimates for NCOG study. Arm A (black curve) has about 2.5 times higher hazard than Arm B (red curve) for all times more than a year after treatment. Standard errors shown at 15 and 30 months.

Homework 3.19 Repeat the Arm A parametric calculations in Figure 3.7, including the estimated standard errors.

The comparison of estimated hazard rate in Figure 3.7 is more informative than the survival curve comparison of Figure 3.6. Both arms show a peak in hazard rates at five months, a swift decline, and then a long slow decline after one year, reflecting to some extent the spline model (3.86). The Arm B hazard is always below that of Arm A by a factor of about 2.5.

3.7 The Proportional Hazards Model

Looking at Table 3.8, the NCOG data, one can suppose that the patients in Arms A and B, each represented by their survival times, differ from each other in important ways: age, gender, weight, stage, etc. Cox (1972) proposed an elegant method for incorporating covariate information into the regression analysis of censored data. As before, we will begin with a condensed version of the theory and an application, before getting to the exponential family connection.

The data for each subject is now a triple[8]

$$(T_i, d_i, x_i), \qquad i = 1, \ldots, N, \tag{3.87}$$

where T_i is a non-negative observed lifetime, x_i is a p-vector of observed covariates, and d_i equals 1 or 0 as subject i was or was not observed to die (so $d_i = 0$ is equivalent to the + in Table 3.8).

If there were no censoring we could do a standard regression analysis of the lifetimes T_i as a function of the covariates x_i. The *proportional hazards model* allows us to proceed in the face of censoring. The model is usually presented in terms of *continuous* hazard rates, rather than the discrete definition (3.76)–(3.78). The lifetime T_i for subject i is assumed to follow a continuous density $f_i(t)$, with survival function $S_i(t)$ and hazard rate $h_i(t)$,

$$S_i(t) = \int_t^\infty f_i(s)\, ds \quad \text{and} \quad h_i(t) = \frac{f_i(t)}{S_i(t)}.$$

Homework 3.20 Show that $S_i(t) = e^{-H_i(t)}$, where

$$H_i(t) = \int_0^t h_i(s)\, ds.$$

How does this relate to formula (3.79)?

The proportional hazards model assumes that each $h_i(t)$ is proportional to a "baseline hazard rate" $h_0(t)$ multiplied by a factor that depends on the covariate vector x_i,

$$h_i(t) = h_0(t) e^{x_i^\top \beta}. \tag{3.88}$$

Here β is an unknown $p \times 1$ parameter vector. It will turn out that $h_0(t)$ does not need to be specified in a proportional hazards analysis, leaving only the regression coefficient vector β to be estimated.

Homework 3.21 Denoting $e^{x_i^\top \beta} = \alpha_i$, show that $S_i(t) = S_0(t)^{\alpha_i}$ (a relationship known as "Lehmann alternatives"), where $S_0(t)$ is the baseline survival function and $S_i(t)$ the survival function for subject i.

[8] Here we are using notation more standard in the survival analysis literature.

The key idea of proportional hazards analysis is to condition the occurrence of each observed event on the *risk set* of subjects under observation just before the event occurred. Let J be the total number of deaths observed; that is, cases with $d_i = 1$, say at times $t_{(1)} < t_{(2)} < \cdots < t_{(j)} < \cdots < t_{(J)}$ (assuming no ties for convenience). The *risk set* $\mathcal{R}_j$ for event j is

$$\mathcal{R}_j = \{\text{subjects under observation at time } t_{(j)}\}.$$

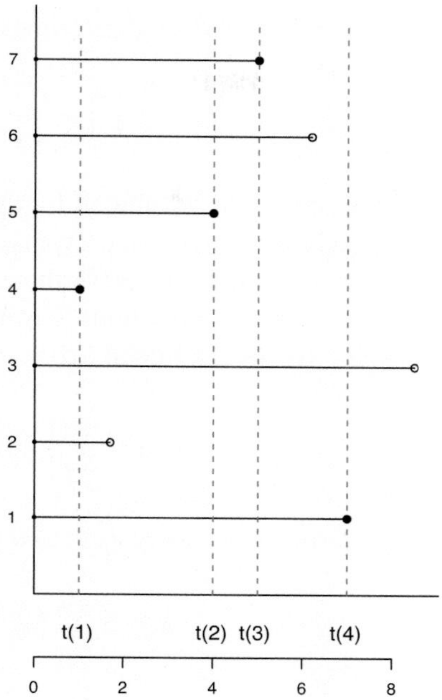

In the example at right there are $N = 7$ subjects, $J = 4$ of whom were observed to die and 3 who were lost to follow-up (the open circles). We also denote

$$i_j = \{\text{index of subject who died at time } t_{(j)}\},$$

$i_1 = 4$, $i_2 = 5$, $i_3 = 7$, $i_4 = 1$ in this case, with risk set $\mathcal{R}_3 = \{1, 3, 6, 7\}$, for example.

A simple but crucial result underlies the proportional hazards method.

Lemma 3.2 *Under the proportional hazards model, the probability that $i_j = i$, i.e., that the death occurred to member i of $\mathcal{R}_j$, is*

$$\pi_i(\beta \mid \mathcal{R}_j) = \frac{e^{x_i^\top \beta}}{\sum_{k \in \mathcal{R}_j} e^{x_k^\top \beta}}. \tag{3.89}$$

The lemma says that subjects with higher values of $x_i^\top \beta$ have greater conditional probabilities of death.

Homework 3.22 Verify Lemma 3.2.

Partial Likelihood

How can we estimate β in the proportional hazards model? Cox (1972, 1975) suggested using the *partial likelihood*, that is, the product of factors

(3.89) over the observed death times,

$$L(\beta) = \prod_{j=1}^{J} \frac{e^{x_{i_j}^\top \beta}}{\sum_{\mathcal{R}_j} e^{x_k^\top \beta}} = \prod_{j=1}^{J} \pi_{i_j}(\beta \mid \mathcal{R}_j) \tag{3.90}$$

as if it were the true likelihood for the unknown parameter β. It is "partial" because it ignores all the non-events, times when nothing happened or there were losses to follow-up. Nevertheless, it can be shown to be quite efficient under reasonable assumptions (Efron, 1977).

The log partial likelihood $l(\beta) = \log L(\beta)$ is

$$l(\beta) = \sum_{j=1}^{J} \left(x_{i_j}^\top \beta - \log \sum_{\mathcal{R}_j} e^{x_k^\top \beta} \right).$$

Taking derivatives with respect to β gives, in the notation of (3.89),

$$\begin{aligned} \underset{p\times 1}{\dot{l}(\beta)} &= \sum_{j=1}^{J} \left(x_{i_j} - E_j(\beta) \right), \\ \text{where} \quad E_j(\beta) &= \sum_{\mathcal{R}_j} x_i \pi_i(\beta \mid \mathcal{R}_j); \\ \text{and} \quad -\underset{p\times p}{\ddot{l}(\beta)} &= \sum_{j=1}^{J} V_j(\beta), \\ \text{where} \quad V_j(\beta) &= \sum_{\mathcal{R}_j} \pi_i(\beta \mid \mathcal{R}_j) \left(x_i - E_j(\beta) \right) \left(x_i - E_j(\beta) \right)^\top . \end{aligned} \tag{3.91}$$

Homework 3.23 (a) Verify (3.91). (b) Show that $l(\beta)$ is a concave function of β.

The partial likelihood estimate of β is defined by

$$\hat{\beta} : \dot{l}\left(\hat{\beta}\right) = \mathbf{0},$$

where $\mathbf{0}$ is a vector of p zeros, with approximate observed information matrix

$$\hat{I} = -\ddot{l}\left(\hat{\beta}\right). \tag{3.92}$$

Considerable theoretical effort has gone into verifying the asymptotic normal approximation

$$\hat{\beta} \dot{\sim} \mathcal{N}_p\left(\beta, \hat{I}^{-1}\right). \tag{3.93}$$

Table 3.9 shows a small portion of the data from the *pediatric abandonment study*. Over a 12-year period, $n = 1620$ children were treated for cancer and other serious diseases at a medical facility in a developing country. The investigators wished to identify the factors impacting abandonment. The survival variable was time, the number of days from entrance to last observation. The response variable was abandonment, playing the role of death in the NCOG study, with further observation ceasing at abandonment. Only one-tenth of the cases were abandoned, which was a good thing for the children but meant that nine-tenths of the data was censored.

Six possible explanatory variables are listed in Table 3.9:

- sex: female = 1; male = 2;
- race: Ladino = 1; Indigina = 2;
- diag: diagnosis leukemia = 1; lymphoma = 2; other = 3;
- age: at admission, in years;
- enter: entry date in days since 01/01/2000;
- far: distance in kilometers from child's home to the medical facility.

Standardized versions of the explanatory variables were used in the analysis, for instance,

$$\text{Age} = (\text{age} - \text{mean(age)}) / \text{sd(age)},$$

and similarly for Sex, Race, Diag, Enter, Far.

An excellent proportional hazards program `coxph` is available in the R package `survival`. Setting

$$S = \texttt{Surv(time,d)},$$

where $d = 1$ or 0 indicated whether or not the child was abandoned, the call

```
coxph(S~Sex+Race+Diag+Age+Enter+Far)
```

gave the results shown in Table 3.10. Sex, Race, and Diag are insignificant as predictors of abandonment. Age is mildly interesting, with two-sided p-value 0.021. The two dramatic predictors are Enter and Far: children entering the study later suffered less abandonment (as indicated by the negative regression coefficient $\hat{\beta}_{\text{Enter}} = -0.469$, which reduces the hazard rate in model (3.88)) while those living farther away had a greater hazard rate for abandonment.

Homework 3.24 Use `coxph` to test the null hypothesis that Arm B is no better than Arm A for the NCOG data listed at the beginning of Section 3.6;

Table 3.9 *40 randomly selected children, pediatric abandonment study: sex (female = 1, male = 2); race (Ladino = 1, Indigina = 2); diag (leuk = 1, lymph = 2, solid = 3); age (at admission); enter (days since 1/1/00); far (kms from home); time (from entry to last observation); d (abandonment observed = 1, not observed = 0).*

child	sex	race	diag	age	enter	far	time	*d*
199	2	1	1	16.8	2742	0	21	0
634	2	2	1	7.5	2949	201	43	0
278	1	1	1	17.2	2174	17	2177	0
1262	1	1	3	0.6	1325	0	3189	0
768	1	2	1	10.3	1347	154	3153	0
668	1	2	1	6.1	3252	165	58	0
844	2	1	2	11.4	1362	102	2662	0
1350	2	2	3	9.8	1158	0	85	0
1085	2	1	2	15.8	1477	120	0	0
305	1	1	1	3.9	1958	12	15	0
1534	1	1	3	1.8	1868	0	286	0
1428	2	1	3	16.2	1957	411	432	0
1266	2	1	3	0.7	2509	130	2067	0
861	2	1	2	6.5	1543	103	35	0
1375	2	2	3	4.7	2721	248	113	0
1669	1	1	3	12.8	3162	9	348	0
101	2	2	1	1.2	965	144	996	0
687	2	1	1	6.2	2911	0	1572	0
709	2	1	1	6.2	2235	284	350	0
1325	1	1	3	0.9	2334	12	2058	0
1733	2	2	3	9.2	1441	155	3	1
358	1	1	1	3.8	2726	283	1918	0
1419	2	2	3	13.1	1757	203	408	0
14	2	2	1	12.3	3076	130	1445	0
264	1	1	1	3.9	730	23	376	1
155	2	1	1	3.7	473	0	3501	0
210	2	1	1	2.4	655	0	3696	0
1064	2	2	2	12.9	2751	165	1	0
546	1	1	1	11.8	3064	240	1093	0
1336	2	2	3	17.9	2781	199	96	0
1025	2	1	2	10.3	920	240	1708	0
845	2	1	2	5.2	2364	85	1939	0
658	2	1	1	6.2	2824	175	1742	0
106	1	2	1	10.7	2008	231	2448	0
832	1	1	2	9.7	1427	134	2884	0
1153	2	1	3	14.7	1921	0	151	0
1269	1	2	3	0.4	3236	24	114	0
613	2	1	1	5.0	3127	154	1502	0
595	1	2	1	5.3	2995	109	1545	0
815	1	1	1	9.0	1271	207	0	1

data is in the file named `ncogdata`. *Hint*: The only explanatory variable is the Arm indicator.

Homework 3.25 In R, run these two calls and comment on your results.

- `coxph(S~Sex+Race+Diag+Age)`
- `coxph(S~Age+Enter+Far)`

Table 3.10 *Proportional hazards analysis, pediatric abandonment data. Six explanatory variables standardized for mean 0, variance 1. High significance noted as in Table 3.1.*

	Coef	St. err	z-value	p-value	Exp(coef)
Sex	−0.015	0.076	−0.196	0.845	0.985
Race	0.110	0.074	1.493	0.135	1.116
Diag	0.146	0.079	1.865	0.062	1.158
Age	−0.197	0.085	−2.315	0.021	0.821
Enter	−0.469	0.077	−6.116	0.000 ***	0.626
Far	0.279	0.069	4.013	0.000 ***	1.321

Proportional Hazards as an Exponential Family Model

The partial likelihood function (3.90) can be written as

$$L(\beta) = \prod_{j=1}^{J} e^{\beta^\top x_{i_j} - \psi_j(\beta)} = e^{\beta^\top \sum_j x_{i_j} - \psi_+(\beta)},$$
$$\text{where } \psi_+(\beta) = \sum_{j=1}^{J} \left(\log \sum_{\mathcal{R}_j} e^{\beta^\top x_k} \right). \tag{3.94}$$

This is the likelihood function of a p-parameter exponential family $f_\beta(\boldsymbol{y})$, having:

- natural parameter β;
- sufficient statistic $\boldsymbol{y} = \sum_{j=1}^{J} x_{i_j}$;
- CGF $\psi_+(\beta)$.

Homework 3.26 (a) Differentiate $\psi_+(\beta)$ to obtain the expectation vector and covariance matrix of $\boldsymbol{y}$.
(b) Apply the general theory of maximum likelihood estimation in (3.94) to get $\dot{l}(\beta)$ and $-\ddot{l}(\beta)$, agreeing with (3.91).

Let $X(j)$ denote the $L_j \times p$ matrix ($L_j = |\mathcal{R}_j|$) having the covariate vectors x_i in the risk set $\mathcal{R}_j$ as rows, and consider the model

$$\eta(j) = X(j)\beta. \tag{3.95}$$

As in Section 2.9, this defines a p-dimensional exponential family of probability vectors $\pi_\beta(j)$ on the L_j-dimensional simplex,

$$\pi_\beta(j) = \frac{e^{\eta(j)}}{\sum_{k=1}^{L_j} e^{\eta_k(j)}}; \tag{3.96}$$

(3.95)–(3.96) can be thought of as a multinomial GLM, constituting a p-parameter subexponential family of the full L_j-parameter unrestricted multinomial family. Partial likelihood analysis (3.90) amounts to considering J notionally independent such subfamilies.

The baseline hazard rate $h_0(t)$ has disappeared from the proportional hazards likelihood (3.94), another example – like Fisher's analysis of 2×2 tables – where conditioning has eliminated nuisance parameters. Nothing in the derivation of (3.94) says that the covariates x_i must stay the same in different risk sets $\mathcal{R}_j$. If x_i refers to a possibly time-varying measurement such as a patient's weight, it may be important to let it vary appropriately in (3.94), though `coxph` then needs modification.

3.8 Overdispersion and Quasi-likelihood

Applications of binomial or Poisson generalized linear models often encounter difficulties with *overdispersion*: after fitting the best GLM we can find, the residual errors are still too large by the standards of binomial or Poisson variability. *Quasi-likelihood* is a simple method for dealing with overdispersion while staying within the GLM framework. A more detailed technique, *double exponential families*, is developed in the next section.

As an example, Table 3.11 reports on the prevalence of toxoplasmosis, an endemic blood infection, in 34 cities of El Salvador (Efron, 1986a). The data consists of triplets (r_i, n_i, s_i), $i = 1, \dots, 34$, where:

- r_i = annual rainfall in city i;
- n_i = number of people sampled;
- s_i = number testing positive for toxoplasmosis.

Let $p_i = s_i/n_i$ be the observed proportion positive in city i. A cubic logistic regression of p_i on r_i was run,

```
glm(p~poly(r,3), binomial, weight=n),        (3.97)
```

with p, r, and n indicating their respective 34-vectors. Part of the output appears in Table 3.12. We see that the cubic regression coefficient 1.3787 is strongly positive, z-value 3.35, two-sided p-value less than 0.001.

The points (r_i, p_i) are shown in Figure 3.8, along with the fitted cubic regression curve. Each p_i is connected to its fitted value $\hat{\pi}_i$ by a dashed line. We will see that the points are too far from the curve according to the standard of binomial variability. This is what overdispersion looks like.

The middle two columns of Table 3.11 show the observed proportions p_i and the fitted values $\hat{\pi}_i$ from the cubic logistic regression. Two measures

Table 3.11 *Toxoplasmosis data: rainfall, # sampled, # positive in 34 cities in El Salvador;* $p = s/n$, $\hat{\pi}$ *fit from cubic logistic regression in rainfall;* R *binomial dev residual,* Rp *Pearson residual;* $\sum(Rp^2)/30 = 1.94$, *estimated overdispersion factor.*

City	r	n	s	p	$\hat{\pi}$	R	Rp
1	1735	4	2	0.500	0.539	−0.16	−0.16
2	1936	10	3	0.300	0.506	−1.32	−1.30
3	2000	5	1	0.200	0.461	−1.22	−1.17
4	1973	10	3	0.300	0.480	−1.16	−1.14
5	1750	2	2	1.000	0.549	1.55	1.28
6	1800	5	3	0.600	0.563	0.17	0.17
7	1750	8	2	0.250	0.549	−1.72	−1.70
8	2077	19	7	0.368	0.422	−0.47	−0.47
9	1920	6	3	0.500	0.517	−0.08	−0.08
10	1800	10	8	0.800	0.563	1.58	1.51
11	2050	24	7	0.292	0.432	−1.42	−1.39
12	1830	1	0	0.000	0.560	−1.28	−1.13
13	1650	30	15	0.500	0.421	0.87	0.88
14	2200	22	4	0.182	0.454	−2.69	−2.57
15	2000	1	0	0.000	0.461	−1.11	−0.92
16	1770	11	6	0.545	0.558	−0.08	−0.08
17	1920	1	0	0.000	0.517	−1.21	−1.03
18	1770	54	33	0.611	0.558	0.79	0.79
19	2240	9	4	0.444	0.506	−0.37	−0.37
20	1620	18	5	0.278	0.353	−0.68	−0.67
21	1756	12	2	0.167	0.552	−2.76	−2.69
22	1650	1	0	0.000	0.421	−1.04	−0.85
23	2250	11	8	0.727	0.523	1.39	1.36
24	1796	77	41	0.532	0.563	−0.54	−0.54
25	1890	51	24	0.471	0.536	−0.93	−0.93
26	1871	16	7	0.438	0.546	−0.87	−0.87
27	2063	82	46	0.561	0.427	2.44	2.46
28	2100	13	9	0.692	0.417	2.00	2.01
29	1918	43	23	0.535	0.518	0.22	0.22
30	1834	75	53	0.707	0.559	2.62	2.57
31	1780	13	8	0.615	0.561	0.40	0.40
32	1900	10	3	0.300	0.530	−1.47	−1.46
33	1976	6	1	0.167	0.477	−1.60	−1.52
34	2292	37	23	0.622	0.611	0.13	0.13

of discrepancy are shown in the last two columns: the binomial deviance

Table 3.12 *Toxoplasmosis data, cubic logistic regression* `glm(formula=p~poly(r,3), family=binomial, weight=n)`. *Null deviance 74.212 on 33 degrees of freedom; residual deviance 62.635 on 30 df. Overdispersion: deviance* 62.635/30 = 2.09; *Pearson 1.94. High significance noted as in Table 3.1.*

Coefficients	Estimate	St. error	z-value	Pr(> $\|z\|$)
(Intercept)	0.0243	0.0769	0.32	0.75240
`poly(r,3)1`	−0.0861	0.4587	−0.19	0.85117
`poly(r,3)2`	−0.1927	0.4674	−0.41	0.68014
`poly(r,3)3`	1.3787	0.4115	3.35	0.00081 ***

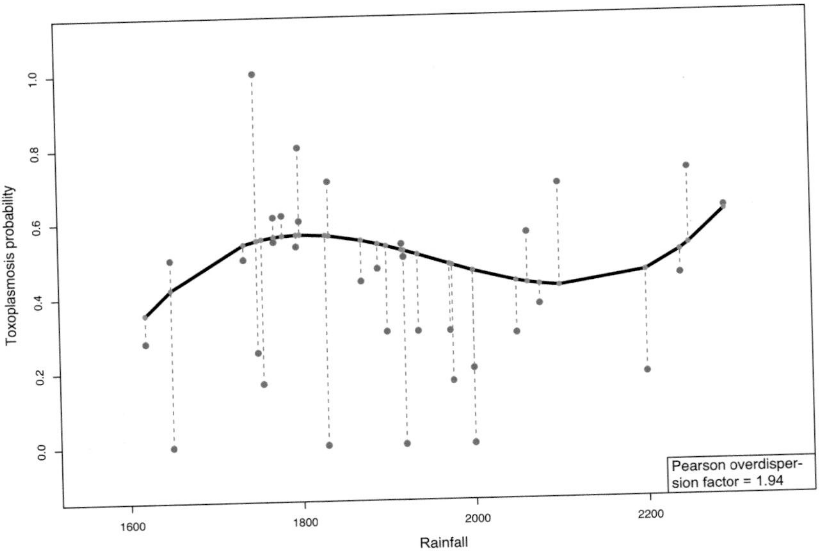

Figure 3.8 Observed proportions of toxoplasmosis, 34 cities in El Salvador; curve is cubic logistic regression.

residual

$$R_i = \text{sign}(p_i - \hat{\pi}_i)\sqrt{n_i D(p_i, \hat{\pi}_i)},$$

as in (1.64) and Homework 1.30, and the Pearson residual

$$Rp_i = \frac{\text{sign}(p_i - \hat{\pi}_i)}{\sqrt{\hat{\pi}_i(1 - \hat{\pi}_i)/n_i}},$$

where $\hat{\pi}_i = (1 + e^{-x_i^\top \hat{\beta}})^{-1}$.

In the absence of overdispersion, we would expect both

$$\sum_1^{34} R_i^2/30 \quad \text{and} \quad \sum_1^{34} Rp_i^2/30$$

to be close to 1 ($30 = 34 - 4$ is the added degrees of freedom in going from cubic regression to the model allowing a separate estimate π_i for each city). Instead we have

$$\sum_1^{34} R_i^2/30 = 2.09 \quad \text{and} \quad \sum_1^{34} Rp_i^2/30 = 1.94; \tag{3.98}$$

the points in Figure 3.8 are about $\sqrt{2}$ farther from the fitted curve than binomial variability suggests.

Homework 3.27 Compute the p-value for Wilks' likelihood ratio test of the null hypothesis that there is no overdispersion around the cubic regression curve, assuming $\sum_1^{34} R_i^2 \sim \sigma^2\chi^2_{30}$.

The toxoplasmosis study comprised 697 subjects. It was originally presented as a 697 by 3 matrix, such as that suggested by Table 3.13, with c_j the city residence for subject j, r_j the rainfall in that subject's city, and z_j either 1 or 0 if the test for toxoplasmosis was positive or not.

Homework 3.28 (a) Create a full version of the 697×3 matrix outlined in Table 3.13; your table should have the correct number of subjects with $z_j = 1$ in each city.

(b) Construct an analysis of deviance table as in (3.4), including these three models:

1. $z \sim 1$ (i.e., only a single constant fit to all cases);
2. $z \sim$ `poly(r,3)`;
3. $z \sim$ `as.factor(city)` (a separate estimate for each city).

(c) Interpret the results.

There is nothing mysterious about overdispersion. Overly large residuals mean that our model is deficient. In the toxoplasmosis example there are certainly other predictors – age, gender, neighborhood, etc. – that would reduce residual error when included in the logistic regression model–if we knew them. We don't, but we can at least assess the degree of overdispersion, and account for its effect on the accuracy of estimates $\hat{\beta}_k$ such as those in Table 3.12. This is what the theory of quasi-likelihood is intended to do.

Table 3.13 *Toxoplasmosis data matrix; c_j city residence for subject j, r_j rainfall in that subject's city, z_j 0 or 1 as positive or negative test.*

	City	Rainfall	Response
1	$\vdots$	$\vdots$	$\vdots$
2	$\vdots$	$\vdots$	$\vdots$
$\vdots$	$\vdots$	$\vdots$	$\vdots$
j	c_j	r_j	z_j
$\vdots$	$\vdots$	$\vdots$	$\vdots$
697	$\vdots$	$\vdots$	$\vdots$

Quasi-likelihood

A normal-theory GLM, that is, ordinary least squares, has no problem with overdispersion. The usual model,

$$\boldsymbol{y} \sim X\beta + \boldsymbol{\epsilon}, \qquad \boldsymbol{\epsilon} \sim \mathcal{N}_N(\mathbf{0}, \sigma^2 I),$$

gives MLE $\hat{\beta} = (X^\top X)^{-1} X^\top \boldsymbol{y}$, with

$$\hat{\beta} \sim \mathcal{N}_p\left(\beta, \sigma^2 (X^\top X)^{-1}\right);$$

σ^2 is estimated from the residual sum of squared errors. A significance test for the kth coefficient is based on $\hat{\beta}_k/\hat{\sigma}$, automatically accounting for dispersion. Notice that the point estimate $\hat{\beta}_k$ doesn't depend on $\hat{\sigma}$, while its variability does.

Homework 3.29 Suppose we observe independent 0/1 random variables $y_1, \dots, y_N$, with unknown expectations $E\{y_i\} = \pi_i$, and wish to estimate $\theta = \sum_1^N \pi_i/N$. An unbiased estimate is $\hat{\theta} = \bar{y}$. What can we learn from

$$\hat{\sigma}^2 = \sum_{i=1}^N (y_i - \bar{y})^2/N?$$

The advantage of the normal-theory GLM

$$y_i \overset{\text{ind}}{\sim} \mathcal{N}(x_i^\top \beta, \sigma^2), \qquad i = 1, \dots, N, \tag{3.99}$$

is that it incorporates a dispersion parameter σ^2 without leaving the world of exponential families. This isn't possible for other GLMs (however, see

Section 3.9). Quasi-likelihood theory says that we can act as if it were possible.

We begin by considering an extension of the GLM structure. As in (3.3)–(3.4), the observations y_i are obtained from possibly different members of a one-parameter exponential family $g_\eta = \exp\{\eta y - \psi(\eta)\}g_0(y)$, with expectation $\mu_i = \dot{\psi}(\eta_i)$ and variance $v_i = \ddot{\psi}(\eta_i)$. (We are reserving our usual notation V for the familiar variance functions, e.g., $V = \mu$ for the Poisson.)

Rather than the GLM stipulation $\boldsymbol{\eta}_\beta = X\beta$, that is $\boldsymbol{\mu}_\beta = \dot{\psi}(X\beta)$, we assume only that $\boldsymbol{\mu}_\beta$ is smoothly defined as a function of a p-dimensional parameter β, having $N \times p$ derivative matrix, say $\boldsymbol{w}_\beta$,

$$\boldsymbol{w}_\beta = \frac{\partial \mu_i}{\partial \beta_j}. \tag{3.100}$$

Lemma 3.3 *The score function and information matrix for an extended GLM family are*

$$\dot{l}_\beta(\boldsymbol{y}) = \boldsymbol{w}_\beta^\top \boldsymbol{v}_\beta^{-1}(\boldsymbol{y} - \boldsymbol{\mu}_\beta) \quad \textit{and} \quad i_\beta = \boldsymbol{w}_\beta^\top \boldsymbol{v}_\beta^{-1} \boldsymbol{w}_\beta, \tag{3.101}$$

where $\boldsymbol{v}_\beta$ is the diagonal matrix with elements $v_i(\beta)$.

The proof of Lemma 3.3 begins by differentiating

$$l_\beta(\boldsymbol{y}) = \boldsymbol{\eta}_\beta^\top \boldsymbol{y} - \sum \psi\left(\eta_i(\beta)\right),$$

using $d\boldsymbol{\eta}_\beta / d\beta = \boldsymbol{v}_\beta^{-1} \boldsymbol{w}_\beta$.

Homework 3.30 (a) Complete the proof. (b) Show that in the GLM case where $\boldsymbol{\eta}_\beta = X\beta$, (3.101) reduces to (3.10) and (3.13).

Under reasonable asymptotic assumptions (Chapter 9 of McCullagh and Nelder, 1989), it can be shown that the solution $\hat{\beta}$ to $\dot{l}_\beta(\boldsymbol{y})$ in (3.101) satisfies the usual MLE approximation

$$\hat{\beta} \mathrel{\dot{\sim}} \mathcal{N}_p(\beta, i_\beta^{-1}).$$

The quasi-likelihood approach to overdispersion is simply to pretend that we are dealing with an extended GLM family having

$$v(\mu) = \sigma^2 V(\mu) \tag{3.102}$$

(for σ^2 an unknown positive constant), where $V(\mu)$ is the variance function in the original family. For instance,

$$v(\mu) = \sigma^2 \mu(1 - \mu) = \sigma^2 \pi(1 - \pi)$$

for the binomial family, or $v(\mu) = \sigma^2\mu$ for the Poisson family. Applied to the original model $\boldsymbol{\eta}_\beta = X\beta$, Lemma 3.3 becomes

$$\dot{l}_\beta(\boldsymbol{y}) = X^\top(\boldsymbol{y} - \boldsymbol{\mu}_\beta)/\sigma^2 \quad \text{and} \quad i_\beta = X^\top V_\beta X/\sigma^2, \tag{3.103}$$

the argument being the same as in part (b) of Homework 3.30. The MLE equation $\dot{l}_\beta(\boldsymbol{y}) = \mathbf{0}$ gives the usual estimate $\hat{\beta}$. However the estimated covariance matrix for $\hat{\beta}$ is now multiplied by σ^2,

$$\hat{\beta} \,\dot{\sim}\, \mathcal{N}_p\left(\beta, \sigma^2(X^\top \boldsymbol{V}_\beta X)^{-1}\right),$$

compared with (3.13) in Section 3.1; σ^2 is obtained from the Pearson residuals, as in (3.98).

The toxoplasmosis data was rerun using a quasi-binomial model, as seen in Table 3.14. It estimated σ^2 as 1.94, the Pearson residual overdispersion estimate from (3.98). Comparing the results with the standard binomial GLM in Table 3.12 we note the following.

- The estimated coefficient vector $\hat{\beta}$ is the same.
- The estimated standard errors are multipled by $1.94^{1/2} = 1.39$.
- The estimated t-values are divided by 1.39.

This last item results in a two-sided p-value for the cubic coefficient of 0.023, compared with 0.00081 previously.

Table 3.14 *Toxoplasmosis data, quasi-binomial logistic regression* `glm(formula=p~poly(r,3), family=quasibinomial, weight=n).`

Coefficients	Estimate	St. error	t-value	Pr(> |t|)
(Intercept)	0.0243	0.1072	0.23	0.822
`poly(r,3)1`	−0.0861	0.6390	−0.13	0.894
`poly(r,3)2`	−0.1927	0.6511	−0.30	0.769
`poly(r,3)3`	1.3787	0.5732	2.41	0.023

3.9 Double Exponential Families

The quasi-likelihood analysis of the toxoplasmosis data proceeded as if the observed proportions p_i were obtained from a one-parameter exponential family with expectation π_i and variance $\sigma^2\pi_i(1 - \pi_i)/n$. There is no such family, but it turns out we can come close using the *double exponential family* construction.

Note These are not to be confused with the Laplace double exponential density $\exp(-|x|)/2$. A close cousin of the double exponential families presented here are *exponential dispersion models* (Jørgensen, 1987). These directly add a dispersion parameter σ to standard exponential families, giving density kernels of the form $\exp\{(\eta^\top y - \psi(y)/\sigma\}$ in our notation. This is shown possible in a range of situations, though not including discrete families such as the Poisson or binomial, of particular interest here.

Forgetting about GLMs for now, suppose we have a single random sample $y_1, \ldots, y_n$ from a one-parameter exponential family $g_\mu(y)$ having expectation μ and variance function $V(\mu)$. The average $\bar{y}$ is then a sufficient statistic, with density say

$$g_{\mu,n}(\bar{y}) = e^{n(\eta\bar{y}-\psi(\eta))} g_{0,n}(\bar{y}),$$

as in Section 1.3, and expectation and variance

$$\bar{y} \sim \left(\mu, \frac{V(\mu)}{n}\right). \tag{3.104}$$

(The change of notation, from $g_\mu^{(n)}(\bar{y})$ to $g_{\mu,n}(\bar{y})$, reflects the role of n in the double theory.) Hoeffding's formula, Section 1.8, expresses $g_{\mu,n}(\bar{y})$ in terms of deviance,

$$g_{\mu,n}(\bar{y}) = g_{\bar{y},n}(\bar{y}) e^{-nD(\bar{y},\mu)/2},$$

with $D(\mu_1, \mu_2)$ the deviance function for $n = 1$.

The double exponential family corresponding to $g_{\mu,n}(\bar{y})$ is the two-parameter family

$$f_{\mu,\theta,n}(\bar{y}) = C\theta^{1/2} g_{\bar{y},n}(\bar{y}) e^{-n\theta D(\bar{y},\mu)/2}, \tag{3.105}$$

with μ in the interval of allowable expectations for $g_{\mu,n}(\bar{y})$, and $\theta > 0$. An important point is that the carrier measure $m(d\bar{y})$ for $f_{\mu,\theta,n}(\bar{y})$, suppressed in our notation, is the same as that for $g_{\mu,n}(\bar{y})$. This is crucial for discrete distributions like the Poisson where the support stays the same – counting measure on $0, 1, 2, \ldots$ – for all choices of θ.

What follows is a list of salient facts concerning $f_{\mu,\theta,n}(\bar{y})$, as verified in Efron (1986a).

Fact 1 The constant $C = C(\mu, \theta, n)$ that makes $f_{\mu,\theta,n}(\bar{y})$ integrate to 1 is close to 1.0. Standard Edgeworth calculations give

$$C(\mu, \theta, n) \doteq 1 + \frac{1}{n}\left(\frac{1-\theta}{\theta} \frac{9\delta_\mu - 15\gamma_\mu^2}{72}\right) + O\left(n^{-2}\right),$$

with γ_u and δ_u the skewness and kurtosis of $g_{\mu,1}(y)$. Taking $C = 1$ in (3.105) is convenient, and usually accurate enough for applications.

Fact 2 Formula (3.105) can also be written as

$$f_{\mu,\theta,n}(\bar{y}) = C\theta^{1/2} g_{\mu,n}(\bar{y})^{\theta} g_{\bar{y},n}(\bar{y})^{1-\theta},$$

which says $\log f_{\mu,\theta,n}(\bar{y})$ is essentially a linear combination of $\log g_{\mu,n}(\bar{y})$ and $\log g_{\bar{y},n}(\bar{y})$.

Homework 3.31 Verify Fact 2.

Fact 3 The expectation and variance of $\bar{y} \sim f_{\mu,\theta,n}$ are, to excellent approximations,

$$\bar{y} \dot{\sim} \left(\mu, \frac{V(\mu)}{n\theta}\right),$$

with errors of order n^{-2} for both terms. Comparison with (3.102) shows that $1/\theta$ measures dispersion,

$$\sigma^2 = 1/\theta.$$

Homework 3.32 Suppose that $g_{\mu,n}(\bar{y})$ represents the normal family $\bar{y} \sim \mathcal{N}(\mu, n^{-1})$. Show that $f_{\mu,\theta,n}(\bar{y})$ has $\bar{y} \sim \mathcal{N}(\mu, (n\theta)^{-1})$.

Homework 3.33 Verify the following as **Fact 4**: $f_{\mu,\theta,n}(\bar{y})$ is a two-parameter exponential family having natural parameter η equal $n(\theta\eta, \theta)$, and sufficient vector y equal $(\bar{y}, -\psi(\bar{\eta}))$, where $\bar{\eta} = \dot{\psi}^{-1}(\bar{y})$. *Hint*: Use Fact 2. Moreover, with θ and n fixed, $f_{\mu,\theta,n}(\bar{y})$ is a one-parameter exponential family with natural parameter $n\theta\eta$ and sufficient statistic $\bar{y}$; with μ and n fixed, $f_{\mu,\theta,n}(\bar{y})$ is a one-parameter exponential family with natural parameter θ and sufficient statistic $-nD(\bar{y}, \mu)/2$.

Together, Facts 3 and 4 say that $f_{\mu,\theta,n}(\bar{y})$, with θ fixed, is a one-parameter exponential family having expectation and variance nearly μ and $V_\mu/(n\theta)$, respectively. It is just what was required for the notional quasi-likelihood families.

Figure 3.9 illustrates the double Poisson distribution: we have taken[9] $n = 1$, $\mu = 10$, and $\theta = 1, 1/2$, or 2 (using $C(\mu, \theta, n)$ from Fact 1). The case $\theta = 1$, which is the standard Poisson distribution, has $\mu = 10$ and $V_\mu = 10^{1/2} = 3.16$. As claimed, the variance doubles for $\theta = 1/2$ and halves

[9] Because the Poisson family is closed under convolutions, the double Poisson family turns out to be essentially the same for any choice of n.

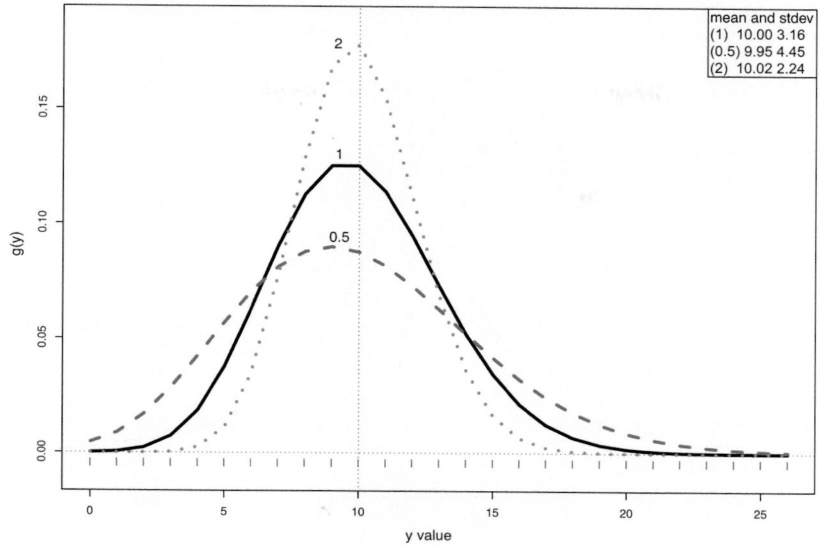

Figure 3.9 Double Poisson densities for $n = 1$, $\mu = 10$, $\theta = 1$ (black solid), $= 1/2$ (red dashed), or $= 2$ (green dotted).

for $\theta = 2$, while μ stays near 10. All three distributions are supported on $0, 1, 2, \ldots$.

Homework 3.34 (a) For the Poisson family ($n = 1$) show that (3.105), with $C = 1$, gives

$$f_{\mu,\theta}(y) = \theta^{1/2} e^{-\theta\mu} \left(\frac{e^{-y} y^y}{y!}\right) \left(\frac{e\mu}{y}\right)^{\theta y} \qquad (y \text{ is } \bar{y} \text{ here}).$$

(b) Compute $f_{\mu,\theta}(y)$ for $\theta = 1/3$ and $\theta = 3$; numerically calculate the expectations and variances.

(c) Use Fact 1 to give an expression for $C(\mu, \theta, n)$.

(d) What are the expectations and variances now using $C(\mu, \theta, n)$ instead of $C = 1$?

Count statistics that appear to be overdispersed Poissons are often modeled by negative binomial distributions (Section 1.5). Figure 3.10 compares $f_{10,.5}(y)$ with the negative binomial density having expectation 10 and variance 20, showing a striking similarity. Negative binomials form a one-parameter family for y, but the auxiliary parameter k cannot be incorporated into a two-parameter exponential family.

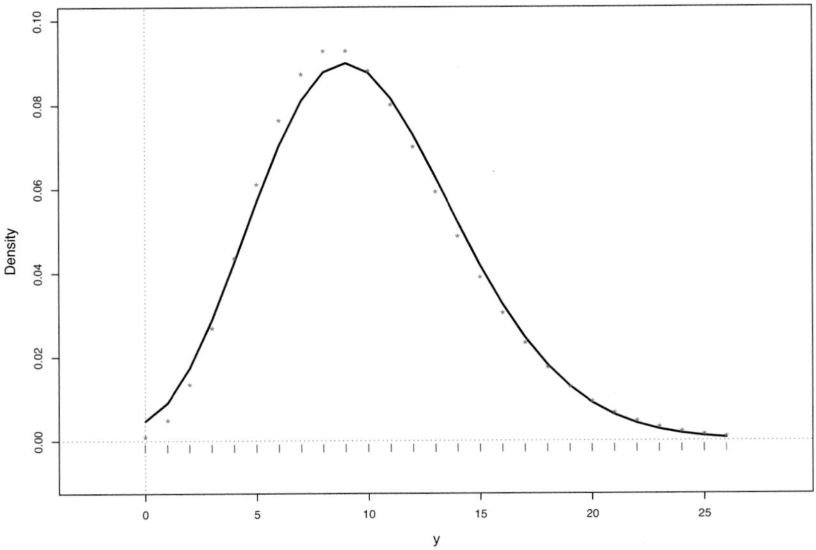

Figure 3.10 Double Poisson density, $\mu = 10$, $\theta = 1/2$ (black solid), compared with negative binomial density, $\mu = 10$, variance $= 20$ (red points).

The fact that $\bar{y} \sim f_{\mu,\theta,n}$ has expectation and variance approximately μ and $V_\mu/(n\theta)$ suggests that the density $f_{\mu,\theta,n}(\bar{y})$ is similar to $g_{\mu,n\theta}(\bar{y})$. Choosing $\theta = 1/2$, say, effectively reduces the sample size in the original family from n to $n/2$. This was exactly true in the normal case of Homework 3.32.

Fact 5 For any interval I of the real line,

$$\int_I f_{\mu,\theta,n}(\bar{y})\, m_n(d\bar{y}) = \int_I g_{\mu,n\theta}(\bar{y})\, m_{n\theta}(d\bar{y}) + O(n^{-1}).$$

Here m_n and $m_{n\theta}$ are the carrying measures in the original family, for sample sizes n and $n\theta$.

For the binomial family $p \sim \text{Bi}(n,\pi)/n$, where we are thinking of $\bar{y} = p$ as the average of n Bernoulli variates $y_i \sim \text{Bi}(1,\pi)$, $m_n(p)$ is counting measure on $0, 1/n, \dots, 1$, while $m_{n\theta}(p)$ is counting measure on $0, 1/n\theta, 2/n\theta, \dots, 1$. This assumes $n\theta$ is an integer, and shows the limitations of Fact 5 in discrete families.

Homework 3.35 In the binomial case, taking $C = 1$, numerically compare the cumulative distribution function of $f_{\mu,\theta,n}(p)$ for $(\mu, \theta, n) = (0.4, 0.5, 16)$ with that of $g_{\mu,n}(p)$, $(\mu, n) = (0.4, 8)$.

Homework 3.36 What would be the comparisons suggested by Fact 5 for the Poisson distributions in Figure 3.9?

Differentiating the log likelihood function $l_{\mu,\theta,n}(\bar{y}) = \log f_{\mu,\theta,n}(\bar{y})$ (ignoring C),

$$l_{\mu,\theta,n}(\bar{y}) \doteq -n\theta\,\frac{D(\bar{y},\mu)}{2} + \frac{1}{2}\log\theta + \log g_{\bar{y},\mu}(\bar{y}),$$

and using $\partial D(\bar{y},\mu)/\partial\mu = 2(\mu - \bar{y})/V_\mu$, gives the next fact.

Fact 6 The score functions for $f_{\mu,\theta,n}(\bar{y})$ are

$$\frac{\partial l_{\mu,\theta,n}(\bar{y})}{\partial\mu} \doteq \frac{\bar{y}-\mu}{V_\mu/(n\theta)} \quad\text{and}\quad \frac{\partial l_{\mu,\theta,n}(\bar{y})}{\partial\theta} \doteq \frac{1}{2\theta} - \frac{nD(\bar{y},\mu)}{2}.$$

Double Family GLMs

Suppose we have a generalized regression setup, observations

$$\bar{y}_i \overset{\text{ind}}{\sim} f_{\mu_i,\theta,n_i}, \qquad \text{for } i = 1,\dots,N, \tag{3.106}$$

with the GLM model $\boldsymbol{\eta}(\beta) = X\beta$ giving $\boldsymbol{\mu}(\beta) = (\cdots \mu_i = \dot{\psi}(\eta_i(\beta))\cdots)^\top$. The approximate score functions for the full data set $\boldsymbol{y} = (\bar{y}_1,\dots,\bar{y}_N)^\top$ are

$$\frac{\partial}{\partial\beta}\, l_{\beta,\theta}(\boldsymbol{y}) = \theta X^\top \operatorname{diag}(\boldsymbol{n})\,(\boldsymbol{y} - \boldsymbol{\mu}(\beta)),$$

$\operatorname{diag}(\boldsymbol{n})$ the diagonal matrix with entries n_i, and

$$\frac{\partial}{\partial\theta}\, l_{\beta,\theta}(\boldsymbol{y}) = \frac{N}{2\theta} - \sum_{i=1}^{N} \frac{n_i D\left(\bar{y}_i, \mu_i(\beta)\right)}{2}.$$

Homework 3.37 Verify the score functions.

We see that the MLE $\hat{\beta}$ does not depend on θ, which only enters the β score function as a constant multiple. The MLE for $\hat{\theta}$ is

$$\hat{\theta} \doteq \frac{N}{\sum_{i=1}^{N} n_i D\left(\bar{y}_i, \mu_i\left(\hat{\beta}\right)\right)}.$$

Homework 3.38 (a) How does this estimate relate to the overdispersion estimates $\hat{\sigma}^2$ for the toxoplasmosis data of Section 3.8?

(b) In the normal case where we begin with $y \sim \mathcal{N}(\mu, \sigma^2)$, show that $1/\hat{\theta}$ is the maximum likelihood estimate for σ^2.

I ran a more ambitious GLM for the toxoplasmosis data, where θ_i as well as p_i was modeled. It used the double binomial model $p_i \sim f_{\pi_i,\theta_i,n_i}$, $i = 1, \dots, 34$. Here p_i and π_i, the observed and true proportion positive in city i, play the roles of y_i and μ_i in (3.105).

The model let π_i be a cubic polynomial function of rainfall, as in Section 3.8; θ_i was modeled as a function of n_i, the number of people sampled in city i. Let

$$\tilde{n}_i = \frac{n_i - \bar{n}}{\text{sd}_n},$$

with $\bar{n}$ and sd_n the mean and standard deviation of the n_i values. I took $\theta_i = 1.25 \cdot (1 + e^{-\lambda_i})$, where $\lambda_i = \gamma_0 + \gamma_1 \tilde{n}_i + \gamma_2 \tilde{n}_i^2$. This allowed the θ_i to range from 1.25 (mild underdispersion) all the way down to zero. All together the model had seven parameters, four for the cubic rainfall regression and three for the θ regression. The seven-parameter MLE was found using the R function nonlinear maximizer `nlm`.

Homework 3.39 What was the function I minimized?

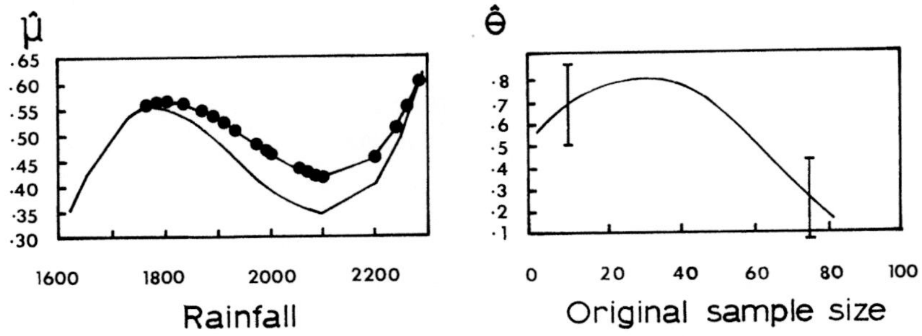

The resulting cubic regression of π_i as a function of r_i (solid curve, above on left) was somewhat more extreme than the original GLM in Table 3.12, the latter shown as the dotted curve. Perhaps more interesting was the fitted regression for the dispersion parameter $\hat{\theta}_i$ as a function of number sampled n_i (above on right). It peaked at about $\hat{\theta}_i = 0.8$ at $n_i = 30$, declining to 0.2 for $n_i = 70$. Rather unintuitively, overdispersion *increased* in the largest samples.

4

Curved Exponential Families, Empirical Bayes, Missing Data, and Stability of the MLE

A probability distribution like the normal or binomial is a powerful and sophisticated extension of basic numerical thinking. Statisticians need *families* of probability distributions for the purposes of statistical inference. The theory of exponential families we have been discussing moves to the next higher level of abstraction: families of families. All of this is to say that we have in our hands an amazingly general tool for statistical theory and practice.

Not fully general, though. Not every problem an be reduced to exponential family form. *Curved exponential families*, our first topic in what follows, directly extends the reach of exponential family theory. This is indicated schematically in Figure 4.1, where a fourth ring has been added to the statistical solar system of Figure 1 in this book's Introduction. Venturing outside of safe Circle 3 allows us to apply exponential family ideas to a wider class of problems, including missing data and empirical Bayes analysis. There is a second benefit: the view from outside makes it clearer what is lost when we go beyond exponential family assumptions, as quantified by the statistical curvature calculations of Section 4.6.

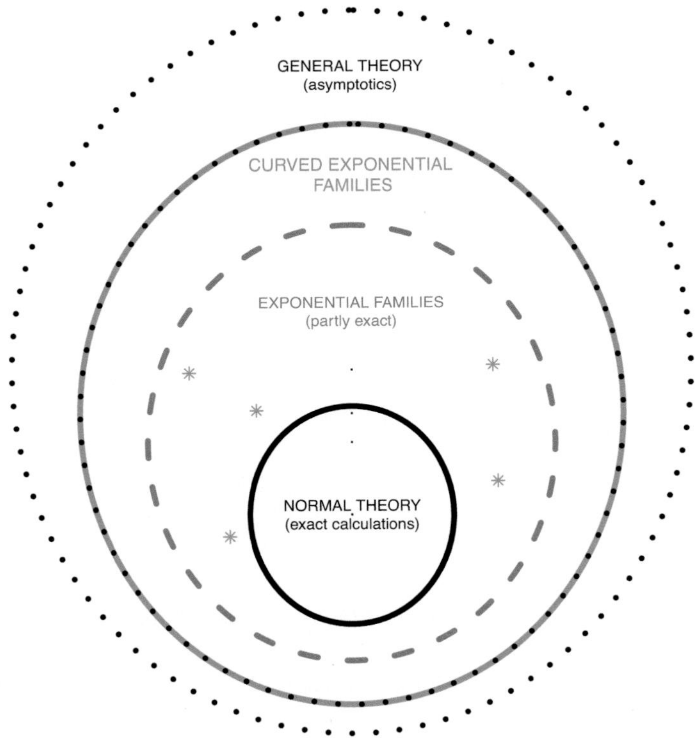

Figure 4.1 Three levels of statistical modeling, now with a fourth added representing curved exponential families.

4.1 Curved Exponential Families: Definitions and First Results

Exponential families, as introduced in Parts 1 and 2, can be thought of as generalizations of normal-theory models that maintain some, if not all, of the latter's exact (non-asymptotic) inference properties. In turn, curved exponential families maintain some, if not all, of the exponential family properties. Most importantly, curved models provide an evocative picture of statistical inference in "smooth" families, in particular for notions of observed and expected information, this being Fisher's original usage.

We begin with a p-parameter exponential family $\mathcal{G}$, as in (2.1),

$$\mathcal{G} = \left\{ g_\eta(y) = e^{\eta^\top y - \psi(\eta)} g_0(y) \text{ for } \eta \in A, y \in \mathcal{Y} \right\}.$$

In some situations though, we may know that η is restricted to lie in a q-parameter curved subspace of A, say

$$\eta = \eta_\theta \text{ for } \theta \in \Theta \subset \mathcal{R}^q \qquad (q < p). \tag{4.1}$$

For the one-parameter subfamilies of Section 2.6, $\eta_\theta = \alpha + \beta\theta$ (2.24), $q = 1$. The mapping $\eta = \eta_\theta$ from Θ into A will be assumed to have continuous second derivatives. We can express the q-parameter subfamily of densities as

$$\mathcal{F} = \left\{ f_\theta(y) = g_{\eta_\theta}(y) \text{ for } \theta \in \Theta \right\}, \tag{4.2}$$

so

$$f_\theta(y) = e^{\eta_\theta^\top y - \psi(\eta_\theta)} g_0(y). \tag{4.3}$$

If the mapping from θ into the natural parameter space A is linear, say $\eta = X\theta$, then

$$f_\theta(y) = e^{\theta^\top X^\top y - \psi(X\theta)} g_0(y) \tag{4.4}$$

is a q-parameter exponential family with natural parameter vector θ, as in the generalized linear models of Part 3. Here we will be interested in *curved exponential familes*, where η_θ is a nonlinear function of θ.

Some Convenient Notation

We will write μ_θ for $\mu(\eta_\theta)$, ψ_θ for $\psi(\eta_\theta)$, and likewise V_θ for $V_{\eta_\theta} = \text{Cov}_\theta(y)$, etc. Derivatives with respect to θ will be indicated by dot notation,

$$\begin{aligned} \underset{p\times q}{\dot\eta_\theta} &= \left(\frac{\partial \eta_i}{\partial \theta_j}\right)_{\eta=\eta_\theta}, \quad \underset{p\times q}{\dot\mu_\theta} = \left(\frac{\partial \mu_i}{\partial \theta_j}\right)_{\mu=\mu_\theta} = V_\theta \dot\eta_\theta, \\ \text{and} \quad \ddot\eta_\theta &= \left(\frac{\partial^2 \eta_i}{\partial \theta_j \partial \theta_k}\right)_{\eta=\eta_\theta}, \end{aligned} \tag{4.5}$$

the last a $p \times q \times q$ array. Family $\mathcal{F}$ has representations in both the η and μ ("A" and "B") spaces,

$$\mathcal{F}_A = \{\eta_\theta, \theta \in \Theta\} \quad \text{and} \quad \mathcal{F}_B = \{\mu_\theta, \theta \in \Theta\}, \tag{4.6}$$

both $\mathcal{F}_A$ and $\mathcal{F}_B$ being q-dimensional manifolds in $\mathcal{R}^p$, usually curved.

The Log Likelihood Function

The log likelihood function is

$$l_\theta(y) = \log f_\theta(y) = \eta_\theta^\top y - \psi(\eta_\theta), \tag{4.7}$$

thought of as a function of θ. Taking derivatives with respect to θ gives the *score function*, a q-dimensional vector,

$$\underset{q\times 1}{\dot{l}_\theta(y)} = \left(\frac{\partial l_\theta}{\partial \theta_j}\right) = \dot{\eta}_\theta^\top (y - \mu_\theta) \tag{4.8}$$

(the large parentheses indicating the q-vector of partial derivatives), where we have used

$$\dot{\psi}_\theta = \left(\frac{\partial \eta_\theta}{\partial \theta}\right)^\top \dot{\psi}(\eta)\bigg|_{\eta_\theta} = \dot{\eta}_\theta^\top \mu_\theta;$$

(4.8) gives the *Fisher information matrix*

$$\underset{q\times q}{i_\theta} = E_\theta\left\{\dot{l}_\theta(y)\dot{l}_\theta(y)^\top\right\} = \dot{\eta}_\theta^\top V_\theta \dot{\eta}_\theta. \tag{4.9}$$

Taking derivatives again yields a simple but important result:

Lemma 4.1 (Second Derivative Lemma) *Minus the second derivative matrix of the log likelihood is*

$$\underset{q\times q}{-\ddot{l}_\theta(y)} = -\left(\frac{\partial^2 l_\theta(y)}{\partial \theta_j \partial \theta_k}\right) = i_\theta - \ddot{\eta}_\theta^\top (y - \mu_\theta), \tag{4.10}$$

where $\ddot{\eta}_\theta^\top (y - \mu_\theta)$ *is the matrix having* (j,k)*th element*

$$\sum_{i=1}^{p} \frac{\partial^2 \eta_i}{\partial \theta_j \partial \theta_k}(y - \mu_\theta)_i. \tag{4.11}$$

In a non-curved exponential family, where η_θ is linear in θ, $\ddot{\eta}_\theta = 0$ and $-\ddot{l}_\theta(y)$ equals the Fisher information matrix i_θ (as in Section 2.6, where η is θ). Curvature makes $-\ddot{l}_\theta(y)$ depend on y, a crucial difference in what follows.

Homework 4.1 (a) Verify the formulas for $\dot{l}_\theta$, $\dot{\psi}_\theta$, i_θ, and $-\ddot{l}_\theta$. (b) Show that $\dot{\mu}_\theta = V_\theta \dot{\eta}_\theta$ and so

$$\dot{\eta}_\theta^\top \dot{\mu}_\theta = \dot{\eta}_\theta^\top V_\theta \dot{\eta}_\theta = i_\theta \geq 0. \tag{4.12}$$

Note In the case $q = 1$, (4.12) shows that the curves η_θ and μ_θ through p-space have directional derivatives within 90° of each other.

4.2 Two Pictures of the MLE

Maximum likelihood estimation in curved exponential families $\mathcal{F}$ (4.2), can be geometrically pictured in two complementary ways, one originally due to Fisher and the other obtained from Hoeffding's formula (2.38).

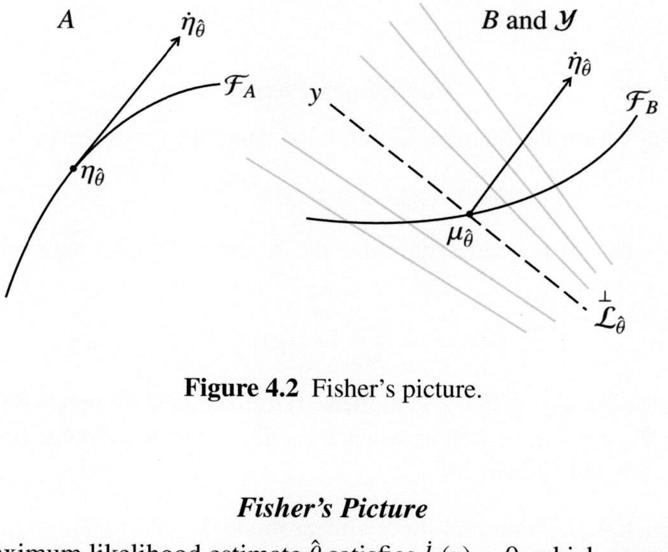

Figure 4.2 Fisher's picture.

Fisher's Picture

The maximum likelihood estimate $\hat{\theta}$ satisfies $\dot{l}_{\hat{\theta}}(y) = 0$, which according to (4.8) is the local orthogonality condition

$$\dot{\eta}_{\hat{\theta}}^\top (y - \mu_{\hat{\theta}}) = 0. \tag{4.13}$$

Here 0 represents a vector of q zeros. We obtain a pleasing geometric visualization of maximum likelihood estimation from Figure 4.2. Notice that:

- The expectation vector $\mu_{\hat{\theta}}$ corresponding to $\hat{\theta}$ is obtained by projecting the data vector y onto $\mathcal{F}_B$ orthogonally to $\dot{\eta}_{\hat{\theta}}$, the tangent directions to $\mathcal{F}_A$ at $\eta_{\hat{\theta}}$.

- If η_θ is linear in θ, say

$$\eta_\theta = a + b\theta,$$

with a $p \times 1$ and b $p \times q$ (that is, if $\mathcal{F}_A$ is a flat space), then $\dot{\eta}_\theta = b$ for all θ, and the projection direction is always orthogonal to b,

$$b^\top (y - \mu_{\hat{\theta}}) = 0. \tag{4.14}$$

- Otherwise, the projection orthogonals $\dot{\eta}_{\hat{\theta}}$ change with $\hat{\theta}$. (Section 4.6 discusses what happens in Figure 4.2 if the flats intersect.)
- The set of y vectors having MLE equaling some particular value $\hat{\theta}$ is a $(p - q)$-dimensional flat space in $\mathcal{Y}$, passing through $\mu_{\hat{\theta}}$ orthogonally to $\dot{\eta}_{\hat{\theta}}$, say

$$\overset{\perp}{\mathcal{L}}_{\hat{\theta}} = \left\{ y : \dot{\eta}_{\hat{\theta}}^\top (y - \mu_{\hat{\theta}}) = 0 \right\}. \tag{4.15}$$

Hoeffding's Picture

Hoeffding's formula, Section 2.7 (now indexing $\mathcal{G}$ by μ rather than η), is

$$f_\theta(y) = g_{\mu_\theta}(y) = g_y(y) e^{-D(y,\mu_\theta)/2}. \tag{4.16}$$

Therefore the MLE $\hat{\theta}$ must minimize the deviance $D(y, \mu_\theta)$ between y and a point μ_θ on $\mathcal{F}_B$,

$$\hat{\theta} : D(y, \mu_{\hat{\theta}}) = \min_{\theta \in \Theta} D(y, \mu_\theta) \equiv D_{\min}(y). \tag{4.17}$$

Another way to say this: as d is increased from 0, the level curves $D(y, \mu) = d$ touch $\mathcal{F}_B$ for the first time when $d = D_{\min}$, the touchpoint being $\mu_{\hat{\theta}}$. Figure 4.3 is the visualization.

Homework 4.2 A normal-theory linear model $y \sim \mathcal{N}_p(\mu, I)$ has

$$\eta_\theta = \mu_\theta = X\theta,$$

for X a known $p \times q$ matrix. What do the two pictures look like? What are their conventional names?

Homework 4.3 Show that the level curves of constant deviance,

$$C_d = \{\mu : D(y, \mu) = d\},$$

become ellipsoidal as $d \downarrow 0$. What determines the shape of the ellipsoids?

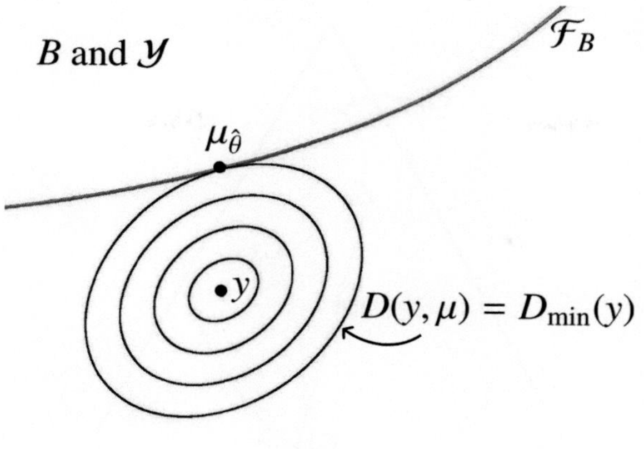

Figure 4.3 Hoeffding's picture.

Fisher never drew "Fisher's picture" but he did carry out an incisive analysis of the multinomial case[1]

$$p \sim \text{Mult}_L(n, \pi_\theta)/n, \tag{4.18}$$

(2.74) with θ one-dimensional, i.e., $q = 1$. This is a curved one-dimensional subfamily of the L-category multinomial family of Section 2.9. In terms of the previous notation, y is now p, the vector of observed proportions, and μ_θ is π_θ, the vector of true probabilities. Figure 4.4 shows $\mathcal{F}_B$ as a curve through the L simplex $\mathcal{S}_L$, with the MLE $\pi_{\hat{\theta}}$ obtained by projecting p onto $\mathcal{F}_B$ orthogonally to $\dot{\eta}_{\hat{\theta}}$.

Homework 4.4 Show that we can take $\dot{\eta}_\theta$ to have components $\dot{\eta}_{\theta j} = \dot{\pi}_{\theta j}/\pi_{\theta j}$, where $\dot{\pi}_{\theta j} = (\partial/\partial\theta_j)\pi_\theta$.

A predecessor to maximum likelihood estimation was "minimum chi-squared", defined by

$$\hat{\theta} = \arg\min_\theta \left\{ \sum_{k=1}^{L} \frac{(p_k - \pi_{\theta k})^2}{\pi_{\theta k}} \right\}. \tag{4.19}$$

Its level surfaces are *curved*, not flat, as indicated by $C_{\hat{\theta}}$ in Figure 4.4. When maximum likelihood was introduced in the 1920s, it was criticized as being more difficult to compute than competitors such as minimum chi-squared. Now that that no longer matters, the MLE has gone on to become a universally favored estimator.

[1] p in (4.18) is a random variable, not the dimension p in (4.1), which is now L.

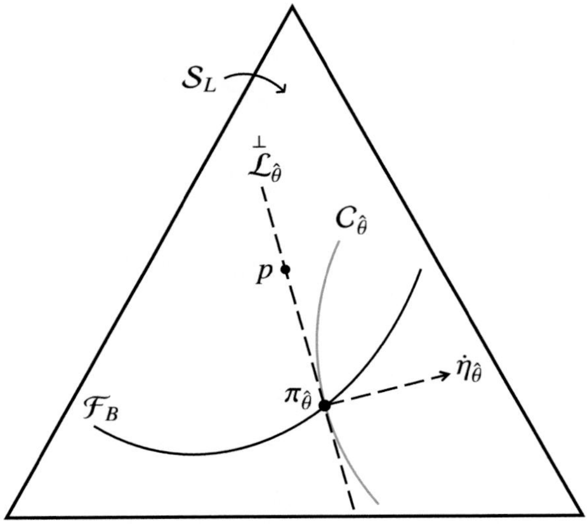

Figure 4.4 Fisher's multinomial picture of maximum likelihood estimation.

Fisher argued the superiority of maximum likelihood over any other smooth estimation method, say $\hat\theta = t(p)$. He made three points:

1. *Consistency* In order for $\hat\theta = t(p)$ to be a consistent estimator of θ as $n \to \infty$, it must satisfy

$$t(\pi_\theta) = \theta \qquad \text{("Fisher consistency").} \tag{4.20}$$

Homework 4.5 What was Fisher's argument?

2. *Efficiency* In order to asymptotically achieve the Fisher information bound for estimating θ, a Fisher consistent estimator must have $C(\hat\theta)$ (the level surface $\{p : t(p) = \hat\theta\}$) intersect $\mathcal{F}_B$ at $\pi_{\hat\theta}$ *orthogonally* to $\dot\eta_\theta$. Such an estimator is said to be *first-order efficient*.

Here is a brief outline of Fisher's efficiency argument. Let θ_0 be the true value of θ, and denote $\pi_0 = \pi_{\theta_0}$ and $\dot\eta_0 = \dot\eta_{\theta_0} = \dot\pi_0/\pi_0$. It then turns out that the Fisher information for estimating θ at θ_0 is

$$i_0 = \sum_{j=1}^{L} \pi_{0j}\dot\eta_{0j}^2, \tag{4.21}$$

and that the local linear approximation for the MLE is

$$\hat{\theta} \doteq \theta_0 + i_0{}^{-1}\dot{\eta}_0^\top (p - \pi_0); \tag{4.22}$$

(4.22) is a special case of Lemma 4.2 in Section 4.3, taking $\bar{y} = p$ and expanding around $p = \pi_0$.

Homework 4.6 Show that $i_0 = \sum_{j=1}^{L} \pi_{0j}\dot{\eta}_{0j}^2$, and that the variance of the local linear approximation to the MLE achieves the Cramér–Rao lower bound i_0^{-1} at θ_0.

Suppose that $\tilde{\theta} = s(p)$ is a Fisher consistent estimator of θ such that the local linear approximation

$$s(p) = \theta_0 + \dot{s}_0^\top (p - \pi_0) \qquad \left[\dot{s}_0 = \left(\cdots \frac{\partial s(p)}{\partial p_j} \cdots\right)^\top\right]$$

crosses $\mathcal{F}_B$ at π_0 *not* orthogonally to $\dot{\eta}_0$ (i.e., with $\dot{s}_0$ *not* along $\dot{\eta}_0$).

Homework 4.7 (a) Show that Fisher consistency implies

$$\sum_{j=1}^{L} \pi_{0j}\dot{s}_{0j}\dot{\eta}_{0j} = 1.$$

(b) Use the preceding to show that the linear approximation to $s(p)$ must have variance $> i_0^{-1}$. *Hint*: This amounts to rederiving the Cramér–Rao lower bound.

Asymptotically, as $n \to \infty$, the MLE shares the good properties of its local linear approximation, achieving the Cramér–Rao lower bound and outperforming any competitor $s(p)$. It turns out that minimum chi-squared estimate $\hat{\theta} = t(p)$ *does* cross $\mathcal{F}_B$ orthogonally to $\dot{\eta}_{\hat{\theta}}$. Fisher's third claim promoted MLE over all competitors at the next higher level of approximation.

3. *Second-order efficiency* If $\mathcal{F}_B$ represents a one-parameter exponential family ($\eta_\theta = a + b\theta$) then the MLE $\hat{\theta}$ is a sufficient statistic. Even if not, $\hat{\theta}$ loses less information than any first-order efficient competitor $\tilde{\theta}$. In other words, the flat level surfaces of maximum likelihood estimation are superior to any curved one of the type suggested in Figure 4.4. This was a controversial claim, but we will see it supported later on.

4.3 Repeated Sampling and the Influence Function of the MLE

Suppose $y_1, \ldots, y_n$ is an i.i.d. sample from some member of a curved family $\mathcal{F} = \{f_\theta(y) = g_{\eta_\theta}(y)\}$. Then $\bar{y}$ is a sufficient statistic in $\mathcal{F}$, since it is so in the larger family $\mathcal{G}$. The family of distributions of $\bar{y}$, $\mathcal{F}_n$, is the curved exponential family[2]

$$\mathcal{F}_n = \left\{ f_{\theta,n}(\bar{y}) = g_{\eta_\theta,n}(\bar{y}) = e^{n(\eta_\theta^\top \bar{y} - \psi_\theta)} g_{0,n}(\bar{y}), \theta \in \Theta \right\}. \tag{4.23}$$

$\mathcal{F}_n$ is essentially the same family as $\mathcal{F}$: B and $\mathcal{F}_B$ stay the same, while $A_n = nA$, $\mathcal{F}_{A,n} = n\mathcal{F}_A$, and $\dot{\eta}_{\theta,n} = n\dot{\eta}_\theta$, as in Section 2.4. Fisher's and Hoeffding's pictures stay exactly as shown, with $\bar{y}$ replacing y. The covariance matrix of $\bar{y}$, $V_{\theta,n}$, equals V_θ/n. This means that typically $\bar{y}$ is closer than y to $\mathcal{F}_B$ by a factor of $n^{-1/2}$, this applying to both pictures. Asymptotic calculations in curved exponential families are greatly simplified by the unchanging geometry.

The log likelihood $l_{\theta,n}(\bar{y})$ in $\mathcal{F}_A$ is

$$l_{\theta,n}(\bar{y}) = n(\eta_\theta^\top \bar{y} - \psi_\theta)$$

(with the case $n = 1$ denoted $l_\theta(y)$ as before). The first and second derivatives with respect to θ are

$$\dot{l}_{\theta,n}(\bar{y}) = n\dot{\eta}_\theta^\top(\bar{y} - \mu_\theta) \quad \text{and} \quad -\ddot{l}_{\theta,n}(\bar{y}) = n\left[i_\theta - \ddot{\eta}_\theta^\top(\bar{y} - \mu_\theta)\right], \tag{4.24}$$

according to (4.8) and (4.10) in Section 4.1.

Looking at Fisher's picture, we can think of the MLE $\hat{\theta}$ as being a function of $\bar{y}$, with $t(\cdot)$ being a mapping from $\mathcal{R}^p$ into $\mathcal{R}^q$:

$$\hat{\theta} = t(\bar{y}).$$

A small change $\bar{y} \to \bar{y} + d\bar{y}$ produces a small change in the MLE, $\hat{\theta} \to \hat{\theta} + d\hat{\theta}$, according to the *influence function* (or derivative matrix)

$$\underset{q\times p}{\frac{dt}{d\bar{y}}} = \frac{\partial \hat{\theta}_i}{\partial \bar{y}_j} \equiv \frac{d\hat{\theta}}{d\bar{y}},$$

which is approximately

$$d\hat{\theta} \doteq \frac{d\hat{\theta}}{d\bar{y}}\, d\bar{y}.$$

The MLE influence function takes a simple but useful form:

2 We might have written $f_\theta^{(n)}(\bar{y})$ etc. as in Section 2.4, but putting n as a subscript is a little more convenient in what follows.

Lemma 4.2 (Influence Lemma) *The influence function of the MLE is*

$$\underset{q\times p}{\frac{d\hat{\theta}}{d\bar{y}}} = \underset{q\times q}{\left(-\ddot{l}_{\hat{\theta}}(\bar{y})\right)^{-1}} \underset{q\times p}{\dot{\eta}_{\hat{\theta}}^{\top}}, \tag{4.25}$$

where

$$-\ddot{l}_{\hat{\theta}}(\bar{y}) = i_{\hat{\theta}} - \ddot{\eta}_{\hat{\theta}}^{\top}(\bar{y} - \mu_{\hat{\theta}}).$$

(Notice that (4.25) uses $-\ddot{l}_{\hat{\theta}}(\bar{y})$, not $-\ddot{l}_{\hat{\theta},n}(\bar{y})$. It can also be expressed as $(-\ddot{l}_{\hat{\theta},n}(\bar{y}))^{-1}\,(n\dot{\eta}_{\hat{\theta}}^{\top})$.)

Homework 4.8 Verify Lemma 4.2. *Hint*: $(\dot{\eta}_{\hat{\theta}+d\hat{\theta}})^{\top}(\bar{y} + d\bar{y} - \mu_{\hat{\theta}+d\hat{\theta}}) = 0$. (See Figure 4.5.)

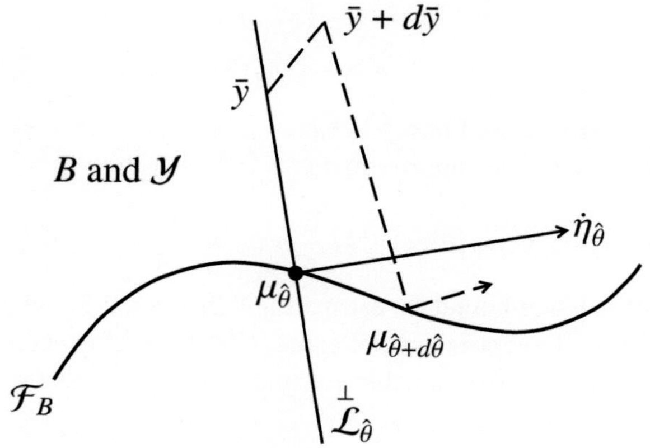

Figure 4.5 Calculating the MLE influence function.

4.4 Variance Calculations for the MLE

Corresponding to a given true value θ in $\mathcal{F}$ is $\overset{\perp}{\mathcal{L}}_{\theta} = \{\bar{y} : \dot{\eta}_{\theta}^{\top}(\bar{y} - \mu_{\theta}) = 0\}$, the set of observations $\bar{y}$ such that the MLE $t(\bar{y})$ equals θ. In particular, $\bar{y} = \mu_{\theta}$ is on $\overset{\perp}{\mathcal{L}}_{\theta}$, i.e., $t(\mu_{\theta}) = \theta$; so the MLE is what is called *Fisher consistent*. Since $\bar{y}$ converges almost surely to μ_{θ} as $n \to \infty$, and $t(\bar{y})$ is a nicely continuous function, Fisher consistency implies ordinary consistency: $\hat{\theta} \to \theta$ a.s.

The symmetric matrix

$$I_n(\bar{y}) = -\ddot{l}_{\hat{\theta},n}(\bar{y}) = n\left[i_{\hat{\theta}} - \ddot{\eta}_{\hat{\theta}}^\top(\bar{y} - \mu_{\hat{\theta}})\right] \tag{4.26}$$

is called the *observed Fisher information* for θ, with the case $n = 1$ denoted $I(y)$. For reasons that will be discussed in Section 4.6, Fisher claimed that the covariance of the MLE is better assessed by $I_n(\bar{y})^{-1}$ rather than $(ni_{\hat{\theta}})^{-1}$. Notice that $I(\mu_\theta) = i_\theta$, so that $I(\bar{y})$ is itself Fisher consistent as an estimate of i_θ,

$$-\ddot{l}_\theta(\mu_\theta) = i_\theta,$$

implying $I_n(\mu_{\hat{\theta}}) = ni_{\hat{\theta}}$, and also $I_n(\bar{y}) \to ni_\theta$ as $n \to \infty$, since $\bar{y} \to \mu_\theta$.

The variance matrix of $\hat{\theta}$ can be approximated using Fisher consistency and the MLE influence function (4.25):

$$\begin{aligned}\hat{\theta} = t(\bar{y}) = t(\mu_\theta + \bar{y} - \mu_\theta) &\doteq t(\mu_\theta) + \left.\frac{dt}{d\bar{y}}\right|_{\mu_\theta}(\bar{y} - \mu_\theta)\\ &\doteq \theta + \left(-\ddot{l}_\theta(\mu_\theta)\right)^{-1}\dot{\eta}_\theta^\top(\bar{y} - \mu_\theta)\\ &\doteq \theta + i_\theta^{-1}\dot{\eta}_\theta^\top(\bar{y} - \mu_\theta),\end{aligned} \tag{4.27}$$

a Taylor series statement of how $\hat{\theta}$ behaves as a function of $\bar{y}$ when $n \to \infty$. This yields the covariance approximation

$$\mathrm{Cov}_\theta\left(\hat{\theta}\right) \doteq i_\theta^{-1}\dot{\eta}_\theta^\top\frac{V_\theta}{n}\dot{\eta}_\theta i_\theta^{-1} = \frac{i_\theta^{-1}}{n}, \tag{4.28}$$

the Cramér–Rao lower bound for estimating θ (Section 2.5), where we have used $\dot{\eta}_\theta^\top V_\theta \dot{\eta}_\theta = i_\theta$. Except in special cases, $\mathrm{Cov}_\theta(\hat{\theta})$ will exceed i_θ^{-1}; nevertheless, $i_{\hat{\theta}}^{-1}$ is usually a reasonable estimate of $\mathrm{Cov}(\hat{\theta})$, as the derivation above suggests.

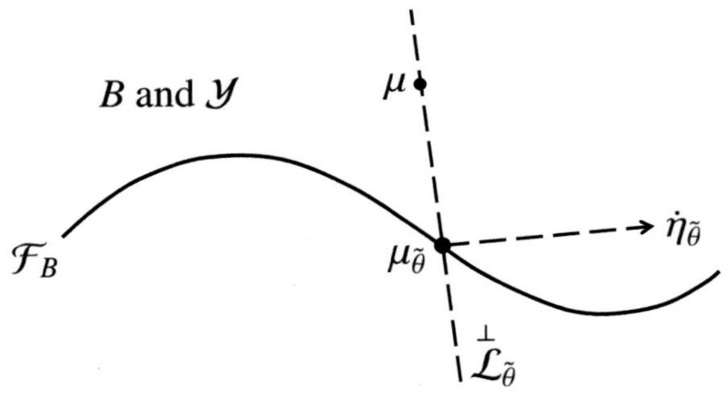

Figure 4.6 Model misspecification: the true mean $\boldsymbol{\mu}$ is not on $\mathcal{F}_B$.

Variance if Our Model is Wrong

Suppose we were wrong in assuming that $\mu \in \mathcal{F}_B$, and that actually μ lies somewhere else in B, as in Figure 4.6. Let

$$\tilde{\theta} = t(\mu),$$

so $\mu_{\tilde{\theta}}$ is the closest point to μ on $\mathcal{F}_B$ in terms of deviance distance (Hoeffding's picture). This means that μ, the true expectation of $\bar{y}$, lies on $\overset{\perp}{\mathcal{L}}_{\tilde{\theta}}$, passing through $\mathcal{F}_B$ at $\mu_{\tilde{\theta}}$, orthogonally to $\dot{\eta}_{\tilde{\theta}}$.

As $n \to \infty$, $\bar{y}$ goes to μ and

$$\hat{\theta} \longrightarrow t(\mu) = \tilde{\theta}.$$

We can still use the MLE influence function (4.25)

$$\left.\frac{d\hat{\theta}}{d\bar{y}}\right|_{\bar{y}=\mu} = \left(-\ddot{l}_{\tilde{\theta}}(\mu)\right)^{-1} \dot{\eta}_{\tilde{\theta}}^{\top} = \left[i_{\tilde{\theta}} - \ddot{\eta}_{\tilde{\theta}}^{\top}(\mu - \mu_{\tilde{\theta}})\right]^{-1} \dot{\eta}_{\tilde{\theta}}^{\top}$$

to approximate the covariance matrix $\operatorname{Cov}_\mu\{\hat{\theta}\}$,

$$\operatorname{Cov}_\mu\left\{\hat{\theta}\right\} \doteq \operatorname{Cov}_\mu\left\{\tilde{\theta} + \left.\frac{d\hat{\theta}}{d\bar{y}}\right|_\mu (\bar{y} - \mu)\right\} = \left.\frac{d\hat{\theta}}{d\bar{y}}\right|_\mu^{\top} \frac{V_\mu}{n} \left.\frac{d\hat{\theta}}{d\bar{y}}\right|_\mu$$

or

$$\operatorname{Cov}_\mu\left\{\hat{\theta}\right\} \doteq \left(-\ddot{l}_{\tilde{\theta}}(\mu)\right)^{-1}\left(\dot{\eta}_{\tilde{\theta}}^{\top}\frac{V_\mu}{n}\dot{\eta}_{\tilde{\theta}}\right)\left(-\ddot{l}_{\tilde{\theta}}(\mu)\right)^{-1}. \tag{4.29}$$

In applied use, we would estimate the covariance of $\hat{\theta}$ by substituting $\bar{y}$ for μ and $\hat{\theta}$ for $\tilde{\theta}$, giving

$$\operatorname{Cov}\left\{\hat{\theta}\right\} \doteq \left(-\ddot{l}_{\hat{\theta}}(\bar{y})\right)^{-1}\left(\dot{\eta}_{\hat{\theta}}^{\top}\frac{V_{\bar{y}}}{n}\dot{\eta}_{\hat{\theta}}\right)\left(-\ddot{l}_{\hat{\theta}}(\bar{y})\right)^{-1}, \tag{4.30}$$

sometimes called "Huber's sandwich estimator".

The covariance matrix $V_{\mu=\bar{y}}$ is the only probabilistic element of the exponential family $\mathcal{G}$ entering into (4.30). Everything else depends only on the geometry of $\mathcal{F}$'s location within $\mathcal{G}$.

Homework 4.9 Suppose $q = 1$, i.e., θ is real-valued, and that both $\mathcal{F}_A$ and $\mathcal{F}_B$ are specified. How might you numerically evaluate $-\ddot{l}_{\hat{\theta}}(\bar{y})$?

We can retreat further from $\mathcal{G}$ by using the nonparametric covariance estimate

$$\overline{V} = \frac{1}{n}\sum_{i=1}^{n}(y_i - \bar{y})(y_i - \bar{y})^{\top}$$

in place of $V_{\bar{y}}$. This results in the *nonparametric delta method* estimate of covariance,

$$\widehat{\text{Cov}}\left(\hat{\theta}\right) = \left(-\ddot{l}_{\hat{\theta}}(\bar{y})\right)^{-1} \left(\frac{\dot{\eta}_{\hat{\theta}}^{\top}\overline{V}\dot{\eta}_{\hat{\theta}}}{n}\right) \left(-\ddot{l}_{\hat{\theta}}(\bar{y})\right)^{-1}. \tag{4.31}$$

Homework 4.10 A nonparametric bootstrap sample $y_1^*, \ldots, y_n^*$ consists of a random sample of size n drawn *with* replacement from $(y_1, \ldots, y_n)$, and gives a bootstrap replication of $\hat{\theta}$,

$$\hat{\theta}^* = t\left(\sum_{i=1}^{n} y_i^*/n\right).$$

The bootstrap estimate of covariance for $\hat{\theta}$ is $\text{Cov}_*\{\hat{\theta}^*\}$, Cov_* indicating the covariance of $\hat{\theta}^*$ under bootstrap sampling, with $(y_1, \ldots, y_n)$ held fixed. Give an argument suggesting that $\text{Cov}_*\{\hat{\theta}^*\}$ is approximately the same as (4.31). *Hint*: $\bar{y}^* \overset{*}{\sim} (\bar{y}, \overline{V}/n)$.

Note In complicated models it is often easiest to evaluate $-\ddot{l}_{\hat{\theta}}(\bar{y})$ and $\dot{\eta}_{\hat{\theta}}$ numerically, by computing the likelihood and η_θ at the $2p$ points $\bar{y} \pm ce_i$, where $e_i = (0, \ldots, 1, 0, \ldots, 0)$, 1 in the ith place. Then the vector $\dot{\eta}_{\hat{\theta}}$ can be approximated by

$$(\eta_{\hat{\theta}+ce_i} - \eta_{\hat{\theta}-ce_i})(2\epsilon)^{-1},$$

and $-\ddot{l}_{\hat{\theta}}(\bar{y})$ by

$$\left(-l_{\hat{\theta}+ce_i}(\bar{y}) + 2l_{\hat{\theta}}(\bar{y}) - l_{\hat{\theta}-ce_i}(\bar{y})\right)(\epsilon^2)^{-1}.$$

Homework 4.11 What is the rationale for these expressions?

Table 4.1 *The spatial test data (NA = not available).*

	A	B		A	B		A	B
1	48	42	10	41	NA	19	32	NA
2	36	33	11	45	NA	20	24	NA
3	20	16	12	14	NA	21	47	NA
4	29	39	13	6	7	22	41	41
5	42	38	14	0	15	23	24	28
6	42	36	15	33	NA	24	26	14
7	20	15	16	28	NA	25	30	NA
8	42	33	17	34	NA	26	41	NA
9	22	NA	18	4	NA			

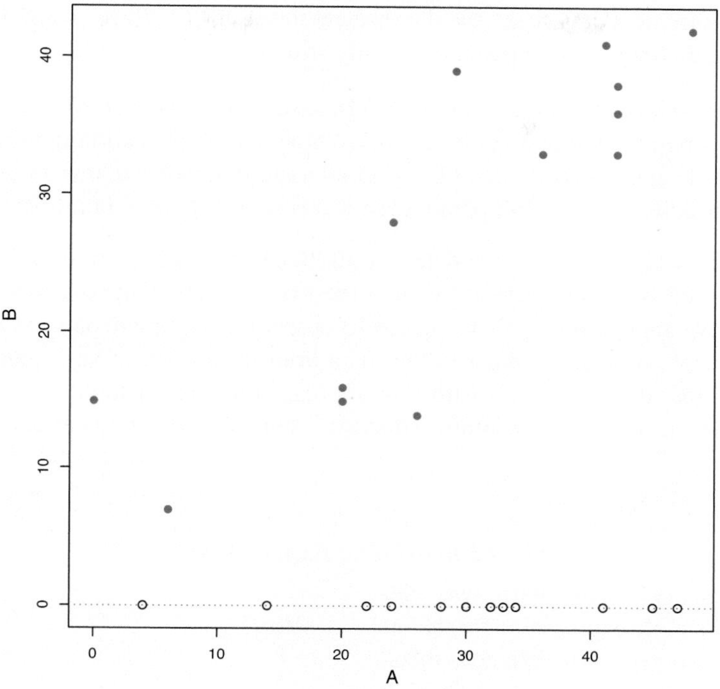

Figure 4.7 Spatial test data; open circles represent 13 children missing B measurement.

4.5 Missing Data and the Fisher–Louis Expressions

Exponential family structure is lost if some of a data set is missing. In the *spatial test data* of Table 4.1 and Figure 4.7, the original data set consisted of $n = 26$ pairs (A_i, B_i), each pair being two different measurements of spatial ability on a neurologically impaired child. However the B score was lost for 13 of the 26 children, as indicated by NA (not available).

Suppose we assume, perhaps not very wisely, that the (A_i, B_i) were originally bivariate normal pairs, obtained independently,

$$\begin{pmatrix} A_i \\ B_i \end{pmatrix} \overset{\text{ind}}{\sim} \mathcal{N}_2(\lambda, \Gamma), \qquad i = 1, \dots, 26. \tag{4.32}$$

This is a five-parameter exponential family (Section 2.8), but with the 13

B measurements missing,[3] the data we get to see, 13 (A, B) pairs and 13 A values, no longer has exponential family structure.

Homework 4.12 In addition to (4.32), assume that $A_i \overset{\text{ind}}{\sim} \mathcal{N}(\lambda_1, \Gamma_{11})$ for the 13 open circle points in Figure 4.7. Show that this situation represents a curved exponential family. (The normal assumptions make this a special case; usually we wouldn't obtain even a curved exponential family.)

In the absence of missing data – if all 26 (A_i, B_i) pairs were observable – it would be straightforward to find the MLE $(\hat{\lambda}, \hat{\Gamma})$. With data missing, the naïve approach, simply using the 13 observable complete pairs, is both inefficient and biased. What follows is a brief discussion of an expansive theory that deals correctly with missing data situations. The theory is not confined to exponential family situations, but takes on a neater form in them.

The Fisher–Louis Expressions

The theory begins with two elegantly simple expressions for the information lost when data goes missing. Let y be a data vector, of which we observe part o, with part u unobserved, and let $\mathcal{Y}(o_1)$ denote the set of y vectors having a certain value o_1 of o,

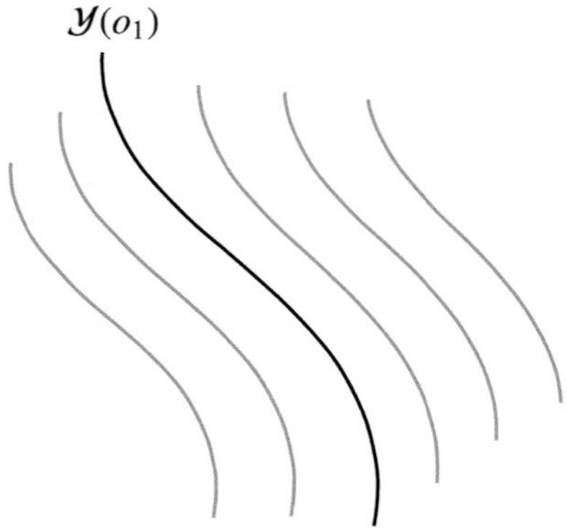

$$\mathcal{Y}(o_1) = \{y = (o, u) : o = o_1\}.$$

Fisher derived a simple but evocative expression for the score function based on observing only o:

Lemma 4.3 (Fisher)

$$\dot{l}_\theta(o) = E_\theta\left\{\dot{l}_\theta(y) \mid o\right\}. \tag{4.33}$$

The diagram above schematically depicts the level surfaces of $y = (o, u)$

3 It is crucial in what follows that missingness should be random, in the sense that the chance B is missing should not depend on the value of A. This is the same point as the "no look ahead" requirement for the Kaplan–Meier estimate of Homework 3.19 in Section 3.6, another missing data situation.

given o. The original score function $\dot{l}_\theta(y)$ is defined at every point on $\mathcal{Y}(o)$; averaging over the *conditional* position of y in $\mathcal{Y}(o_1)$ according to the conditional density given $o = o_1$ provides the missing data score function $\dot{l}_\theta(o)$.

Fisher's lemma applies to all smoothly defined q-parameter families, not just exponential families. Its proof is particularly transparent when the sample space $\mathcal{Y}$ of y is discrete. The discrete density of o is then

$$f_\theta(o) = \sum_{\mathcal{Y}(o)} f_\theta(y).$$

Letting $\dot{f}_\theta$ indicate the gradient $(\partial f_\theta/\partial\theta_j)$, a q-vector,

$$\begin{aligned} \dot{f}_\theta(o) &= \sum_{\mathcal{Y}(o)} \dot{f}_\theta(y) = \sum_{\mathcal{Y}(o)} \frac{\dot{f}_\theta(y)}{f_\theta(y)} f_\theta(y) \\ \text{or} \quad \frac{\dot{f}_\theta(o)}{f_\theta(o)} &= \sum_{\mathcal{Y}(o)} \frac{\dot{f}_\theta(y)}{f_\theta(y)} \frac{f_\theta(y)}{f_\theta(o)}, \end{aligned} \tag{4.34}$$

which is (4.33) since $f_\theta(y)/f_\theta(o)$ is the conditional density of y given o.

There is also an expression for $-\ddot{l}_\theta(o)$, attributed in its multiparameter version to Tom Louis:

Lemma 4.4 (Louis)

$$-\ddot{l}_\theta(o) = E_\theta\left\{-\ddot{l}_\theta(y) \mid o\right\} - \mathrm{Cov}_\theta\left\{\dot{l}_\theta(y) \mid o\right\}. \tag{4.35}$$

The covariance matrix of the MLE $\hat{\theta}$ based on y is approximately $-\ddot{l}_\theta(y)^{-1}$, and likewise $-\ddot{l}_\theta(o)^{-1}$ for the MLE covariance based on o. Lemma 4.4 says that we lose some of the minus-second-derivative matrix in going from y to o, and should expect a corresponding increase in the MLE's covariance matrix.

Homework 4.13 What happens in the Fisher–Louis expressions if o is sufficient for y?

Homework 4.14 Prove (4.35) (assuming discreteness if you wish).

The Fisher–Louis lemmas take on a more specific form in exponential families. Suppose now that $\boldsymbol{y} = (y_1, \ldots, y_n)$ is an i.i.d. sample from a curved exponential family

$$f_\theta(y_i) = g_{\eta_\theta}(y_i) = e^{\eta_\theta^\top y_i - \psi(\eta_\theta)} g_0(y_i),$$

but we only observe part o_i of each y_i, $\boldsymbol{o} = (o_1, \ldots, o_n)$. Let $y_i(\theta)$ and $V_i(\theta)$ indicate the conditional expectation and covariance of y_i given o_i,

$$y_i(\theta) = E_\theta\{y_i \mid o_i\} \quad \text{and} \quad V_i(\theta) = \text{Cov}_\theta\{y_i \mid o_i\}. \tag{4.36}$$

Since

$$\dot{l}_\theta(y_i) = \dot{\eta}_\theta^\top(y_i - \mu_\theta) \quad \text{and} \quad -\ddot{l}_\theta(y_i) = i_\theta - \ddot{\eta}_\theta^\top(y_i - \mu_\theta),$$

the Fisher–Louis expressions become

$$\begin{aligned} \dot{l}_\theta(o_i) &= \dot{\eta}_\theta^\top\left(y_i(\theta) - \mu_\theta\right), \\ -\ddot{l}_\theta(o_i) &= i_\theta - \ddot{\eta}_\theta^\top\left(y_i(\theta) - \mu_\theta\right) - \dot{\eta}_\theta^\top V_i(\theta)\dot{\eta}_\theta. \end{aligned} \tag{4.37}$$

If we also assume that the missing data mechanism $y_i \to o_i$ operates independently for $i = 1, \ldots, n$, then the o_i are independent, $\dot{l}_\theta(\boldsymbol{o}) = \sum_1^n \dot{l}_\theta(o_i)$ and $\ddot{l}_\theta(\boldsymbol{o}) = \sum_1^n \ddot{l}_\theta(o_i)$, putting the Fisher–Louis expressions for $\boldsymbol{o}$ into compact form:

$$\begin{aligned} \dot{l}_\theta(\boldsymbol{o}) &= n\left[\dot{\eta}_\theta^\top\left(\bar{y}(\theta) - \mu_\theta\right)\right] && \left(\bar{y}(\theta) = \sum_{i=1}^n \frac{y_i(\theta)}{n}\right), \\ -\ddot{l}_\theta(\boldsymbol{o}) &= n\left[i_\theta - \ddot{\eta}_\theta^\top\left(\bar{y}(\theta) - \mu_\theta\right) - \dot{\eta}_\theta^\top \overline{V}(\theta)\dot{\eta}_\theta\right] && \left(\overline{V}(\theta) = \sum_{i=1}^n \frac{V_i(\theta)}{n}\right). \end{aligned} \tag{4.38}$$

Fisher's Picture with Missing Data

The MLE equation is now

$$\hat{\theta} : \dot{\eta}_{\hat{\theta}}^\top\left(\bar{y}(\hat{\theta}) - \mu_{\hat{\theta}}\right) = 0.$$

Fisher's picture, Figure 4.8, is the same as in Figure 4.2 (with $\bar{y}$ for y) *except* that $\bar{y}$ is now $\bar{y}(\theta)$, itself a function of the unknown parameter θ. This induces extra variability in $\hat{\theta}$, as suggested by the decrease of amount $n\dot{\eta}_{\hat{\theta}}^\top \overline{V}(\hat{\theta})\dot{\eta}_{\hat{\theta}}$ in the observed Fisher information.

Homework 4.15 Suppose $o_i = y_i$ for $i = 1, \ldots, n_1$, and $o_i =$ NA for $i = n_{1+1}, n_{1+2}, \ldots, n$. What does Figure 4.8 look like?

Homework 4.16 Assuming the normal model (4.32), use the nonlinear maximizer `nlm` to find the maximum likelihood estimate of λ and Γ.

In missing data problems with more complicated structure it can be arduous to directly calculate maximum likelihood estimates. The EM algorithm, a clever computational device, can often lessen the computational burden.

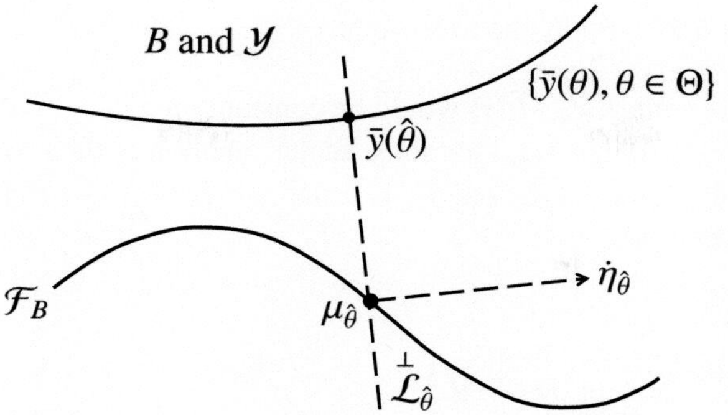

Figure 4.8 Fisher's picture from Figure 4.2 with $\bar{y}$ now $\bar{y}(\theta)$.

Reference Dempster et al. (1977), "Maximum likelihood from incomplete data via the EM algorithm", *JRSS-B* 1–38.

4.6 Statistical Curvature

Exponential families of distributions enjoy desirable inferential properties such as sufficiency of the MLE. Some of the ideal is lost when we move to curved exponential families. The theory of statistical curvature provides a quantitative description of the loss, in particular giving a striking picture of the MLE's insufficiency in curved families.[4]

We begin with a one-dimensional curved exponential family, $q = 1$, where $\mathcal{F}_A$ and $\mathcal{F}_B$ are curves through $\mathcal{R}^p$ and with sample size $n = 1$, so $\bar{y} = y$; this last being no restriction since the geometry in Fisher's and Hoeffding's pictures stays the same for all n.

If $\mathcal{F}$ were an (uncurved) one-parameter subfamily of $\mathcal{G}$ then the MLE $\hat{\theta}$ would be a sufficient statistic for θ. This is not the case for curved families, where different values of y on $\overset{\perp}{\mathcal{L}}_{\hat{\theta}}$ yield the same MLE $\hat{\theta} = t(y)$ but different amounts of observed information,

$$I(y) = -\ddot{l}_{\hat{\theta}}(y) = i_{\hat{\theta}} - \ddot{\eta}_{\hat{\theta}}^{\top}(y - \mu_{\hat{\theta}}). \tag{4.39}$$

This happens because $\hat{\theta}$ is *not* sufficient in curved families. Initially, Fisher believed in the sufficiency of the MLE, but notional pictures such as Figure 4.2 led him to the theory of observed information and a deeper understanding of maximum likelihood estimation.

[4] Derivations and details for the material in this section appear in Efron (1975, 1978, 2018); Efron and Hinkley (1978).

Homework 4.17 Show the following:

(a) If $\hat{\theta}$ were sufficient, $-\ddot{l}_{\hat{\theta}}(y)$ would be constant on $\overset{\perp}{\mathcal{L}}_{\hat{\theta}}$.
(b) If $\ddot{\eta}_\theta = c_\theta \dot{\eta}_\theta$ for all θ and for some scalar function c_θ then $-\ddot{l}_{\hat{\theta}}(y) \equiv i_{\hat{\theta}}$.
(c) Under the same condition, $\mathcal{F}_A$ is a flat subset of A (so $\mathcal{F}$ is an exponential family).

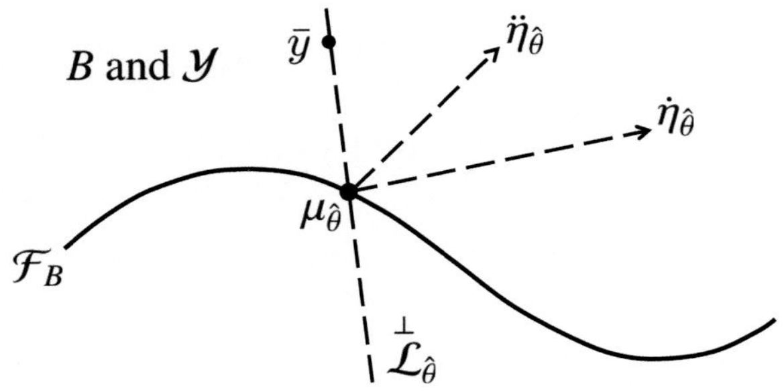

Figure 4.9 Curved exponential family, $q = 1$.

When $q = 1$, $\ddot{\eta}_{\hat{\theta}}$ as well as $\dot{\eta}_{\hat{\theta}}$ is a p-dimensional vector, and we can draw a helpful schematic diagram like that shown in Figure 4.9. Fisher argued for $I(y)^{-1}$ rather than $i_{\hat{\theta}}^{-1}$ as an estimate of $\text{Cov}(\hat{\theta})$. From Figure 4.9 we see $I(y) = i_{\hat{\theta}} - \ddot{\eta}_{\hat{\theta}}^\top(y - \mu_{\hat{\theta}})$ is *less* than $i_{\hat{\theta}}$ for y above $\mu_{\hat{\theta}}$ ("above" meaning in the direction best aligned with $\ddot{\eta}_{\hat{\theta}}$), presumably making $\hat{\theta}$ *more* variable; and conversely for y below $\mu_{\hat{\theta}}$. Curvature theory is intended to quantify this argument.

For a given value of θ, denote the 2×2 covariance matrix of $\dot{l}_\theta(y) = \dot{\eta}_\theta^\top(y - \mu_\theta)$ and $\ddot{l}_\theta(y) = \ddot{\eta}_\theta^\top(y - \mu_\theta) - i_\theta$ by

$$\begin{pmatrix} \nu_{11\theta} & \nu_{12\theta} \\ \nu_{21\theta} & \nu_{22\theta} \end{pmatrix} = \begin{pmatrix} \dot{\eta}_\theta^\top V_\theta \dot{\eta}_\theta & \dot{\eta}_\theta^\top V_\theta \ddot{\eta}_\theta \\ \ddot{\eta}_\theta^\top V_\theta \dot{\eta}_\theta & \ddot{\eta}_\theta^\top V_\theta \ddot{\eta}_\theta \end{pmatrix} \tag{4.40}$$

(so $\nu_{11\theta} = i_\theta = E_\theta\{-\ddot{l}_\theta(y)\}$). The residual of $\ddot{l}_\theta(y)$ after linear regression on $\dot{l}_\theta(y)$,

$$\overset{\perp}{l}_\theta(y) = \ddot{l}_\theta(y) - \frac{\nu_{12\theta}}{\nu_{11\theta}} \dot{l}_\theta(y),$$

has variance

$$\text{Var}_\theta\left\{\overset{\perp}{l}(y)\right\} = \nu_{22\theta} - \frac{\nu_{12\theta}^2}{\nu_{11\theta}}.$$

Definition The *statistical curvature* γ_θ of $\mathcal{F}$ at θ is

$$\gamma_\theta = \left(\frac{\nu_{22\theta}}{\nu_{11\theta}^2} - \frac{\nu_{21\theta}^2}{\nu_{11\theta}^3} \right)^{1/2} = \frac{\text{sd}_\theta \left\{ -\overset{\perp}{\ddot{l}}_\theta(y) \right\}}{E_\theta \left\{ -\ddot{l}_\theta(y) \right\}}, \tag{4.41}$$

using $E_\theta\{-\ddot{l}_\theta(y)\} = \nu_{11\theta}$. (In classical differential geometry terms, γ_θ is the curvature of $\mathcal{F}_A$ in the metric V_θ.)

Curvature can be computed for general one-parameter families $\mathcal{F} = \{f_\theta(y), \theta \in \Theta\}$, not necessarily of exponential family form. Having calculated

$$\dot{l}_\theta(y) = \frac{\partial}{\partial\theta} \log f_\theta(y) \quad \text{and} \quad \ddot{l}_\theta(y) = \frac{\partial^2}{\partial\theta^2} \log f_\theta(y),$$

we define the 2×2 matrix (4.40) as the covariance matrix of $(\dot{l}_\theta(y), \ddot{l}_\theta(y))$ and then compute γ_θ as in (4.41).

Some important properties are derived in Efron (1975):

- The value of the curvature is invariant under smooth monotonic transformations of θ and y.
- $\gamma_\theta = 0$ for all $\theta \in \Theta$ if and only if $\mathcal{F}$ is a one-parameter exponential family.
- Large values of γ_θ indicate a breakdown of exponential family properties; for instance, locally most powerful tests of $H_0 : \theta = \theta_0$ won't be globally most powerful.
- In repeated sampling situations, $\gamma_{\theta,n} = \gamma_\theta n^{-1/2}$ (so increased sample sizes make $\mathcal{F}_n$ more exponential family-like).
- In repeated sampling situations, with θ the true value,

$$\frac{-\ddot{l}_{\hat{\theta},n}(\bar{y})}{i_{\hat{\theta},n}} = \frac{-\ddot{l}_{\hat{\theta}}(\bar{y})}{i_{\hat{\theta}}} \longrightarrow \mathcal{N}(1, \gamma_\theta^2/n). \tag{4.42}$$

This says that $\gamma_{\theta,n} = \gamma_\theta n^{-1/2}$ determines the variability of the observed Fisher information $I_n(\bar{y})$ around the expected Fisher information $i_{\hat{\theta},n}$.

Efron and Hinkley (1978) show that there is reason to take $I_n(\bar{y})^{-1/2}$ rather than $i_{\hat{\theta},n}^{-1/2}$ as the estimate of $\text{sd}\{\hat{\theta}\}$. If $\gamma_{\hat{\theta},n}^2$ is large, say $\geq 1/8$, then the two estimates can differ substantially. In uncurved families, having $\gamma_\theta = 0$, $I_n(\bar{y})$ always equals $i_{\hat{\theta}}$, but the greater the value of γ_θ the more the two information measures can diverge.

Homework 4.18 Suppose $\gamma_{\hat{\theta},n} = 1/8$. What is a rough 68% interval for the ratio of the two estimates of $\text{sd}\{\hat{\theta}\}$?

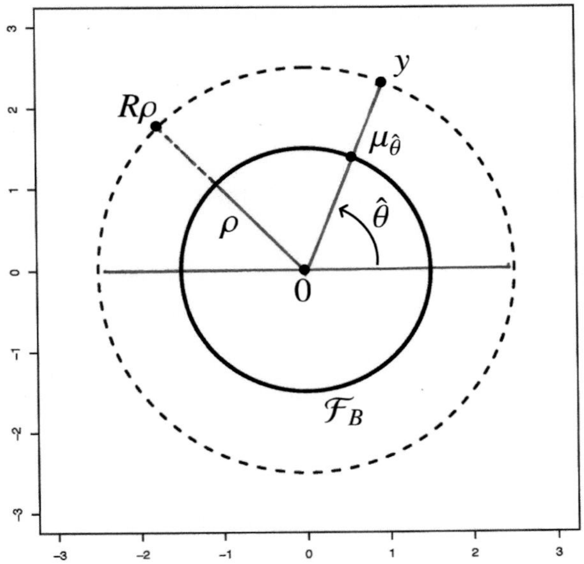

Figure 4.10 Fisher's circle model.

Fisher's Circle Model

Fisher provided a salient example of what we have been calling a curved exponential family, designed to show why $I(y)^{-1/2}$ is better than $i_{\hat\theta}^{-1/2}$ as an assessment of sd$\{\hat\theta\}$. The example has $p = 2$, and y bivariate normal with identity covariance matrix $y \sim \mathcal{N}_2(\mu, I)$, where μ lies on a circle of known radius ρ:

$$\mathcal{F}_B = \left\{ \mu_\theta = \rho \begin{pmatrix} \cos\theta \\ \sin\theta \end{pmatrix}, -\pi < \theta \leq \pi \right\}.$$

Homework 4.19 Show that:

(a) $i_\theta = \rho^2$ for all θ;
(b) $\gamma_\theta = \rho^{-1}$ for all θ;
(c) $\mu_{\hat\theta}$ is the point on $\mathcal{F}_B$ nearest y;
(d) $I(y) = -\ddot{l}_{\hat\theta}(y) = Ri_\theta$ if y lies on a circle of radius $R\rho$ (i.e., if $\|y\| = R\rho$).

Finally, give a heuristic argument supporting Fisher's preference for $I(y)$ in place of $i_{\hat\theta}$. *Hint*: Consider the case illustrated in Figure 4.10.

R is an *ancillary statistic*: a random variable whose distribution does not depend on θ, but whose value determines the accuracy of $\hat{\theta}$ as an estimate of θ.

Homework 4.20 What is the distribution of R?

Bayesians have criticized frequentist calculations averaging over possible data sets that are different from the one actually observed: $\mathrm{sd}(\hat{\theta}) = i_{\hat{\theta}}^{-1/2}$ is accurate on average, but is misleading for R very large or very small. Conditioning on ancillary statistics was Fisher's method for reconciling frequentist and Bayesian calculations (though he wouldn't have said it that way). Like the MLE itself, $I(y)$ depends only on the observed likelihood function and not its frequentist framework, moving it closer to the Bayesian perspective. From a practical point of view, $I(y)$ can be numerically calculated directly from the likelihood, obviating the need for theoretical derivation of i_θ.

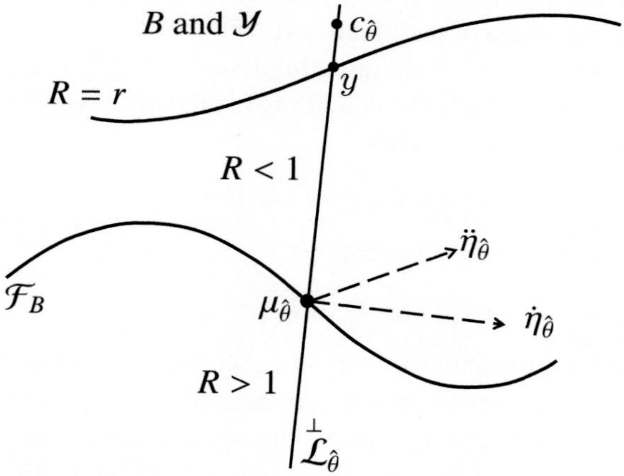

Figure 4.11 Approximate ancillarity of statistic R.

Approximate versions of ancillarity hold more generally. This is illustrated in Figure 4.11, drawn for the situation $p = 2$ and $q = 1$. From $I(y) = i_{\hat{\theta}} - \ddot{\eta}_{\hat{\theta}}^\top (y - \mu_{\hat{\theta}}) = -\ddot{l}_{\hat{\theta}}(y)$, we see that there must be a critical point $c_{\hat{\theta}}$ above $\mu_{\hat{\theta}}$ on $\overset{\perp}{\mathcal{L}}_{\hat{\theta}}$ such that

$$I(c_{\hat{\theta}}) = -\ddot{l}_{\hat{\theta}}(c_{\hat{\theta}}) = 0. \tag{4.43}$$

Homework 4.21 Show that $i_{\hat{\theta}} = \ddot{\eta}_{\hat{\theta}}^\top (c_{\hat{\theta}} - \mu_{\hat{\theta}})$.

Let $\rho_{\hat{\theta}}$ be the *radius of curvature* at $\hat{\theta}$, $\rho_{\hat{\theta}} = \gamma_{\hat{\theta}}^{-1}$. Some results from Efron and Hinkley (1978) are:

- $\rho_{\hat{\theta}} = [(c_{\hat{\theta}} - \mu_{\hat{\theta}})^\top V_{\hat{\theta}}^{-1}(c_{\hat{\theta}} - \mu_{\hat{\theta}})]^{1/2}$, that is, $\rho_{\hat{\theta}}$ equals the Mahalanobis distance from $c_{\hat{\theta}}$ to $\mu_{\hat{\theta}}$.
- Let R be the proportional distance of y from $c_{\hat{\theta}}$ to $\mu_{\hat{\theta}}$,

$$R = \frac{||c_{\hat{\theta}} - y||}{||c_{\hat{\theta}} - \mu_{\hat{\theta}}||} \tag{4.44}$$

(in any metric since $\overset{\perp}{\mathcal{L}}_{\hat{\theta}}$ is one-dimensional). Then

$$R = \frac{-\ddot{l}_{\hat{\theta}}(y)}{i_{\hat{\theta}}} = \frac{I(y)}{i_{\hat{\theta}}},$$

with the repeated sampling version of R asymptotically $\mathcal{N}(1, \gamma_{\theta}^2/n)$. R is > 1 below $\mathcal{F}_B$ and < 1 above it.
- Above the critical point $c_{\hat{\theta}}$, R is less than 0 and $\hat{\theta}$ is a local *minimum* rather than *maximum* of the likelihood.
- Let C_r be the curve in $\mathcal{Y}$ representing those y having R equal to some particular value r. Then the conditional standard deviation of $\hat{\theta}$ given that y is on C_r is approximately

$$\text{sd}\left\{\hat{\theta} \mid R = r\right\} \doteq \sqrt{r i_{\theta}}^{-1}, \tag{4.45}$$

leading to the data-based estimate

$$\text{sd}\left\{\hat{\theta} \mid R\right\} \doteq \sqrt{R i_{\hat{\theta}}}^{-1} = \sqrt{I(y)}^{-1}. \tag{4.46}$$

R is approximately ancillary for θ – i.e., its distribution (almost) doesn't depend on θ – strengthening the rationale for (4.46).[5] Note that $I_n(\bar{y})/i_{\hat{\theta},n} = I(\bar{y})/i_{\hat{\theta}}$, so the curve C_r stays the same under repeated sampling: the difference being that $\bar{y}$ tends closer to $\mathcal{F}_B$, making R closer to 1.

Homework 4.22 (a) In Figure 4.11, identify the elements of Fisher's circle model, $c_{\hat{\theta}}$, $\dot{\eta}_{\hat{\theta}}$, etc.
(b) Give a heuristic argument justifying (4.46) in this case.

Large values of the curvature $\gamma_{\hat{\theta}}$ make $\rho_{\hat{\theta}} = \gamma_{\hat{\theta}}^{-1}$ smaller; since $\rho_{\hat{\theta}}$ is the Mahalanobis distance of $c_{\hat{\theta}}$ from $\mu_{\hat{\theta}}$, this moves the critical point $c_{\hat{\theta}}$ nearer to $\mathcal{F}_B$, potentially destabilizing the MLE. In addition to increased variance,

[5] A better approximate ancillary is $\tilde{R} = (R - 1)\gamma_{\hat{\theta}}^{-1}$, better in the sense of having its distribution depend less on θ, but conditioning on $\tilde{R}$ lends again to $I(y)^{-1/2}$ as the preferred estimate of sd$\{\hat{\theta}\}$. The underlying rationale for (4.46) is geometrical, as exemplified by Fisher's circle model.

there is the possibility of false roots, the probability of falling on the far side of the critical boundary being approximately

$$\Pr_\theta\{R < 0\} \doteq \Phi(-\gamma_\theta^{-1}), \tag{4.47}$$

Φ the standard normal CDF.

Curvature theory also has something to say about Fisher's claim that the MLE is a superior estimate of θ. As in Section 4.2, suppose that $\tilde{\theta} = t(\bar{y})$ is a locally unbiased and first-order efficient estimator of the true value θ, so that asymptotically

$$\lim_{n\to\infty} n \operatorname{Var}_{\theta,n}\left\{\tilde{\theta}\right\} = i_\theta.$$

The theorem in Section 10 of Efron (1975) says that

$$\operatorname{Var}_\theta\left\{\tilde{\theta}\right\} \doteq \frac{1}{ni_\theta} + \frac{1}{n^2 i_\theta}\{\gamma_\theta^2 + C_1 + C_2\}, \tag{4.48}$$

where

C_1 is the same positive constant for all choices of the first-order efficient estimator $\tilde{\theta} = t(\bar{y})$,

while

$C_2 \geq 0$, equaling zero only for the MLE.

This result demonstrates the "second-order efficiency of the MLE": using an estimator other than the MLE increases asymptotic variance at the $1/n^2$ term.

Some Examples of One-parameter Curved Families

- **Normal with Known Coefficient of Variation** Suppose $x \sim \mathcal{N}(\theta, \Gamma)$ as in Section 2.8, but with $\Gamma = \tau\theta^2$, τ known (so the coefficient of variation of x is $\sqrt{\Gamma}(|\theta|)^{-1} = \sqrt{\tau}$). This is a one-parameter curved family in $p = 2$ dimensions, with $y = (x, x^2)$ and

$$\eta_\theta = \frac{1}{\tau}\left(\frac{1}{\theta}, -\frac{1}{2\theta^2}\right)^\top.$$

Homework 4.23 (a) For this model, augment Figure 2.6 to include $\mathcal{F}_A$ and $\mathcal{F}_B$. (b) What is γ_θ?

- **Autoregressive Process** $x_0, x_1, \ldots, x_T$ are observations of an autoregressive process, with $x_0 = u_0$ and

$$x_{t+1} = \theta x_t + (1 - \theta^2)^{1/2} u_{t+1} \qquad (t = 1, \ldots, T), \tag{4.49}$$

where $u_t \overset{\text{ind}}{\sim} \mathcal{N}(0, 1)$ for $t = 0, 1, \ldots, T$, and $\Theta = (-1, 1)$.

Homework 4.24 Show that the preceding is a one-parameter curved family in $p = 3$ dimensions, and give expressions for y and η_θ. (Efron 1975 shows that $\gamma_0^2 = (8T - 6)/T^2$.)

- **A Genetics Linkage Model** (Dempster et al., 1977) In an animal genetics study, four phenotypes occur in proportions

$$\pi_\theta = \left(\frac{1}{2} + \frac{\theta}{4}, \frac{1}{4} - \frac{\theta}{4}, \frac{1}{4} - \frac{\theta}{4}, \frac{\theta}{4}\right), \tag{4.50}$$

where θ is an unknown parameter between 0 and 1. A sample of $n = 197$ animals gave observed proportions

$$p = (125, 18, 20, 34)/197,$$

with $p \sim \text{Mult}_4(n, \pi_\theta)/n$ as in Section 2.9. Model (4.50) is linear in the B space, as shown in Figure 4.12, but curved in the A space, making it a one-parameter curved family: with η_θ the curve $\log \pi_\theta$, $\mathcal{F}_B$ is the line

$$\pi_\theta = a + b\theta \qquad \text{where} \begin{cases} a = \left(\frac{1}{2}, \frac{1}{4}, \frac{1}{4}, 0\right) \\ b = \left(\frac{1}{4}, -\frac{1}{4}, -\frac{1}{4}, \frac{1}{4}\right), \end{cases}$$

Homework 4.25 (a) Write an iterative program to find the MLE $\hat{\theta}$.
(b) Provide an estimated standard error of $\hat{\theta}$.
(c) Show, numerically, that $\gamma_{\hat{\theta},n} = 0.0632$.

- **Cauchy Translation Family** A Cauchy observation y has density

$$g_\theta(y) = \frac{1}{\pi} \frac{1}{1 + (y - \theta)^2} \qquad (y \text{ and } \theta \in \mathcal{R}^1), \tag{4.51}$$

the translation parameter θ being the centerpoint of the symmetric density. Though *not* an exponential family, we can differentiate the log likelihood function $l_\theta(y) = -\log[1 + (y - \theta)^2] - \log \pi$ to get $\dot{l}_\theta(y)$ and $\ddot{l}_\theta(y)$, and then compute their covariance matrix (4.40) to get the curvature (4.41), $\gamma_\theta = \sqrt{2.5}$ (the same for all θ because this is a translation family). A Cauchy sample of say $n = 10$ observations $y_1, \ldots, y_n$ would have

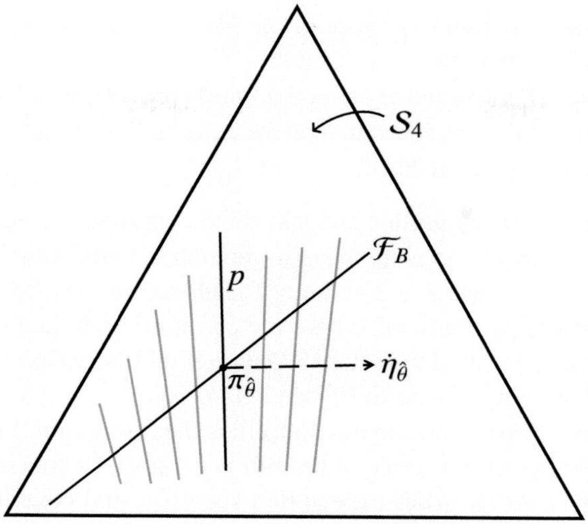

Figure 4.12 Genetics linkage model.

$\gamma_{\theta,n} = \sqrt{0.25} = 0.50$: a large curvature implying, correctly, difficulties with multiple roots to the MLE equation $\dot{l}_{\hat{\theta}}(\mathbf{y}) = 0$.

For θ near 0 we can expand $l_\theta(y)$ in a Taylor series,

$$l_\theta(y) = l_0(y) + \dot{l}_0(y)\theta + \ddot{l}_0(y)\theta^2/2 + \dddot{l}_0(y)\theta^3/6 + \cdots, \tag{4.52}$$

and think of this as a one-parameter curved exponential family having sufficient vector $(\dot{l}_0(y), \ddot{l}_0(y), \dddot{l}_0(y), \dots)$ and natural parameter vector

$$\eta_\theta = (\theta, \theta^2/2, \theta^3/6, \dots). \tag{4.53}$$

The point here is that curved exponential families can be thought of as an approximating framework for all smoothly defined parametric families, with the amount of curvature indicating favorable or unfavorable properties of the MLE. Section 4.7 discusses an example of a multiparameter curved family ($q > 1$) and generalizations of Figure 4.11.

4.7 Regions of Stability for the MLE

Fisher made three bold claims for maximum likelihood estimation, all more or less still accepted, at least in low-parameter families:

1. That maximum likelihood is a superior general method for estimation.

2. That the unconditional variance of the MLE is approximately one over the Fisher information.
3. That outside of uncurved exponential family models, the observed Fisher information is more relevant than the expected information for assessing the accuracy of the MLE.

Curved exponential families and statistical curvature illustrate and verify the three claims. The ideas have a geometric foundation, pictured in Figure 4.11, for the case $p = 2$ and $q = 1$. This section extends the discussion to general (p, q) families, where the geometry is basically the same, though more complicated to picture.[6] Only a brief discussion is developed here, the full story appearing in Efron (2011).

A transformation of coordinates simplifies the discussion. Let η_0 be any point in A, the parameter space of the full p-parameter exponential family $\mathcal{G}$ (Section 4.1). At η_0, y has expectation vector μ_0 and covariance matrix V_0. Also let M be a symmetric $p \times p$ square-root matrix of V_0, $M^2 = V_0$. (If V_0 has eigenvalue-eigenvector representation $V_0 = \Gamma D \Gamma^\top$, D the diagonal matrix of eigenvalues, we can set $M = \Gamma D^{1/2} \Gamma^\top$.)

Transform η and y to

$$\tilde{\eta} = M\eta \quad \text{and} \quad \tilde{y} = M^{-1}(y - \mu_0). \tag{4.54}$$

Then $\tilde{y}$ has expectation 0 and covariance matrix

$$M^{-1} V_0 M^{-1} = I_p,$$

the $p \times p$ identity matrix at $\eta = \eta_0$. The family $\mathcal{G}$ transforms into an equivalent exponential family,

$$\widetilde{\mathcal{G}} = \left\{ \tilde{g}_{\tilde{\eta}}(\tilde{y}) = e^{\tilde{\eta}^\top \tilde{y} - \tilde{\psi}(\tilde{\eta})} \tilde{g}_0(\tilde{y}), \tilde{\eta} \in \tilde{A}, \tilde{y} \in \widetilde{\mathcal{Y}} \right\}, \tag{4.55}$$

with $\tilde{y} \sim (0, I_p)$ at $\tilde{\eta}_0 = M\eta_0$.

Homework 4.26 Verify (4.55).

In what follows, we will assume that (η, y) have *already been transformed*, with η_0 chosen to be the MLE point $\eta_{\hat{\theta}}$, so that

$$\mu_{\hat{\theta}} = 0 \quad \text{and} \quad V_{\hat{\theta}} = I_p \tag{4.56}$$

in the curved family $\mathcal{F}$. All the calculations will be conditional on $\theta = \hat{\theta}$, making (4.56) legitimate for the purpose of easy discussion.

[6] This section contains some more specialized material, though the discussion of penalized maximum likelihood estimation is of general interest.

Returning to one-parameter curved families as in Section 4.6, let $\hat{\nu}_{11} = \nu_{11\hat{\theta}}$, $\hat{\nu}_{12} = \nu_{12\hat{\theta}}$, and $\hat{\nu}_{22} = \nu_{22\hat{\theta}}$ in (4.40), and define

$$\overset{\perp}{\eta}_{\hat{\theta}} = \ddot{\eta}_{\hat{\theta}} - \frac{\hat{\nu}_{12}}{\hat{\nu}_{11}} \dot{\eta}_{\hat{\theta}}, \tag{4.57}$$

the component of $\ddot{\eta}_{\hat{\theta}}$ orthogonal to $\dot{\eta}_{\hat{\theta}}$.

Figure 4.13 generalizes Figure 4.11 to $p \geq 2$ (still keeping $q = 1$ for the sake of illustration). Now $\overset{\perp}{\mathcal{L}}_{\hat{\theta}} = \{y : \text{MLE} = \hat{\theta}\}$ is a flat space of dimension $p - 1$. $\mathcal{B}_{\hat{\theta}}$ is the $(p-2)$-dimensional subspace of $\overset{\perp}{\mathcal{L}}_{\hat{\theta}}$ for which the observed information $I(y) = -\ddot{l}_{\hat{\theta}}(y)$ equals 0; $c_{\hat{\theta}}$ is the nearest point to $\mu_{\hat{\theta}}$ in $\mathcal{B}_{\hat{\theta}}$ and has Mahalanobis distance $\gamma_{\hat{\theta}}^{-1}$ from $\mu_{\hat{\theta}}$. All of this is verified in Homework 4.27 and Homework 4.28 which follow.

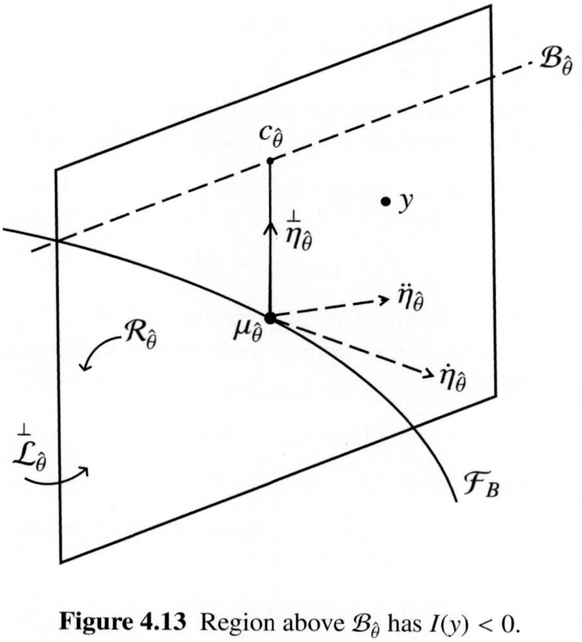

Figure 4.13 Region above $\mathcal{B}_{\hat{\theta}}$ has $I(y) < 0$.

Homework 4.27 Show that, in terms of the examples described in Section 4.6, (a) $\|\overset{\perp}{\eta}_{\hat{\theta}}\| = i_{\hat{\theta}}\gamma_{\hat{\theta}}$ and (b) $-\ddot{l}_{\hat{\theta}}(y) = i_{\hat{\theta}} - \overset{\perp}{\eta}{}_{\hat{\theta}}^{\top}(y - \mu_{\hat{\theta}})$.

Homework 4.28 Let $v_{\hat{\theta}} = \overset{\perp}{\eta}_{\hat{\theta}} / i_{\hat{\theta}}\gamma_{\hat{\theta}}$ (so $\|v_{\hat{\theta}}\| = 1$). Show that:

(a) $c_{\hat{\theta}} = v_{\hat{\theta}}/\gamma_{\hat{\theta}}$;

(b) $I(y) = i_{\hat\theta}(1 - b\gamma_{\hat\theta})$, where $y = \mu_{\hat\theta} + bv_{\hat\theta} + r$, with r in $\overset{\perp}{\mathcal{L}}_{\hat\theta}$ and $v_{\hat\theta}^\top r = 0$.

The critical boundary

$$\mathcal{B}_{\hat\theta} = \{y = \mu_{\hat\theta} + v_{\hat\theta}/\gamma_{\hat\theta} + r\} \tag{4.58}$$

separates $\overset{\perp}{\mathcal{L}}_{\hat\theta}$ into halves having $I(y) > 0$ or < 0. By definition, the *stable region* $\mathcal{R}_{\hat\theta}$ is the good half,

$$\mathcal{R}_{\hat\theta} = \{y : y = \mu_{\hat\theta} + bv_{\hat\theta} + r, b < \gamma_{\hat\theta}^{-1}\}. \tag{4.59}$$

Large values of the curvature move $\mathcal{B}_{\hat\theta}$ closer to $\mu_{\hat\theta}$, destabilizing the MLE, allowing $I(y)$ to vary more from $i_{\hat\theta}$ and even to go negative. Since $\mathrm{Cov}_{\hat\theta}(y) = I_p$ in our transformed coordinates, we have the approximation

$$\Pr_{\hat\theta}\{I(y) < 0\} \doteq \Phi(-\gamma_{\hat\theta}^{-1}), \tag{4.60}$$

Φ the standard normal CDF.

What happens in multiparameter curved exponential families, that is, those with $q > 1$? A schematic answer appears in Figure 4.14. The stable region $\mathcal{R}_{\hat\theta}$ is now a convex subset of $\overset{\perp}{\mathcal{L}}_{\hat\theta}$. For a given unit vector u in p-dimensions, let

$$\mathcal{F}_u = \{f_{\hat\theta+\lambda u}, \lambda \in \Lambda\} \tag{4.61}$$

be a one-parameter subfamily of $\mathcal{F}$, as in Section 2.6. Here Λ is an interval of R^1 containing 0 as an interior point. $\mathcal{F}_u$ determines Fisher information i_u and curvature γ_u at $\lambda = 0$, and also direction vector $\overset{\perp}{\eta}_u/i_u\gamma_u$.

Finally, let a_u be the angle between $\overset{\perp}{\eta}_u$ and $\overset{\perp}{\mathcal{L}}_{\hat\theta}$ ($a_u = 0$ if $q = 1$), and w_u the unit projection vector of $\overset{\perp}{\eta}_u$ into $\overset{\perp}{\mathcal{L}}_{\hat\theta}$. The $q = 1$ theory yields a half-space $\mathcal{R}_u$ of $\overset{\perp}{\mathcal{L}}_{\hat\theta}$ as the stable region for family $\mathcal{F}_u$. Its boundary $\mathcal{B}_u$ lies at distance d_u from $\mu_{\hat\theta}$ and it can be shown that

$$d_u = [\cos(a_u) \cdot \gamma_u]^{-1}. \tag{4.62}$$

Theorem 4.5 *The observed Fisher information matrix* $I(y) = -\ddot{l}_{\hat\theta}(y)$ *is positive definite if and only if* y *is in the* stable region

$$\mathcal{R}_{\hat\theta} = \bigcap_u \mathcal{R}_u, \tag{4.63}$$

where the intersection is over all p-dimensional unit vectors. (The boundary of $\mathcal{R}_{\hat\theta}$ *is labeled "*$\mathcal{B}_{\hat\theta}$*" in Figure 4.14.)*

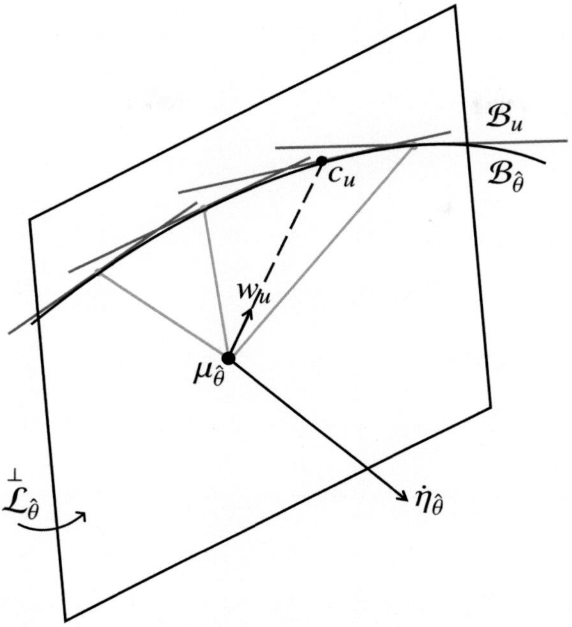

Figure 4.14 $\mathcal{R}_{\hat{\theta}}$ of Theorem 4.5 is region below boundary $\mathcal{B}_{\hat{\theta}}$.

All of this is verified in Efron (2018). An example having $p = 37$ and $q = 8$ (dimension $\overset{\perp}{\mathcal{L}}_{\hat{\theta}} = 37 - 8 = 29$) appears here in Section 4.8.

Homework 4.29 In family $\mathcal{F}_u$, show that, at $\lambda = 0$,

(a) the expected Fisher information $i_u = u^\top i_{\hat{\theta}} u$;
(b) the observed Fisher information $I_u(y) = u^\top I(y) u$;
(c) $u^\top I(y) u = u^\top i_{\hat{\theta}} u \cdot (1 - b/d_u)$, where $y = \mu_{\hat{\theta}} + b w_u + r$, with r in $\overset{\perp}{\mathcal{L}}_{\hat{\theta}}$ and orthogonal to w_u.

Once again, large curvatures move the critical boundary $\mathcal{B}_{\hat{\theta}}$ closer to $\mu_{\hat{\theta}}$, increasing conditioning effects and potentially destabilizing the MLE. Note that not all choices of u contribute to $\mathcal{B}_{\hat{\theta}}$. In our examples, some choices give enormous values of d_u, so that $\mathcal{B}_u$ lies far beyond $\mathcal{B}_{\hat{\theta}}$.

Figure 4.15 shows a toy example from Efron (2018). The observed data y comprises seven independent Poisson observations,

$$y_i \overset{\text{ind}}{\sim} \text{Poi}(\mu_i), \qquad i = 1, \dots, 7, \tag{4.64}$$

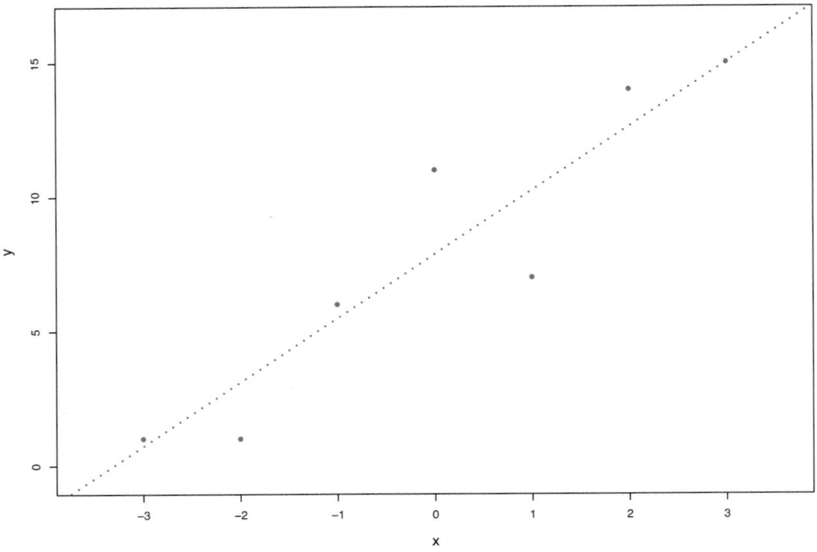

Figure 4.15 Toy example: $y_i \sim \text{Poi}(\mu_i)$ for $i = 1, \ldots, 7$; curved model $\mu_i = \theta_0 + \theta_1 x_i$; MLE line $\hat{\theta} = (7.86, 2.38)$.

$y = (1, 1, 6, 11, 7, 14, 15)$. A hypothesized model supposes that

$$\mu_i = \theta_1 + \theta_2 x_i, \qquad \text{for } x_i = -3, -2, -1, 0, 1, 2, 3. \tag{4.65}$$

In this case, $\mathcal{G}$ is a $(p = 7)$-parameter exponential family, having $\mathcal{F}$ as a two-parameter curved subfamily. (If instead we had taken $\log \mu_i = \theta_1 + \theta_2 x_i$, $\mathcal{F}$ would be a two-parameter GLM.)

Homework 4.30 Write an explicit expression of $\mathcal{F}$ for this $\mathcal{G}$.

Direct numerical calculation gave MLE $\hat{\theta} = (\hat{\theta}_1, \hat{\theta}_2) = (7.86, 2.38)$ in model (4.64)–(4.65), illustrated by the straight line fit $\hat{\theta}_1 + \hat{\theta}_2 x_i$ in Figure 4.15. The observed and expected Fisher information matrices were calculated to be

$$i_{\hat{\theta}} = \begin{pmatrix} 2.63 & -5.75 \\ -5.75 & 14.98 \end{pmatrix} \quad \text{and} \quad I(y) = \begin{pmatrix} 3.00 & -6.97 \\ -6.97 & 23.00 \end{pmatrix}.$$

Model (4.64)–(4.65) is a $(p, q) = (7, 2)$ curved family, making $\overset{\perp}{\mathcal{L}}_{\hat{\theta}}$ in Figure 4.14 a five-dimensional flat space. Direction/distance pairs (w_u, d_u) were calculated for 101 choices of u, $u(t) = (\cos t, \sin t)$, $t = k\pi/100$ for

$k = 0, 1, \ldots, 100$. Somewhat surprisingly, w_u was always the same,

$$w = (-0.390, 0.712, 0.392, 0.145, -0.049, -0.210, -0.340); \qquad (4.66)$$

d_u took on mininum value $d_{\min} = 2.85$ at $u = (0, 1)$, with the other d_u values ranging up to 1133. In this case, the stable region $\mathcal{R}_u$ was a half-space of $\overset{\perp}{\mathcal{L}}_{\hat{\theta}}$, with boundary $\mathcal{B}_{\hat{\theta}}$ minimum distance 2.85 from $\mu_{\hat{\theta}}$ (after transformation to standard coordinates (4.56)).

The observed information matrix $I(bw)$, w as in (4.66), decreases toward singularity as b increases; it becomes singular at $b = 2.85$, at which point its lower right corner equals 0. Further increases of b reduce other quadratic forms $u(t)^\top I(bw)u(t)$ to zero, as in Homework 4.29(c).

Homework 4.31 (a) Why was it the lower right corner? (b) Given a list of $d_{u(t)}$ versus t, which choice $u(t)^\top I(bw)u(t)$ would give the second zero?

From $y \sim (0, I_p)$ in our transformed coordinates, we get

$$b = (y - \mu_{\hat{\theta}})^\top w \sim (0, 1)$$

for the projection of y along w. Using Homework 4.29(a), applied to $u = (0, 1)$, the lower right corners of $I(y)$ and $i_{\hat{\theta}}$ have

$$I_{11}(y) i_{\hat{\theta}11}^{-1} \sim (1, 2.85^{-2}) = (1, 0.35^2), \qquad (4.67)$$

so we can expect conditioning effects on the order of 35%.

Penalized Maximum Likelihood Estimation

The performance of maximum likelihood estimates can often be improved by regularization, that is, by adding a penalty term to the log likelihood in order to tamp down volatile behavior of $\hat{\theta}$, this being especially true when the number of parameters p is large, as famously the case in large-scale prediction algorithms like *deep learning*.

We define the penalized log likelihood function $m_\theta(y)$ to be

$$m_\theta(y) = l_\theta(y) - s_\theta, \qquad (4.68)$$

where s_θ is a non-negative *penalty function*. Two familiar choices are $s_\theta = c \sum_1^p \theta_j^2$ ("ridge regression") and $s_\theta = c \sum_1^p |\theta_j|$ ("the lasso"), c a positive constant. The "g-modeling" example of the next section uses $c(\sum_1^p \theta_j^2)^{1/2}$. Larger values of c pull the penalized maximum likelihood estimate (pMLE) more strongly toward 0 (when using definitions such that $\theta = 0$ is a plausible choice in the absence of much data).

By definition, the pMLE is the value of θ maximizing $m_\theta(y)$. In a (p,q) curved exponential family $\mathcal{F}$, the equivalent of the score function condition $\dot{l}_\theta(y) = 0$ is $\dot{m}_\theta(y) = 0$ or, as in (4.13),

$$\dot{\eta}_{\hat\theta}^\top (y - \mu_{\hat\theta}) - \dot{s}_{\hat\theta} = 0,$$

where $\dot{s}_\theta$ is the q-dimensional gradient vector $(\partial s/\partial\theta_j)$. For a given value of $\hat\theta$, the set of y vectors satisfying this is the $(p-q)$-dimensional hyperplane $\overset{\perp}{\mathcal{M}}_{\hat\theta}$,

$$\overset{\perp}{\mathcal{M}}_{\hat\theta} = \{y : \dot{\eta}_{\hat\theta}^\top (y - \mu_{\hat\theta}) = \dot{s}_{\hat\theta}\}. \tag{4.69}$$

$\overset{\perp}{\mathcal{M}}_{\hat\theta}$ lies parallel to $\overset{\perp}{\mathcal{L}}_{\hat\theta}$ (4.15), but intersecting $\mathcal{F}_B$ at an offset from $\mu_{\hat\theta}$.

In Euclidean distance, the nearest point in $\overset{\perp}{\mathcal{M}}_{\hat\theta}$ to $\mu_{\hat\theta}$ is

$$\nu_{\hat\theta} = \mu_{\hat\theta} + \dot{\eta}_{\hat\theta}(\dot{\eta}_{\hat\theta}^\top \dot{\eta}_{\hat\theta})^{-1} \dot{s}_{\hat\theta}, \tag{4.70}$$

having squared distance from $\mu_{\hat\theta}$

$$\|\nu_{\hat\theta} - \mu_{\hat\theta}\|^2 = \dot{s}_{\hat\theta}^\top (\dot{\eta}_{\hat\theta}^\top \dot{\eta}_{\hat\theta})^{-1} \dot{s}_{\hat\theta}, \tag{4.71}$$

these being standard projection calculations.

Homework 4.32 Draw the pMLE equivalent of Figure 4.2.

By analogy with the observed information matrix $I(y) = -\ddot{l}_{\hat\theta}(y)$, we define

$$J(y) = -\ddot{m}_{\hat\theta}(y) = I(y) + \ddot{s}_{\hat\theta},$$

where $\ddot{s}$ is the $q \times q$ matrix $(\partial^2 s/\partial\theta_j\partial\theta_k)$. $J(y)$ plays a central role in the accuracy and stability of the pMLE, as discussed in Efron (2018). For instance, the influence function (4.25) becomes

$$\frac{\partial\hat\theta}{\partial y} = J(y)^{-1}\dot{\eta}_{\hat\theta}^\top.$$

Versions of Figure 4.13 and Figure 4.14 apply to the pMLE as well, with $J(y)$ and $\nu_{\hat\theta}$ playing the roles of $I(y)$ and $\mu_{\hat\theta}$.

4.8 Empirical Bayes Estimation Strategies: f-modeling and g-modeling

Empirical Bayes methods, introduced in Section 1.7, aim to use the data from parallel inference problems to carry out Bayesian estimation without the need to specify a prior distribution.[7] This section brings exponential family theory to bear on the two main empirical Bayes inferential strategies, *f-modeling* and *g-modeling*. After some initial discussion, we'll begin with f-modeling, the strategy used on the prostate cancer data of Figure 1.5.

A typical empirical Bayes estimation problem begins with a collection $\theta_1, \ldots, \theta_N$ of real-valued unobserved parameters sampled from an unknown probability density,

$$\theta_i \overset{\text{ind}}{\sim} g(\theta), \qquad \text{for } i = 1, \ldots, N. \tag{4.72}$$

Each θ_i independently produces an observation z_i according to a known probability density function $p(z \mid \theta)$,

$$z_i \mid \theta_i \sim p(z_i \mid \theta_i), \tag{4.73}$$

for instance $z_i \sim \mathcal{N}(\theta_i, 1)$ or $z_i \sim \text{Poi}(\theta_i)$.

We wish to estimate the θ_i, perhaps a specific one of them or perhaps all. If the prior density $g(\theta)$ were known, Bayes rule would yield ideal inferences based on the posterior density $g(\theta_i \mid z_i)$. It came as a pleasant surprise to statisticians in the 1950s[8] that, in situation (4.72)–(4.73), it is often possible to do nearly as well *without* knowledge of $g(\cdot)$.

How this can be done is the subject of empirical Bayes theory. A key element is the *marginal density* $f(z)$ (defined with respect to Lebesgue measure),

$$f(z) = \int_{\mathcal{T}} p(z \mid \theta) g(\theta)\, d\theta, \tag{4.74}$$

with $\mathcal{T}$ the sample space for the θs. The observed data $\boldsymbol{z} = (z_1, \ldots, z_N)$ is a random sample from $f(z)$,

$$z_i \overset{\text{iid}}{\sim} f(\cdot), \qquad i = 1, \ldots, N,$$

which is all the statistician gets to see.

In what follows, we will bin the data as in Section 3.4, partitioning the z-axis into K bins $\mathcal{Z}_k$, and letting $y_k = \#\{z_i \in \mathcal{Z}_k\}$ be the count in bin k. The count vector $\boldsymbol{y} = (y_1, \ldots, y_K)$ is a sufficient statistic for the discretized

[7] The notation in this section is specialized to empirical Bayes calculations, and differs from that in Section 1.7; the main references again are Efron and Hastie (2016) and Efron (2018).

[8] This work was pioneered by Herbert Robbins in key papers in the early 1950s; he coined the name "empirical Bayes".

data. It has a multinomial distribution (Section 2.9)

$$\boldsymbol{y} \sim \text{Mult}_K(N, \boldsymbol{f}),$$

where $\boldsymbol{f} = (f_1, \ldots, f_K)$ is a discretized version of $f(z)$, with

$$f_k = \int_{\mathcal{Z}_k} f(z)\, dz. \tag{4.75}$$

Here is a schematic diagram of the empirical Bayes model (4.72)–(4.75):

$$g \longrightarrow \boldsymbol{f} \longrightarrow \boldsymbol{y} \sim \text{Mult}_K(N, \boldsymbol{f}). \tag{4.76}$$

We observe $\boldsymbol{y}$ and wish to estimate various functions of g, perhaps

$$E\{\theta \mid z\} = \int_{\mathcal{T}} \theta p(z \mid \theta) g(\theta)\, d\theta / f(z). \tag{4.77}$$

Efficient estimation requires some form of parametric modeling. There are two basic choices, modeling g or modeling f, each with advantages and disadvantages.

We begin with an example of f-modeling. A diffusion tensor imaging (DTI) study compared six dyslexic children with six normal controls at 15,443 brain locations, or voxels. Here we will analyze only the $N = 477$ voxels located at the extreme back of the brain. Each voxel produced a statistic z_i comparing dyslexics with normals; a reasonable model for this is

$$z_i \sim \mathcal{N}(\theta_i, 1), \qquad i = 1, \ldots, N = 477, \tag{4.78}$$

with θ_i the true effect size for voxel i. The investigators were hoping to pinpoint voxels having θ_i far away from the "null" value 0, either in a positive or negative direction.

The left panel of Figure 4.16 shows a histogram of the 477 z-values, and a smooth fitted curve obtained from a Poisson GLM regression, as in Section 3.4. Letting c_k be the centerpoint of bin $\mathcal{Z}_k$ and $\boldsymbol{c} = (c_1, \ldots, c_K)$,

$$\hat{\boldsymbol{f}} = \texttt{glm}(\boldsymbol{y} \sim \texttt{poly}(\boldsymbol{c}, 5), \texttt{Poisson})\$\texttt{fit}/N$$

estimates the marginal density $f(z)$; the smooth curve is $N \cdot \hat{\boldsymbol{f}}$.

Some familiar empirical Bayes results can be computed directly from $\hat{f}$ without any need to estimate the prior density $g(\theta)$. Two of these appear in the right panel of Figure 4.16: the solid black curve shows Tweedie's estimate (Section 1.6)

$$\widehat{E}\{\theta \mid z\} = z + \frac{d}{dz} \log \hat{f}(z),$$

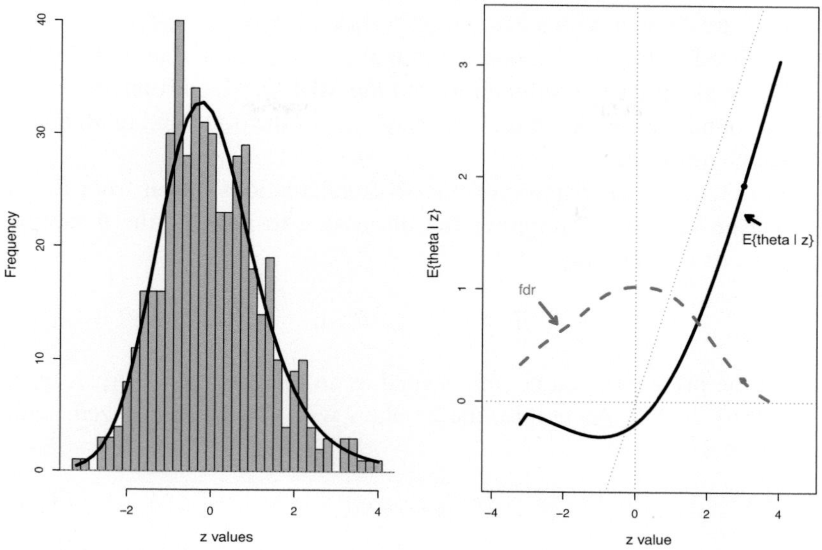

Figure 4.16 *Left*: DTI data, $N = 477$; `poly(df=5)` fit. *Right*: Tweedie estimate $E\{\theta \mid z\}$ at $z = 3$; $E\{\theta \mid z\} = 1.93$, fdr $= 0.20$.

while the dashed red curve is the local false discovery rate (Section 1.7)

$$\widehat{\text{fdr}}(z) = \widehat{\Pr}\{\theta = 0 \mid z\} \doteq \frac{\phi(z)}{\hat{f}(z)},$$

with $\phi(z) = \exp\{-z^2/2\}(2\pi)^{-1/2}$.

Homework 4.33 What is an estimate of $E\{\theta \mid z = 3 \textit{ and } \theta \neq 0\}$?

But what if we need to estimate something *not* in the small catalog of those having direct expressions in terms of $\hat{f}$? Reverting to the (θ_i, z_i) notation of (4.72)–(4.73), some examples might be

$$\Pr\{|\theta| > 2\} \quad \text{or} \quad E\{e^{-\theta} \mid z\}. \tag{4.79}$$

This is where g-modeling becomes essential. We assume that the prior density $g(\theta)$ belongs to a multiparameter exponential family, say $g_\beta(\theta)$, with β is an unknown p-dimensional natural parameter vector. Corresponding to each choice of β is a marginal density $f_\beta(z)$ (4.74). The schematic diagram (4.76) becomes

$$\beta \longrightarrow g_\beta \longrightarrow \boldsymbol{f}_\beta \longrightarrow \boldsymbol{y} \sim \text{Mult}_K(N, \boldsymbol{f}_\beta). \tag{4.80}$$

Now we are dealing with a *hidden exponential family*, $\theta \sim g_\beta(\cdot)$.

The disadvantage of g-modeling is that $z \sim f_\beta(\cdot)$ is *not* an exponential family,[9] making it more difficult to find the MLE $\hat{\beta}$. The advantage is that having found $\hat{\beta}$ we have a direct estimate $g_{\hat{\beta}}(\cdot)$ of the prior, and of all quantities such as (4.79).

Here is a brief description of the MLE calculations, taken from Efron and Hastie (2016). It simplifies the discussion to take $\mathcal{T}$, the θ sample space, to be discrete, say

$$\mathcal{T} = \{\theta_{(1)}, \dots, \theta_{(m)}\},$$

so that the prior $g(\theta)$ can be represented as an m-vector $\boldsymbol{g} = (g_1, \dots, g_m)$, $g_j = \Pr\{\theta = \theta_{(j)}\}$. As before, the z values will also be discretized, with sample space

$$\{z_{(1)}, \dots, z_{(K)}\} \tag{4.81}$$

and marginal density $\boldsymbol{f} = (f_1, \dots, f_K)$. Defining the $K \times m$ matrix $\boldsymbol{P}$,

$$\boldsymbol{P} = (p_{kj}), \qquad p_{kj} = \Pr\{z = z_{(k)} \mid \theta = \theta_{(j)}\}, \tag{4.82}$$

we have $f_k = \sum_j p_{kj} g_j$ or $\boldsymbol{f} = \boldsymbol{P}\boldsymbol{g}$.

The p-parameter exponential family for $\boldsymbol{g}_\beta$ is written as

$$\boldsymbol{g}_\beta = e^{\boldsymbol{Q}\beta - \phi(\beta)}, \tag{4.83}$$

where $\boldsymbol{Q}$ is an $m \times p$ matrix having jth row $q_j^\top$; that is,

$$g_j = e^{q_j^\top \beta - \phi(\beta)} \qquad \left(\phi(\beta) = \log \sum_{l=1}^{m} e^{q_l^\top \beta}\right). \tag{4.84}$$

Then $\boldsymbol{y} \sim \text{Mult}_K(N, \boldsymbol{f}_\beta)$ will be a (K, p) *curved* exponential family,[10] with $\boldsymbol{y}$ in the simplex $\mathcal{S}_K$, as in Figure 4.4.

Model (4.80)–(4.83) yields simple expressions for the score function and Fisher information. Define $w_{kj}(\beta) = g_j\{p_{kj}/f_k - 1\}$ and let $W_k(\beta)$ be the m-vector

$$W_k(\beta) = (w_{k1}(\beta), \dots, w_{km}(\beta))^\top,$$

[9] Except in the case where g_β represents a normal regression model and $z \mid \theta$ is also normal.

[10] Notice the change of notation to (K, p) from (p, q) used previously.

with $W_+(\beta) = \sum_1^K W_k(\beta)$. It turns out that

$$\dot{l}_\beta(\boldsymbol{y}) = \boldsymbol{Q}^\top W_+(\beta)\boldsymbol{Q},$$

$$I(\boldsymbol{y}) = -\ddot{l}_{\hat{\beta}}(\boldsymbol{y}) = \boldsymbol{Q}^\top \left(\sum_{k=1}^K W_k\left(\hat{\beta}\right) y_k W_k\left(\hat{\beta}\right)^\top - \operatorname{diag}\left\{ W_+\left(\hat{\beta}\right)\right\} \right) \boldsymbol{Q}, \tag{4.85}$$

$$\text{and} \qquad i_\beta = \boldsymbol{Q}^\top \left(\sum_{k=1}^K W_k(\beta)\{N f_k\} W_k(\beta)^\top \right) \boldsymbol{Q},$$

with $\operatorname{diag}\{W_+(\hat{\beta})\}$ denoting the diagonal matrix having entries $W_{+j}(\hat{\beta})$.

Homework 4.34 Show that $w_{kj}(\beta) = g(\theta = \theta_{(j)} \mid z = z_{(k)}) - g(\theta = \theta_{(j)})$.

The MLE $\hat{\beta}$ can be found by direct numerical maximization of the log likelihood, employing say, the R algorithm `nlm`. (Making use of the previous expression for $\dot{l}_\beta(\boldsymbol{y})$ accelerates the convergence rate of `nlm`.) Because $\boldsymbol{y} \sim \text{Mult}_K(N, \boldsymbol{f}_\beta)$ is not an exponential family there is the possibility of multiple local maxima. Regularization, as in Section 4.7, greatly improves the performance of $\hat{\beta}$. In what follows, the penalty function for the pMLE will be

$$s_\beta = \left(\sum_{j=1}^P \beta_j^2 \right)^{1/2}. \tag{4.86}$$

Homework 4.35 Show that

$$\dot{s}_\beta = \frac{\beta}{\|\beta\|} \quad \text{and} \quad \ddot{s}_\beta = \frac{I - \frac{\beta\beta^\top}{\|\beta\|^2}}{\|\beta\|}. \tag{4.87}$$

The g-model of (4.83) was applied to the DTI data[11] of Figure 4.16, $N = 477$, taking the θ sample space to be $\mathcal{T} = (-2.4, -2.2, \ldots, 3.6)$, $m = 31$. The structure matrix $\boldsymbol{Q}$ for the hidden GLM had $p = 8$ degrees of freedom,

$$\boldsymbol{Q} = [\delta_0, \texttt{poly}(\mathcal{T}, 7)], \tag{4.88}$$

δ_0 indicating a delta function at $\theta = 0$ (that is, vector $(0, \ldots, 1, 0, \ldots, 0)$ having 1 in the 16th place) and $\texttt{poly}(\mathcal{T}, 7)$ the $m \times 7$ matrix provided by the R function `poly`.

Specification (4.88) represents a "spike and slab" prior $g(\theta)$. It allows a spike of null cases at $\theta = 0$, i.e., zero difference betwen dyslexics and controls, and a smooth log polynomial distribution for the non-null cases.

[11] This example is carried out in more detail in Efron (2018).

The MLE $\hat{\beta}$ put weight 0.644 on the prior probability of 0, and estimated a mildly long-tailed density $\hat{g}(\theta)$ to the right of zero. This is indicated by the solid curve in panel A of Figure 4.17, pictured against a histogram of the z-values; $\hat{g}(\theta)$ gave for instance $\widehat{E}\{\theta \mid z = 3\} = 2.23$ and $\hat{P}\{\theta \geq 2.5\} = 0.029$.

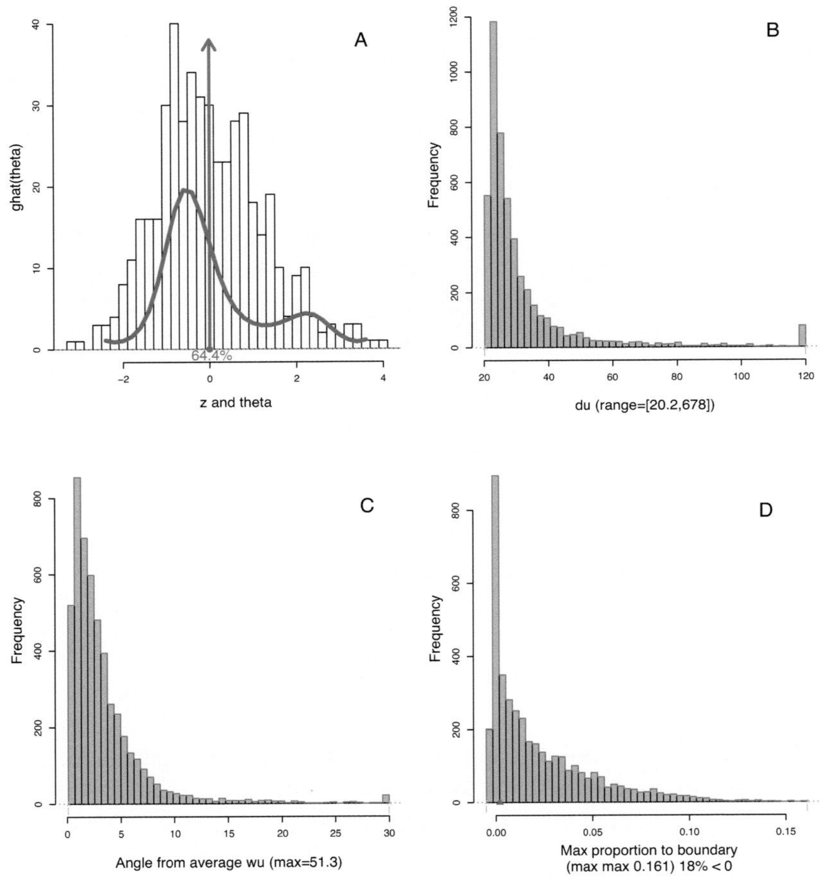

Figure 4.17 Empirical Bayes g-modeling analysis of DTI data, Figure 4.16. Panel A indicates fitted slab and spike prior in red. Panels B, C, and D refer to stability calculations for g-model fitting algorithm as described in the text.

An important question: Is the pMLE estimation process stable for g-modeling? To this end, a calculation of the stable region $\mathcal{R}_{\hat{\beta}}$ was carried out for the DTI data. The computations were done after transformation to standardized coordinates (4.56),

$$\mu_{\hat{\beta}} = 0 \quad \text{and} \quad V_{\hat{\beta}} = I_8. \tag{4.89}$$

We wish to compute the boundary $\mathcal{B}_{\hat{\beta}}$ as in Figure 4.14, now in $\overset{\perp}{\mathcal{M}}_{\hat{\beta}}$ (4.69), a 29-dimensional hyperplane. ($K = 37$, the number of bins in the panel A histogram, minus $p = 8$ equals 29.)

With $p = 8$, as opposed to $p = 2$ in the toy example of Figure 4.15, choosing the vectors u for the one-parameter bounding families $\mathcal{F}_u$ becomes challenging. For this analysis, 5000 u vectors were chosen randomly and uniformly from the surface of the unit sphere $\mathcal{S}^8$ in $\mathcal{R}^8$. Each u yielded a direction vector w_u and a bounding distance d_u (4.62). Panel B of Figure 4.17 is a truncated histogram of the 5000 d_u values,

$$20.2 \leq d_u \leq 678. \tag{4.90}$$

The minimum of d_u over *all* vectors on $\mathcal{S}^8$ – found by direct numerical minimization – was 20.02, suggesting that our 5000 random choices are enough to provide a good estimate of $\mathcal{B}_{\hat{\beta}}$.

Even though the u vectors pointed in all directions on $\mathcal{S}^8$, the direction vectors w_u indicated in Figure 4.14 clustered closely around their average vector $\bar{w}$. Panel C shows the histogram of angular distances in degrees between the 5000 w_us and $\bar{w}$: 95% of them are less than 10 degrees.[12]

The stable region $\mathcal{R}_{\hat{\beta}}$ has its boundary more than 20 standard Mahalanobis units from $\mu_{\hat{\beta}}$. Is this sufficiently distant to rule out unstable behavior? As a check, 4000 parametric bootstrap Y^* replications were generated,

$$Y_i^* \sim \text{Mult}_{37}(477, f_{\hat{\beta}}), \qquad i = 1, \ldots, 4000, \tag{4.91}$$

and then standardized and projected into vectors y_i^* in the 29-dimensional space $\overset{\perp}{\mathcal{M}}_{\hat{\beta}}$. Each y_i^* yielded a worst-case value,

$$m_i^* = \max_h \{ y_i^{*\top} w_{uh} d_{uh}^{-1}, h = 1, \ldots, 5000 \}; \tag{4.92}$$

$m_i^* > 1$ would indicate that y_i^* fell outside the stable region $\mathcal{R}_{\hat{\beta}}$.

Homework 4.36 Why is that last statement true?

In fact, none of the 4000 m_i^*s exceeded 0.16, so none of the y_i^*s fell anywhere near the boundary of $\mathcal{R}_{\hat{\beta}}$. For the actual observation $\boldsymbol{y}$, m equaled

[12] A spherical cap of radius 10° on the surface of a 29-dimensional sphere covers only proportion 3.9×10^{-23} of the "area" of the full sphere.

0.002, locating it quite close to $\mu_{\hat{\beta}}$. Observed and expected Fisher information are almost the same, and probably would not vary much for other possible observation vectors y^*.

Panel D of Figure 4.17 shows the histogram of the 4000 m_i^* values; 18% of them had $m_i^* < 0$. These represent y_i^* vectors having negative correlation with all 5000 w_{uh} direction vectors, implying that $\mathcal{R}_{\hat{\beta}}$ is open (going off to infinity) in some directions, just as suggested in Figure 4.14.

Homework 4.37 Draw a schematic picture supporting this last conclusion.

5

Bootstrap Confidence Intervals

5.1 Introduction

REFERENCE Efron (1987), "Better bootstrap confidence intervals", *JASA* 171–200.
REFERENCE DiCiccio and Efron (1996), "Bootstrap confidence intervals", *Statist. Sci.* 189–228.

Among the most useful, and most used, statistical constructions are the *standard intervals*

$$\hat{\theta} \pm z^{(\alpha)}\hat{\sigma}, \tag{5.1}$$

giving approximate confidence limits for a real-valued parameter of interest θ. Here $\hat{\theta}$ is a point estimate of θ, $\hat{\sigma}$ an estimate of its standard error, and $z^{(\alpha)}$ the αth quantile of a standard normal distribution. With $\alpha = 0.975$, the standard two-sided interval $\hat{\theta} \pm 1.96\hat{\sigma}$ has approximate coverage 0.95, with approximate noncoverage probability 0.025 in each tail.

The prime virtues of the standard intervals are the ease and universality of their construction. In principle, a single program can be written that automatically produces intervals (5.1), say with $\hat{\theta}$ a maximum likelihood estimate and $\hat{\sigma}$ its bootstrap standard error.

Their prime defect concerns accuracy: the actual coverage of $\hat{\theta} \pm 1.96\hat{\sigma}$ may only poorly approximate 0.95. Worse, from the point of view of what "confidence" is supposed to mean, the claimed tail-area noncoverage probabilities may be misleading, typically too large in one direction and too small in the other.

Here is an example. Table 5.1 shows the scores of $n = 22$ students, each of whom has taken five tests: mechanics, vectors, algebra, analysis, and statistics.[1] Suppose we are interested in the correlation between the mechanics and vectors scores. The usual Pearson correlation coefficient is calculated to be $\hat{\theta} = 0.498$. Suppose we are willing to assume that the (mechanics, vectors) scores follow a two-dimensional normal distribution. What is a 95% central interval for the true correlation θ?

The top row of Table 5.2 gives $(0.159, 0.837)$ as the the standard interval $\hat{\theta} \pm 1.96\hat{\sigma}$, using the familiar estimate

$$\hat{\sigma} = \frac{1 - \hat{\theta}^2}{(n-3)^{1/2}} = 0.173.$$

This is a poor approximation of the exact interval $(0.093, 0.751)$. The standard endpoints give noncoverage probabilities much too large on the left

[1] The 22 students were randomly selected from a larger set of 88 in Mardia et al. (1979).

Table 5.1 *Student score data; 22 students received scores from tests in five subjects.*

	Mechanics	Vectors	Algebra	Analysis	Statistics
1	7	51	43	17	22
2	44	69	53	53	53
3	49	41	61	49	64
4	59	70	68	62	56
5	34	42	50	47	29
6	46	40	47	29	17
7	0	40	21	9	14
8	32	45	49	57	64
9	49	57	47	39	26
10	52	64	60	63	54
11	44	61	52	62	46
12	36	59	51	45	51
13	42	60	54	49	33
14	5	30	44	36	18
15	22	58	53	56	41
16	18	51	40	56	30
17	41	63	49	46	34
18	48	38	41	44	33
19	31	42	48	54	68
20	42	69	61	55	45
21	46	49	53	59	37
22	63	63	65	70	63

Table 5.2 *Two-sided 95% CI endpoints for correlation between mechanics and vectors scores, Table 5.1; "noncoverage" is probability of falling on opposite side of* $\hat{\theta} = 0.498$ *when* θ *equals endpoint value, supposed to be 0.025. (Assumes a bivariate normal distribution for the two scores.) Fisher, bca, abc intervals explained in succeeding sections.*

	Endpoints		Noncoverages	
	Left	Right	Left	Right
Standard	0.159	0.837	0.0475	0.0015
Exact	0.093	0.751	0.0250	0.0250
Fisher	0.097	0.760	0.0261	0.0201
bca	0.083	0.754	0.0225	0.0230
abc	0.087	0.752	0.0236	0.0242

and too small on the right. The statistician gets a distorted picture of which θ values are or are not plausible.

Exact confidence intervals are unavailable in most situations. This accounts for the popularity of standard intervals but often leaves the applied statistician sailing in uncharted waters. The theory of bootstrap intervals was developed to fill the gap, using modern computing power to provide highly accurate intervals on a routine basis. The theory, outlined in Section 5.3 and Section 5.4, doesn't require exponential family structure, but this is misleading. It performs best on exponential family problems, and, more to the point, computer algorithms for its day-to-day application depend on exponential family relationships, as developed in Section 5.5 to Section 5.8. Section 5.9 concludes with a version of the bootstrap theory that applies to *fiducial inference*, Fishers' competitor to confidence intervals.

5.2 Exact Confidence Intervals

Suppose we observe data $\boldsymbol{x}$ from a parametric family of densities, and wish to construct a confidence interval for a real-valued parameter of interest θ. As the amount of information in $\boldsymbol{x}$ grows large – for instance, by repeated sampling – we can usually expect asymptotic normality to set in, with the maximum likelihood estimate $\hat{\theta}$ approaching its asymptotic normal distribution

$$\hat{\theta} \dot{\sim} \mathcal{N}(\theta, \sigma^2). \tag{5.2}$$

The standard intervals (5.1) take (5.2) literally: $\hat{\sigma}$, an estimate of σ, is obtained in some way from $\boldsymbol{x}$, giving the α-level one-sided standard upper confidence limit

$$\hat{\theta}_{\text{stand}}[\alpha] = \hat{\theta} + z^{(\alpha)}\hat{\sigma};$$

(5.1) is the two-sided version.

In some special situations, the density of $\hat{\theta}$ may depend only on θ, free of any nuisance parameters, say with densities $f_\theta(\hat{\theta})$. This is the case for the correlation example in Table 5.2 if we assume bivariate normality for the (mechanics,vectors) scores. Fisher's first statistics paper, "Frequency distribution of the values of the correlation coefficient in samples from an indefinitely large population" which appeared in *Biometrika* in 1915, derived the density formula

$$f_\theta\left(\hat{\theta}\right) = \frac{(n-2)(1-\theta^2)^{1/2(n-1)}(1-\hat{\theta}^2)^{1/2(n-4)}}{\pi} \times \int_1^\infty \frac{dx}{(x-\theta\hat{\theta})^{n-1}(x^2-1)^{1/2}}. \tag{5.3}$$

(Other expressions, equally formidable, are available; see Chapter 32 of Johnson and Kotz, 1970b.)

The solid black curve in Figure 5.1 shows (5.3) when θ equals 0.498, the MLE based on the $n = 22$ students in Table 5.1. With a long tail to the left, it is not very normal looking. Larger values of n would give nicer results, but evidently $n = 22$ is not big enough to induce normality in this situation. It's not surprising that the standard intervals perform poorly here.

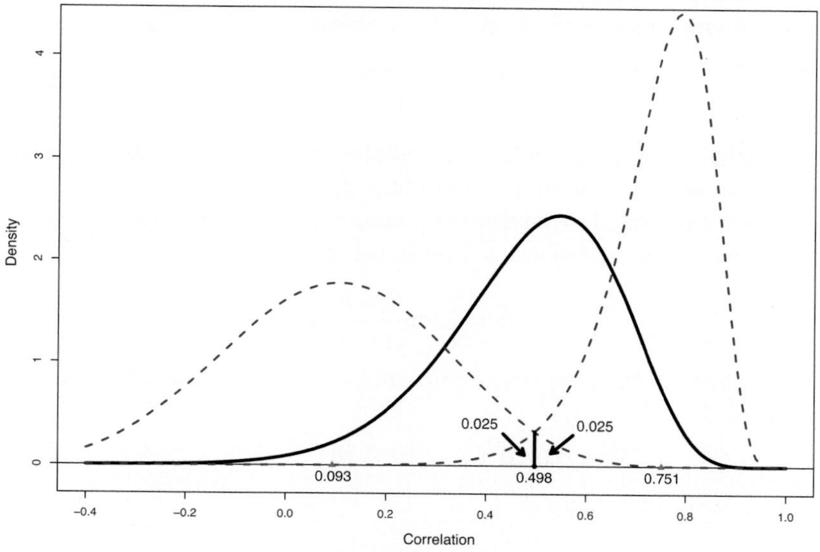

Figure 5.1 Solid curve: density (5.3) for correlation coefficient if $\theta = 0.498$, MLE $\hat{\theta}$. Dashed curves: $f_\theta(\cdot)$ for θ = exact endpoints 0.093, 0.751. These put probability 0.025 on opposite sides of $\hat{\theta}$ in accordance with Neyman's construction of exact intervals.

Under mild conditions, we can calculate *exact* confidence intervals for θ for a one-parameter family of densities $f_\theta(\hat{\theta})$. Figure 5.1 illustrates *Neyman's construction* for doing so: the lower exact limit $\hat{\theta}_{\text{exact}}[0.025] = 0.093$ is the smallest value of θ that puts probability at least 0.025 above the MLE $\hat{\theta} = 0.498$; that is, it's the smallest value of θ not rejected by a one-tailed level 0.025 test. The upper exact limit $\hat{\theta}_{\text{exact}}[0.975] = 0.751$ is the largest value of θ not rejected by a left-sided one-tailed 0.025 test. With θ fixed at its true value and $\hat{\theta}$ distributed according to $f_\theta(\hat{\theta})$, Neyman's construction yields exact 0.95 confidence intervals having 0.025 noncoverage in both tails, for any choice of θ.

Modern computers make short work of formula (5.3), but calculating the exact limits in Figure 5.1 would have been a major project in 1915. Fisher proposed a typically ingenious approximation. It relies on the fact that confidence intervals, by their nature, are *transformation invariant*: if $\hat{\theta}[\alpha]$ represents an upper one-sided level α confidence limit[2] for θ,

$$\Pr\left\{\theta \le \hat{\theta}[\alpha]\right\} = \alpha, \qquad \text{for all choices of } \theta$$

(the probability being calculated with θ the true value), and if $\phi = m(\theta)$ is a continuous monotonic increasing function of θ, then $\hat{\phi}[\alpha] = m(\hat{\theta}[\alpha])$ satisfies

$$\Pr\left\{\phi \le \hat{\phi}[\alpha]\right\} = \alpha, \qquad \text{for all } \phi;$$

that is, $\hat{\phi}[\alpha] = m(\hat{\theta}[\alpha])$ is a level α confidence limit for ϕ. This only says that the set $\{\theta \le \hat{\theta}[\alpha]\}$ is equivalent to $\{\phi \le \hat{\phi}[\alpha]\}$.

Using ideas related to the theory of Section 1.10, Fisher showed that the transformations $\phi = m(\theta)$ and $\hat{\phi} = m(\hat{\theta})$, using

$$\hat{\phi} = \frac{1}{2}\log\frac{1+\hat{\theta}}{1-\hat{\theta}}, \tag{5.4}$$

result in a much better normal approximation than (5.2), and nearly constant σ:

$$\hat{\phi} \mathrel{\dot{\sim}} \mathcal{N}(\phi, \sigma^2) \qquad \left[\sigma \doteq (n-3)^{-1/2}\right]. \tag{5.5}$$

The standard endpoint for ϕ,

$$\hat{\phi}[\alpha] = \hat{\phi} + z^{(\alpha)}\sigma,$$

can then be transformed back to $\hat{\theta}[\alpha] = m^{-1}(\hat{\phi}[\alpha])$ via the inverse transformation

$$\hat{\theta} = \frac{e^{2\hat{\phi}} - 1}{e^{2\hat{\phi}} + 1},$$

giving reasonably accurate confidence limits for θ, with very little computation required.

The third row of Table 5.2 shows Fisher's method giving good, if not perfect, results for the student score correlation problem. The more striking fact is that Fisher's transformation (5.4) does a beautiful job of transforming the highly non-normal family of densities (5.3) into something close to a *normal translation family* (5.5). Figure 5.2 shows the actual densities $f_\phi(\hat{\phi})$: the solid curve is for $\phi = m(0.498)$ while the dashed curves are for

[2] Here α, the intended coverage, will typically be a number near 1.0, such as 0.95, with two-sided noncoverage $1 - \alpha$.

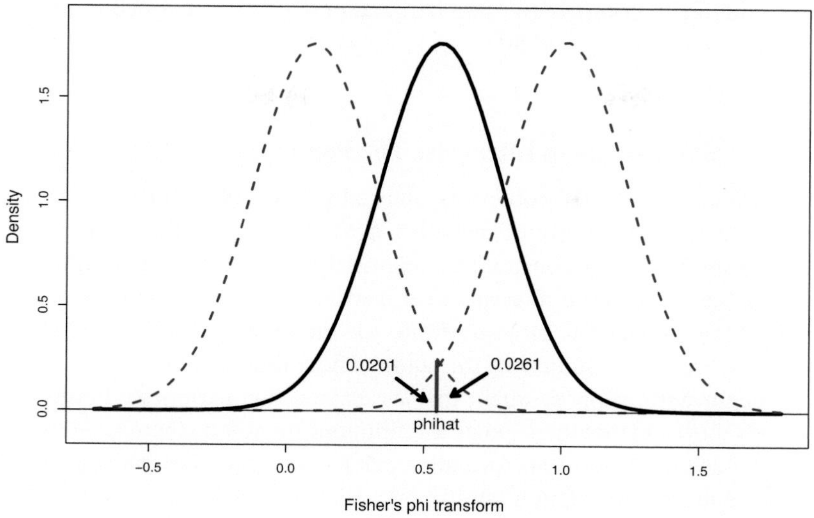

Figure 5.2 Fisher's transformation (5.4) gives $\hat{\phi}$ a nearly normal translation family of densities (5.3); the three curves correspond to those in Figure 5.1.

$\phi = m(0.097)$ and $\phi = m(0.760)$ (Fisher's endpoint values in Table 5.2), all three looking nicely normal. It is much easier to understand and work with model (5.5) than (5.2).

From an applied point of view, it is asking too much of the working statistician to derive appropriate transformations for each new situation. In what follows we will use bootstrap methods to implement an automatic version of Fisher's transformation, leading to highly accurate approximate confidence intervals in exponential families. The price is thousand-fold increases in computation, but that's an easy price to pay these days.

Homework 5.1 We observe $\hat{\theta} \sim \theta G_N/N$, where G_N is a gamma distribution with N degrees of freedom, as in Section 1.5.

(a) Give a formulaic expression for $\hat{\theta}_{\text{stand}}[\alpha] = \hat{\theta} + z^{(\alpha)}\hat{\sigma}$.

(b) Show that the exact level α endpoint is

$$\hat{\theta}_{\text{exact}}[\alpha] = N\hat{\theta}/G_N^{(1-\alpha)},$$

where $G_N^{(1-\alpha)}$ is the $(1-\alpha)$ quantile of a G_N distribution.

(c) Numerically compare the standard and exact 95% two-sided intervals for $N = 10, 20, 40$, and 80.

5.3 Bootstrap Intervals: The Percentile Method

The normal correlation problem is unusually tractable in that it reduces to a one-parameter family of densities $f_\theta(\hat{\theta})$. Most often this is impossible, making Neyman's construction unavailable. Bootstrap theory aims to produce highly accurate intervals over a wide variety of situations, using large-scale computation as a substitute for theoretical calculations. How this is possible is the subject of this and the next four sections.

To begin with, suppose again that we have a one-parameter family of densities $f_\theta(\hat{\theta})$ and that there exists a continuous monotone increasing function $m(\cdot)$ such that the transformations $\phi = m(\theta)$ and $\hat{\phi} = m(\hat{\theta})$ exactly produce a normal translation family[3]

$$\hat{\phi} \sim \mathcal{N}(\phi, 1), \qquad \text{for all } \phi. \tag{5.6}$$

It is easy to see that in this case exact confidence intervals exist and satisfy

$$\hat{\phi}_{\text{exact}}[\alpha] = \hat{\phi} + z^{(\alpha)}, \quad \text{and so} \quad \hat{\theta}_{\text{exact}}[\alpha] = m^{-1}\left(\hat{\phi} + z^{(\alpha)}\right). \tag{5.7}$$

Homework 5.2 Verify (5.7).

Now let $\hat{\theta}$ be the MLE of θ and $G(\cdot)$ be the cumulative distribution function of the density $f_{\hat{\theta}}(\cdot)$,

$$G(t) = \int_{-\infty}^{t} f_{\hat{\theta}}(t)\, dt.$$

Likewise, let $H(t)$ be the CDF of $\hat{\phi}^* \sim \mathcal{N}(\hat{\phi}, 1)$, with $\hat{\phi}$ the MLE of ϕ, $\hat{\phi} = m(\hat{\theta})$. Monotonicity implies that

$$G\left(m^{-1}\left(\hat{\phi}^*\right)\right) = H\left(\hat{\phi}^*\right),$$

for all $\hat{\phi}^*$.

By (5.6), (5.7), and the definition of $z^{(\alpha)}$,

$$H\left(\hat{\phi}_{\text{exact}}[\alpha]\right) = \alpha. \tag{5.8}$$

But by the transformation invariance of confidence intervals and CDFs,

$$G\left(\theta_{\text{exact}}[\alpha]\right) = H\left(\phi_{\text{exact}}[\alpha]\right) = \alpha.$$

[3] If $\hat{\phi} \sim \mathcal{N}(\phi, \sigma^2)$ for some constant σ then $\tilde{m}(\theta) = m(\theta)/\sigma$ gives (5.6), so we can assume to begin with that $\sigma = 1$.

This yields an interesting result:

Lemma 5.1 *If a one-parameter family $f_\theta(\hat{\theta})$ can be monotonically transformed into a normal translation family* (5.6)*, then*

$$\hat{\theta}_{\text{exact}}[\alpha] = G^{-1}(\alpha), \tag{5.9}$$

the αth quantile of the density $f_{\hat{\theta}}(\cdot)$.

Homework 5.3 Give a careful derivation of Lemma 5.1.

Notice that the actual transformation $\phi = m(\theta)$ is *not* needed in (5.9); only the assumption that it exists is required. This opens the door to a computer-driven confidence interval algorithm:

1. Draw a random sample of size B from $f_{\hat{\theta}}(\cdot)$, say

$$\hat{\theta}^*(1), \ldots, \hat{\theta}^*(B) \overset{\text{iid}}{\sim} f_{\hat{\theta}}(\cdot).$$

   (This is called a *parametric bootstrap* sample, with the star notation intended to differentiate the resampled values $\hat{\theta}^*$ from the original $\hat{\theta}$. In Figure 5.1, the $\hat{\theta}^*$s would be drawn according to the heavy black density curve. Size B might be 1000 or 2000, as discussed later.)
2. Calculate $\widehat{G}$, the empirical CDF of the $\hat{\theta}^*$s,

$$\widehat{G}(t) = \#\left\{\hat{\theta}^*(i) \leq t\right\}/B.$$

3. Approximate $\hat{\theta}_{\text{exact}}[\alpha]$ by the *percentile method* endpoint,

$$\hat{\theta}_{\text{pct}}[\alpha] = \widehat{G}^{-1}(\alpha).$$

   In other words, $\hat{\theta}_{\text{pct}}[\alpha]$ is the empirical 100αth percentile of the $\hat{\theta}^*$s.

It seems we might be stating the obvious here: that the α level confidence limit is the αth percentile of the bootstrap distribution, but there is more to the story as Section 5.4 will show.

Figure 5.3 shows the histogram of $B = 10{,}000$ bootstrap replications drawn from $f_{0.498}(\cdot)$ in Figure 5.1. They gave $[0.113, 0.766]$ as the 95% percentile interval for the student score correlation coefficient. $B = 10{,}000$ replications is more than usually needed. Moreover, in this case we have an analytic formula for $f_{0.498}(\cdot)$ that allows us to directly calculate the ideal $B = \infty$ percentile limits,

$$\hat{\theta}_{\text{pct}}[\alpha] = [0.104, 0.760].$$

Homework 5.4 Describe the calculation.

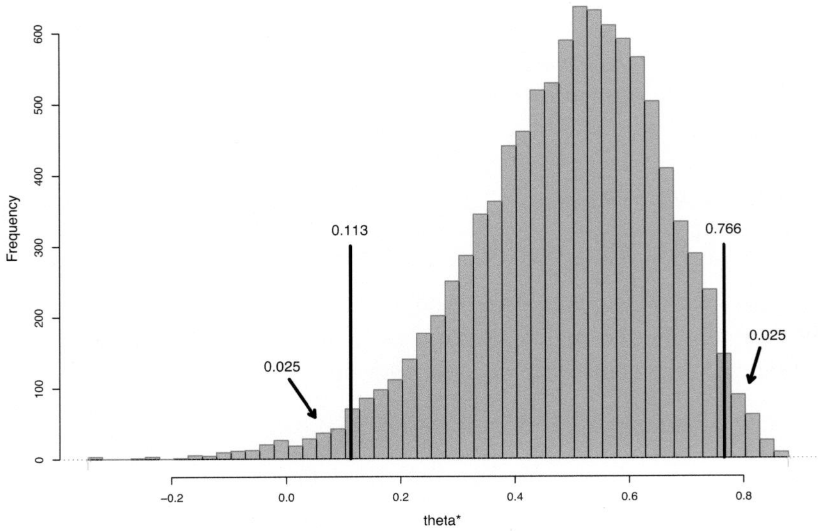

Figure 5.3 10,000 parametric bootstrap resamples from heavy solid curve in Figure 5.1; 95% percentile interval $[0.113, 0.766]$.

Table 5.3 appends the percentile results to those in Table 5.2, showing them doing better than Standard but not as well as Fisher, bca, or abc. The last two make further corrections to the standard intervals, to be discussed in Section 5.4 and Section 5.5.

Table 5.3 *Table 5.2 for student score correlation, now including percentile method interval.*

	Endpoints		Noncoverages	
	Left	Right	Left	Right
Standard	0.159	0.837	0.0475	0.0015
Exact	0.093	0.751	0.0250	0.0250
Fisher	0.097	0.760	0.0261	0.0201
bca	0.083	0.754	0.0225	0.0230
abc	0.087	0.752	0.0236	0.0242
Percentile	0.104	0.760	0.0281	0.0199

An important property of the percentile intervals is their *transformation invariance*: for any monotone transformation $\phi = m(\theta)$, the percentile endpoints transform correctly,

$$\phi_{\text{pct}}[\alpha] = m\left(\theta_{\text{pct}}[\alpha]\right). \tag{5.10}$$

This isn't true for the standard intervals, which can perform poorly on a "wrong" scale. Over the years, applied statisticians developed special transformations, like (5.4), to improve the standard method. One can describe the percentile method as a device for automatically selecting the best scale, a device that only the advent of electronic computation made practical.

Homework 5.5 Verify the transformation invariance of the percentile intervals.

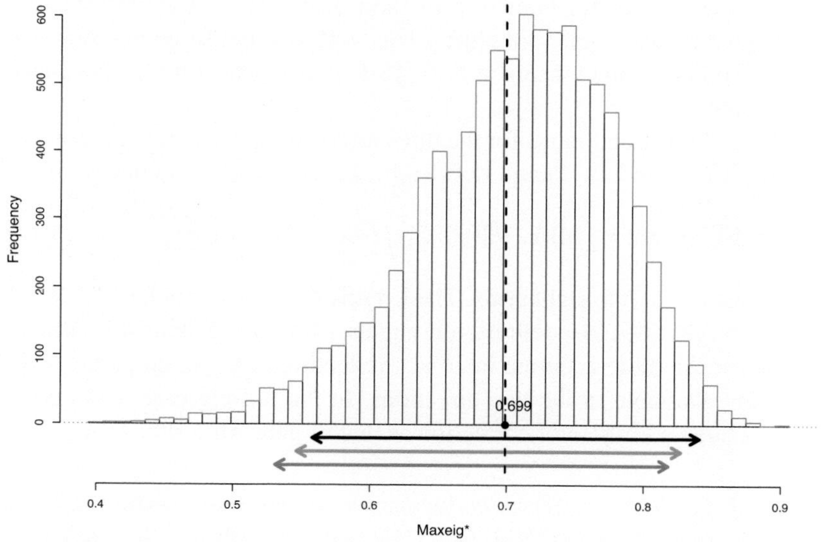

Figure 5.4 10,000 normal-theory bootstrap replications of maxeig; (0.025, 0.975) limits: standard black, percentile green, bca red.

The advantage of bootstrap methods is that they routinely provide highly accurate confidence intervals for that great majority of situations where exact intervals don't exist. Figure 5.4 illustrates an example. The student score data of Table 5.1 is a 22 × 5 matrix, say $\boldsymbol{X}$. Let $\widehat{\boldsymbol{C}}$ be the sample covariance matrix of $\boldsymbol{X}$ and $\hat{\boldsymbol{e}}$ its vector of five eigenvalues (ordered from largest to smallest), and define the *maxeig* statistic as

$$\hat{\theta} = \hat{e}_1 \left(\sum_1^5 \hat{e}_j \right)^{-1}; \tag{5.11}$$

$\hat{\theta} = 0.699$ for the student score data.[4] What is a confidence interval for the true value θ?

Ten thousand parametric bootstrap replications $\hat{\theta}^*$ were generated, assuming a five-dimensional multivariate normal distribution for the student score vectors. (See Section 5.4 for details.) The histogram of the 10,000 $\hat{\theta}^*$ values is markedly long-tailed to the left. We can be certain that the standard interval will be inaccurate, though there are no exact endpoints for comparison here. Figure 5.4 shows the percentile endpoints shifted to the left; bca, the better bootstrap method discussed in the next section, makes an additional leftward shift. (This will turn out to compensate for an upward bias in the maxeig statistic (5.11), as revealed by the bootstrap replications.)

Table 5.4 gives endpoints for the three intervals, and also two descriptive statistics for each: Length and Left/Right, a measure of asymmetry,

$$\text{Left/Right} = \left(\text{MLE} - \hat{\theta}[0.025]\right) \Big/ \left(\hat{\theta}[0.975] - \text{MLE}\right), \tag{5.12}$$

i.e., the ratio of left to right arms. The lengths don't vary much in Table 5.4 but Left/Right does. By definition it equals 1.00 for the standard method, but not for the more accurate intervals, reflecting different amounts of information available in the two directions. In the maxeig case, bca makes twice as large a Left/Right correction as that of percentile.

Table 5.4 *Confidence interval endpoints for student score maxeig statistic (normal model), MLE* = 0.699. *Left/Right is distance from 0.025 endpoint to MLE divided by distance from 0.975 endpoint to MLE.*

	0.025	0.975	Length	Left/Right
Standard	0.558	0.841	0.283	1.00
Percentile	0.546	0.827	0.281	1.19
bca	0.531	0.818	0.287	1.42

[4] This says that about 70% of the variability can be explained by a single linear combination of the five scores–what might be interpreted as a quantitative IQ measure.

5.4 The Bca Intervals

Fisher's transformation (5.4) for the normal correlation does not produce, exactly, $\hat{\phi} \sim \mathcal{N}(\phi, \sigma^2)$. If it did we would have $\Pr\{\hat{\phi} \leq \phi\} = 0.50$ and hence, by monotonicity, $\Pr\{\hat{\theta} \leq \theta\} = 0.50$. But for $\theta = 0.498$, integrating $f_\theta(\hat{\theta})$ in (5.3) gives

$$\Pr_{\theta=0.498}\left\{\hat{\theta} \leq \theta\right\} = 0.478. \tag{5.13}$$

This indicates that $\hat{\theta}$ is a little upwardly biased as an estimate of θ.

Bias can be a more serious problem for other estimators, as we will see. Another flaw in the model $\hat{\phi} \sim \mathcal{N}(\phi, \sigma^2)$ that motivates the percentile method is the assumption that the variance σ^2 does not depend on ϕ. The *bca method* is an improved version of the percentile method that takes both bias and changing variance into account.

The motivating model for bca is an extension of that for the percentile method: we assume the existence of a monotone transformation $\phi = m(\theta)$ and $\hat{\phi} = m(\hat{\theta})$, *and* two constants z_0 and a such that

$$\hat{\phi} \sim \mathcal{N}(\phi - z_0\sigma_\phi, \sigma_\phi^2), \qquad \text{with } \sigma_\phi = 1 + a\phi. \tag{5.14}$$

Definition z_0 is the *bias corrector* and a is the *acceleration*.

The percentile method sets $z_0 = a = 0$, while (5.14) allows for bias and changing variance[5] in ϕ.

Model (5.14) produces exact confidence endpoints for θ, called here $\hat{\theta}_{\text{bca}}[\alpha]$, with "bca" abbreviating *bias corrected and accelerated*:

Theorem 5.2 *If a one-parameter family $f_\theta(\hat{\theta})$ can be monotonically transformed by $\phi = m(\theta)$ and $\hat{\phi} = m(\hat{\theta})$ into model* (5.14), *then an exact level α confidence limit for θ given $\hat{\theta}$ is*

$$\hat{\theta}_{\text{bca}}[\alpha] = G^{-1}\left(\Phi\left(z_0 + \frac{z_0 + z^{(\alpha)}}{1 - a(z_0 + z^{(\alpha)})}\right)\right); \tag{5.15}$$

as before, G is the CDF of $f_\theta(\cdot)$ if $\theta = \hat{\theta}$, and $z^{(\alpha)} = \Phi^{-1}(\alpha)$.

The asymptotic approximation (5.2) – the basis of the standard intervals – portrays the normal translation family $\hat{\theta} \,\dot{\sim}\, \mathcal{N}(\theta, \sigma^2)$. Convergence to (5.2) can be slow, degrading coverage accuracy. The peculiar-looking bca formula (5.15) allows for three data-based corrections to (5.2): a monotone transformation of scale, a bias correction, and an adjustment for changing standard error. Taken together, the bca formula speeds convergence of

5 We could set $\sigma_\phi = \sigma_{\phi_0} + a(\phi - \phi_0)$ for some value ϕ_0 but, by modifying the function $m(\phi)$, we can always suppose $\phi_0 = 0$ and $\sigma_0 = 1$.

asymptotic accuracy by an order of magnitude. Multivariate exponential families, combined with modern computation, permit routine use of bca intervals, as discussed in the next few sections.

Proof Model (5.14) can be expressed as

$$\hat{\phi} = \phi - z_0\sigma_\phi + \sigma_\phi Z,$$

where Z is a $\mathcal{N}(0, 1)$ random variable. This can be rewritten as

$$\left(1 + a\hat{\phi}\right) = (1 + a\phi)\left[1 + a(Z - z_0)\right].$$

Taking logarithms gives

$$\hat{\zeta} = \zeta + W, \tag{5.16}$$

where $\hat{\zeta} = \log(1 + a\hat{\phi})$, $\zeta = \log(1 + a\phi)$, and

$$W = \log\left[1 + a(Z - z_0)\right];$$

statement (5.16) is a *translation family*, with the parameter ζ shifting the known distribution of W along the real line. The obvious exact interval for a translation family has endpoint

$$\hat{\zeta}_{\text{exact}}[\alpha] = \hat{\zeta} - W^{(1-\alpha)}, \tag{5.17}$$

with $W^{(1-\alpha)}$ the $1 - \alpha$ quantile of the distribution of W. Parameter ζ is a monotone function of ϕ, which is a monotone function of the parameter of interest θ. Some algebra shows that (5.17) is equivalent to (5.15). ■

Homework 5.6 Justify (5.17).

Homework 5.7 Complete the algebra that shows (5.17) is equivalent to (5.15).

The preceding proof assumed that $|a|$ was not too big, so $1-a(z_0+z^{(\alpha)}) > 0$ in (5.15), and that $a \neq 0$. If $a = 0$, we get what are called the "bc" endpoints,

$$\hat{\theta}_{\text{bc}}[\alpha] = G^{-1}\Phi\left(2z_0 + z^{(\alpha)}\right). \tag{5.18}$$

This is just $\hat{\theta}_{\text{bca}}[\alpha]$ assuming $a = 0$. The acceleration a can be difficult to estimate (as discussed later), making the bc intervals, sometimes, an attractive option in spite of being theoretically less desirable.

Table 5.5 concerns the example of Homework 5.1,

$$\hat{\theta} \sim \theta G_N/N, \tag{5.19}$$

where G_N is Gamma with N degrees of freedom; $N = 4$ in the table. Because this is a scale family, all confidence endpoints are proportional to $\hat{\theta}$; $\hat{\theta} = 1$ in the table. We can see the approximate endpoints approaching exactness as we go from standard to percentile to bc to bca, the last being nearly exact.

Table 5.5 *Increasingly accurate confidence interval endpoints for gamma scale problem* (5.19) *utilizing standard, percentile, bc, bca (bc is bca assuming acceleration* $a = 0$*), as compared to exact;* $N = 4$ *df,* $\hat{\theta} = 1$.

α	0.025	0.05	0.1	0.9	0.95	0.975
Stand	0.02	0.18	0.36	1.64	1.82	1.98
Pct	0.27	0.34	0.44	1.67	1.94	2.19
bc	0.35	0.43	0.54	1.92	2.21	2.48
bca	0.46	0.52	0.60	2.29	2.93	3.69
Exact	0.46	0.52	0.60	2.29	2.93	3.67

Homework 5.8 Recalculate Table 5.5 for $N = 8$, 16, and 32. *Note*: No simulation is required. In this situation you can set $a = z_0$; see below.

Theorem 5.2 suggests a computational method for approximating the bca intervals, even in very complicated situations. We suppose that the observed data $\boldsymbol{x}$ follows a density function $g_\eta(\boldsymbol{x})$ from a known parametric family $\mathcal{G}$,

$$\mathcal{G} = \left\{g_\eta(\boldsymbol{x}), \eta \in A \subset \mathcal{R}^p\right\}, \tag{5.20}$$

η an unknown p-dimensional vector, and that we wish to set confidence intervals for a real-valued parameter $\theta = s(\eta)$. For the student score correlation problem, $\boldsymbol{x}$ is the 22×5 matrix in Table 5.1, $g_\eta(\boldsymbol{x})$ is the density for 22 independent replications $x_1, \dots, x_{22} \sim \mathcal{N}_5(\lambda, \Gamma)$ (so η is 20-dimensional as in Section 2.8), and $\theta = \Gamma_{12}(\Gamma_{11}\Gamma_{22})^{-1/2}$.

The bca algorithm begins by finding the MLE $\hat{\eta}$ of η, and then resampling some large number B of parametric bootstrap replications $\boldsymbol{x}^*$ from $g_{\hat{\eta}}(\cdot)$,

$$\boldsymbol{x}^*(1), \boldsymbol{x}^*(2), \dots, \boldsymbol{x}^*(B) \overset{\text{iid}}{\sim} g_{\hat{\eta}}(\cdot).$$

Each $\boldsymbol{x}^*$ yields an MLE $\hat{\eta}^*$ and a bootstrap replication $\hat{\theta}^* = s(\hat{\eta}^*)$,

$$\hat{\theta}^*(1), \dots, \hat{\theta}^*(B). \tag{5.21}$$

In the student score correlation example, each $\boldsymbol{x}^*$ consisted of 22 independent draws from a $\mathcal{N}_5(\hat{\lambda}, \widehat{\Gamma})$ distribution, and $\hat{\theta}^*$ was the sample correlation of the first two columns of $\boldsymbol{x}^*$. (This gave the same results as sampling $\hat{\theta}^*$ directly from the solid curve in Figure 5.1, but didn't require knowing Fisher's formula (5.3). The ability to avoid theoretical calculations is what makes bootstrap methods most useful in applied settings.) For the maxeig example of Figure 5.4, each $\boldsymbol{x}^*$ gave covariance matrix $\widehat{\boldsymbol{C}}^*$, eigenvalues $\hat{\boldsymbol{e}}^*$, and $\hat{\theta}^* = \hat{e}_1^* / \sum_1^5 \hat{e}_j^*$.

The empirical CDF of the bootstrap replications $\hat{g}$ (5.2) is an estimate of G in the bca formula (5.15) (exactly equaling G when $B \to \infty$). Moreover, $\widehat{G}$ provides an estimate of the bias corrector z_0:

Lemma 5.3 *Under the assumptions of Theorem 5.2 we have*

$$z_0 = \Phi^{-1} G\left(\hat{\theta}\right),$$

giving

$$\hat{z}_0 = \Phi^{-1} \widehat{G}\left(\hat{\theta}\right) = \Phi^{-1}\left(\# \left\{\hat{\theta}^*(i) \leq \hat{\theta}\right\} / B\right) \tag{5.22}$$

as an estimate of z_0.

Homework 5.9 Verify Lemma 5.3.

For the maxeig example of Figure 5.4, 4259 of the 10,000 bootstrap replications were less than $\hat{\theta} = 0.699$, so $\hat{z}_0 = \Phi^{-1}(0.4259) = -0.187$. That is, the fact that less than half of the $\hat{\theta}^*$s were less than $\hat{\theta}$ indicated that the MLE was biased upward, so bias correction was in the negative direction.

Two of the three elements of the bca formula (5.15), G and z_0, have now been estimated from the bootstrap replications (5.21). The acceleration a is less straightforward. One situation where a is easily available is in a one-parameter exponential family (parameterized with its expectation parameter μ),

$$g_\mu(\boldsymbol{x}) = e^{\eta y - \psi(\eta)} g_0(y),$$

where the real-valued sufficient statistic y is a function of $\boldsymbol{x}$, $\mu = \dot{\psi}(\eta)$ is the expectation parameter, and θ is a monotonic function of μ, say $\theta = t(\mu)$.

Lemma 5.4 *In a one-parameter exponential family, the estimated acceleration* $\hat{a}$ *is one-sixth the estimated skewness,*

$$\hat{a} = \frac{1}{6}\hat{\gamma}. \tag{5.23}$$

Proof According to Section 1.10, the normalizing transformation $\phi = m(\mu)$ can be taken to satisfy

$$\frac{d\phi}{d\mu} = V(\mu)^{-1/3}, \tag{5.24}$$

so that the delta method gives standard deviation

$$\sigma\{\hat{\phi}\} \doteq \sigma\{\hat{\mu}\}\widehat{V}^{-1/3} = \widehat{V}^{1/6},$$

where $\widehat{V} = V(\hat{\mu})$, and

$$\frac{d\sigma\{\hat{\phi}\}}{d\mu} = \frac{1}{6}\widehat{V}^{-5/6}\widehat{V}'.$$

But since $d\phi/d\mu = V^{-1/3}$ according to (5.24), we get

$$\left.\frac{d\sigma\{\hat{\phi}\}}{d\phi}\right|_{\hat{\phi}} = \frac{1}{6}\frac{\widehat{V}^{-5/6}\widehat{V}'}{\widehat{V}^{-1/3}} = \frac{1}{6}\frac{\widehat{V}'}{\widehat{V}^{1/2}} = \frac{1}{6}\hat{\gamma},$$

using the "useful result" (1.20) in Section 1.2. Since $a = d\sigma\{\phi\}/d\phi$ according to (5.14), this verifies Lemma 5.4. ■

A two-term Edgeworth expansion gives the approximation

$$\widehat{G}\left(\hat{\theta}\right) \doteq 0.50 + \frac{1}{6\sqrt{2\pi}}\hat{\gamma},$$

which then yields

$$\hat{z}_0 \doteq \frac{\hat{\gamma}}{6} = \hat{a}. \tag{5.25}$$

In other words, $\hat{z}_0$ and $\hat{a}$ are about the same in one-parameter exponential family situations. Unfortunately this isn't true in multiparameter problems, as discussed in Section 5.6.

Homework 5.10 How does (5.25) follow from the expression for $\widehat{G}(\hat{\theta})$?

If $\boldsymbol{x}$ represents an i.i.d. sample of size n, then $\gamma = \gamma^{(n)} = \gamma^{(1)}n^{-1/2}$, as in Section 1.3. Generally, both $\hat{z}_0$ and $\hat{a}$ are order of magnitude $O_p(n^{-1/2})$ under repeated sampling. In practice, the bca corrections to the standard intervals can be quite large compared with, say student-t effects; see Section 5.5.

5.5 Confidence Intervals in Multiparameter Exponential Families

The bca method of Section 5.4 was motivated by calculations in one-parameter families $f_\theta(\hat{\theta})$. This is where it is least necessary. The real need for accurate approximate confidence intervals arises in multiparameter contexts, where inference about the parameter of interest is complicated by nuisance parameters, perhaps many of them.

Here we will assume that the observed data set $\boldsymbol{x}$ comes from a p-parameter exponential family, with densities

$$g_\eta(\boldsymbol{x}) = e^{\eta^\top y - \psi(\eta)} g_0(\boldsymbol{x}), \tag{5.26}$$

as in (2.1), having p-dimensional sufficient vector y, say $y = Y(\boldsymbol{x})$, and expectation parameter vector $\mu = E_\eta\{y\}$. We have a real-valued parameter of interest θ which can be expressed as a function of either η or μ,

$$\theta = s(\eta) = t(\mu). \tag{5.27}$$

If $\hat{\eta}$ and $\hat{\mu}$ are the maximum likelihood estimates of η and μ then the MLE of θ is

$$\hat{\theta} = s(\hat{\eta}) = t(\hat{\mu}).$$

We wish to set accurate confidence limits for θ having observed $\boldsymbol{x}$.

Parametric bootstrapping proceeds as at (5.21): resampling from $g_{\hat{\eta}}(\cdot)$ gives

$$\boldsymbol{x}^*(1), \dots, \boldsymbol{x}^*(B) \overset{\text{iid}}{\sim} g_{\hat{\eta}}(\cdot) \tag{5.28}$$

and

$$\hat{\theta}^*(1), \dots, \hat{\theta}^*(B),$$

where $\hat{\theta}^*(i) = s(\hat{\eta}^*(i))$, with $\hat{\eta}^*(i)$ the MLE of η for data set $\boldsymbol{x}^*(i)$. As before, we can estimate two of the three elements of the bca formula (5.15), G and z_0, directly from the bootstrap replications (5.28): $\widehat{G}$ their empirical CDF and $\hat{z}_0$ (5.22). That leaves the acceleration a.

As a device for getting back to the one-parameter family situation – where we know how to estimate a – we invoke Stein's *least favorable family* (Section 2.6), applied at η_0 equal the MLE $\hat{\eta}$. That is, we suppose that $\boldsymbol{x}$ has been sampled from the one-parameter exponential family

$$\hat{f}_\lambda(\boldsymbol{x}) = g_{\hat{\eta}+\hat{t}'\lambda}(\boldsymbol{x}), \qquad \hat{t}' = \begin{pmatrix} \vdots \\ \partial t(\mu)/\partial \mu_j \\ \vdots \end{pmatrix}_{\hat{\mu}}, \tag{5.29}$$

where $\hat{\eta}$ and $\hat{t}'$ are considered fixed at their observed value and λ is the unknown real-valued parameter. As in Section 2.6, (5.29) has cumulant generating function

$$\hat{\phi}(\lambda) = \psi\left(\hat{\eta} + \hat{t}'\lambda\right).$$

Numerical differentiation of $\hat{\phi}(\lambda)$,

$$\hat{k}_2 = \left.\frac{\partial^2}{\partial\lambda^2}\hat{\phi}(\lambda)\right|_0 \quad \text{and} \quad \hat{k}_3 = \left.\frac{\partial^3}{\partial\lambda^3}\hat{\phi}(\lambda)\right|_0, \tag{5.30}$$

provides estimated skewness $\hat{\gamma} = \hat{k}_3\hat{k}_2^{-3/2}$ and acceleration

$$\hat{a} = \hat{\gamma}/6, \tag{5.31}$$

from Lemma 5.4.

Homework 5.11 The applicability of (5.31) to the acceleration for parameter θ depends on $\theta(\lambda) = s(\hat{\eta} + \hat{t}'\lambda)$ being a monotone increasing function of λ, at least for λ near 0. Show that

$$\left.\frac{d\theta(\lambda)}{d\lambda}\right|_0 = \hat{t}'^{\top} V_{\hat{\eta}} \hat{t}',$$

the delta method estimate of variance for $\hat{\theta} = t(\bar{y})$, and so is positive.

The standard intervals $\hat{\theta}_{\text{stand}}[\alpha] = \hat{\theta} + z^{(\alpha)}\hat{\sigma}$ are *first-order accurate*, that is, their actual coverage $\hat{\alpha}$ approaches the claimed value α at rate $O(n^{-1/2})$ in repeated sampling situations. The estimated bca interval endpoints

$$\hat{\theta}_{\text{bca}}[\alpha] = \widehat{G}^{-1}\Phi\left(\hat{z}_0 + \frac{\hat{z}_0 + z^{(\alpha)}}{1 - \hat{a}(\hat{z}_0 + z^{(\alpha)})}\right) \tag{5.32}$$

are *second-order accurate*, with $\hat{\alpha} - \alpha$ going to zero at rate $O(n^{-1})$ in the sample size n (DiCiccio and Efron, 1996). This can produce the kind of improvements seen in Table 5.5.

One can ask whether second-order corrections are worthwhile in practice. The answer depends on the importance attached to the value of the parameter of interest θ. As a point of comparison, $\hat{\theta}_{\text{student}}[\alpha] = \hat{\theta} + t_n^{(\alpha)}\hat{\sigma}$, the usual student-$t$ endpoint – with $t_n^{(\alpha)}$ the student-t αth quantile for degrees of freedom n – makes corrections only of order $O(\hat{\sigma}n^{-1})$ to $\hat{\theta}_{\text{stand}}[\alpha]$, while the bca corrections are $O(\hat{\sigma}n^{-1/2})$, an order of magnitude larger. The corrections are typically asymmetric, reflecting the different amounts of information to the left and right of the point estimate $\hat{\theta}$.

The bca method makes three corrections to the standard intervals:

1. For non-normality of $\hat{\theta}$ (by using $\widehat{G}$ in place of the normal CDF Φ).

2. For the bias of $\hat{\theta}$ (through the bias corrector $\hat{z}_0$).
3. For acceleration, the nonconstant variance of $\hat{\theta}$ (through $\hat{a}$).

All three corrections can be necessary for second-order accuracy, as seen in the gamma scale example of Table 5.5. Only the last of these makes specific use of exponential family structure, the real need for which concerns practical computation, as will be discussed in the next section.

Bca intervals are transformation invariant: if $\phi = m(\theta)$ with $m(\cdot)$ monotone increasing, then $\hat{\phi}_{\text{bca}}[\alpha] = m(\hat{\theta}_{\text{bca}}[\alpha])$. A considerable practical and theoretical advantage of transformation invariance is that the statistician can never be working on the wrong scale.

Homework 5.12 Verify the transformation invariance of (5.32).

The bca method can be recommended on three counts:

- It gives second-order accurate intervals.
- It is transformation invariant.
- It is exactly correct under a model (5.14) that substantially extends the standard interval model $\hat{\theta} \sim \mathcal{N}(\theta, \sigma^2)$.

There are other second-order methods in the literature, some with further theoretical advantages. The fourth recommendation for bca intervals is that they are (relatively) easy to compute in a wide variety of applied situations, as discussed next.

5.6 Computing the Bca Intervals

In practice, statistical methods aren't used much unless they can be invoked automatically, that is, without requiring the statistician to perform substantial theoretical work before each individual application. The standard intervals (5.1) can be fully automated, especially so in our era of speedy calculation: given an estimate $\hat{\theta} = T(\boldsymbol{x})$, bootstrap replications – either parametric or nonparametric – produce $\hat{\sigma}$ and $\hat{\theta}_{\text{stand}}[\alpha] = \hat{\theta} + z^{(\alpha)}\hat{\sigma}$.

Automaticity will turn out to hold for nonparametric bca intervals (Section 5.7) and, to a lesser extent, parametric bca intervals, as discussed next. We will be employing the R package `bcaboot`, available from CRAN. Computations in this section use the parametric bca program `bcapar` from `bcaboot` (which also provided the maxeig results in Figure 5.4 and Table 5.4).

As in (5.28), parametric bootstrap resamples $\boldsymbol{x}^*(1), \ldots, \boldsymbol{x}^*(B)$ are generated, giving bootstrap replications

$$\hat{\theta}^*(i) = T\left(\boldsymbol{x}^*(i)\right), \qquad i = 1, \ldots, B, \tag{5.33}$$

where $\hat{\theta} = T(\boldsymbol{x})$ is the estimated parameter of interest, now expressed directly as a function of $\boldsymbol{x}$. In the maxeig example of Figure 5.4, the 22×5 resampled student score matrix $\boldsymbol{x}^*(i)$ gives covariance matrix $\widehat{\boldsymbol{C}}^*$, eigenvalues $\hat{\boldsymbol{e}}^*$, and bootstrap replications $\hat{\theta}^*(i) = \hat{e}_i^* / \sum_1^5 \hat{e}_j^*$ as at (5.11). An important practical point is that we do *not* have to explicitly express $\hat{\theta}^*$ as a function $s(\hat{\eta}^*)$ or $t(\hat{\mu}^*)$ (which would often involve unwanted theoretical calculations that undermine the automatic application of `bcapar`). All we have to do is reapply the original function $T(\cdot)$.

Each replication $\boldsymbol{x}^*(i)$ yields a p-dimensional sufficient vector, say

$$\boldsymbol{y}^*(i) = S\left(\boldsymbol{x}^*(i)\right), \qquad i = 1, \ldots, B.$$

In the student score maxeig example, $\boldsymbol{y}^*(i)$ has $p = 20$ components, the column means of $\boldsymbol{x}^*(i)$, and the column means of the squares and cross-products of the scores, as detailed in Section 2.8. In addition to the value of the point estimate $\hat{\theta}$ and the vector $\hat{\boldsymbol{\theta}}^*$ of bootstrap replications,

$$\hat{\boldsymbol{\theta}}^* = \left(\hat{\theta}^*(1), \ldots, \hat{\theta}^*(B)\right), \tag{5.34}$$

`bcapar` requires the $B \times p$ matrix having rows $\boldsymbol{y}^*(i)^\top$, say

$$\boldsymbol{Y}^* = \begin{pmatrix} \vdots \\ \boldsymbol{y}^*(i)^\top \\ \vdots \end{pmatrix}. \tag{5.35}$$

An example follows next.

The *neonate data* comprises the records of 812 very sick babies admitted for treatment at an African medical facility, of whom 605 lived and 207 died. For each baby, a vector of 11 baseline variables was recorded, say m_j for baby j, with outcome y_j,

$$y_j = \begin{cases} 1 & \text{baby died} \\ 0 & \text{baby lived.} \end{cases}$$

The data set $\boldsymbol{x}$ in this case is[6]

$$\boldsymbol{x} = \{\boldsymbol{M}, \boldsymbol{y}\}, \tag{5.36}$$

[6] Notice that $\boldsymbol{y}$ is the response vector for the entire neonate data set, length 812, rather than $\boldsymbol{y}^*(i)$ the sufficient vector for a single bootstrap replication, denoted as z^* in what follows.

where $\boldsymbol{y}$ is the 812-vector $(\cdots y_j \cdots)^\top$ and $\boldsymbol{M}$ is the 812×12 matrix having jth row $(1, m_j)^\top$ (the 1 allows for an intercept term); $\boldsymbol{M}$ had its other 11 columns standardized to mean 0, variance 1. See the data set file *neonate-data*.

We assume a logistic regression model for y_j as a function of m_j, as in Section 3.2,

$$\boldsymbol{\eta} = \boldsymbol{M}\beta,$$

with $\eta_j = \log(\pi_j/(1-\pi_j))$, $\pi_j = \Pr\{y_j = 1\}$. This is a $(p = 12)$-parameter exponential family having natural parameter vector β and sufficient vector

$$z = \boldsymbol{M}^\top \boldsymbol{y} \tag{5.37}$$

(as in (3.6), now with $\boldsymbol{M}$ playing the role of $\boldsymbol{X}$).

Running the logistic regression program `glm(`$\boldsymbol{y} \sim \boldsymbol{M}$`, binomial)` gave MLE $\hat{\beta}$ whose component values are shown in the first column of Table 5.6, along with standard errors (from normal approximation (3.15)),[7] z-values, and two-sided p-values. Four of the 11 predictor variables were highly significant: gestational age, apgar score, respiratory difficulty, and "cpap", an airway blockage measure.

Table 5.6 *Logistic regression analysis of neonate data. Predictor variables were standardized; four are highly significant (***), that is, with p-values* < 0.005.

	Estimate	Std error	z-value	p-value
(Intercept)	−1.598	0.122	−13.094	0.000 ***
gest	−0.610	0.192	−3.177	0.001 ***
ap	−0.671	0.108	−6.236	0.000 ***
bwei	−0.295	0.189	−1.559	0.119
gen	−0.017	0.107	−0.157	0.875
resp	0.896	0.135	6.649	0.000 ***
head	0.141	0.112	1.254	0.210
hr	−0.002	0.103	−0.015	0.988
cpap	0.357	0.120	2.974	0.003 ***
age	0.138	0.122	1.124	0.261
temp	−0.056	0.133	−0.418	0.676
size	−0.247	0.142	−1.743	0.081

The MLE for the 812-vector $\boldsymbol{\pi}$, $\pi_j = \Pr\{y_j = 1\}$, is

$$\hat{\boldsymbol{\pi}} = \mathbf{1} \Big/ \left(\mathbf{1} + \exp\left\{-\boldsymbol{M}\hat{\beta}\right\}\right).$$

[7] As a check, bootstrap standard errors for the estimates in column 1 were computed and showed close agreement with the values in column 2.

A parametric bootstrap resample of $\boldsymbol{y}$ is generated as

$$y_j^* = \begin{cases} 1 & \Pr\{\hat{\pi}_j\} \\ 0 & \Pr\{(1-\hat{\pi}_j)\}, \end{cases} \qquad \text{independently for } j = 1, \dots, 812. \tag{5.38}$$

(Because $\boldsymbol{M}$ is considered fixed in logistic regression, it stays constant in the resampling process.) Then

$$\texttt{glm(}\boldsymbol{y}^* \sim \boldsymbol{M}\texttt{, binomial)} \tag{5.39}$$

gives $\hat{\boldsymbol{\beta}}^*$ and bootstrap sufficient vector

$$z^* = \boldsymbol{M}^\top \boldsymbol{y}^*. \tag{5.40}$$

If $\hat{\theta} = s(\hat{\beta})$ is the parameter of interest, we also get

$$\hat{\theta}^* = s\left(\hat{\beta}^*\right), \tag{5.41}$$

but (5.33), which doesn't require the function $s(\cdot)$, is usually handier.

Independently rerunning (5.38)–(5.41) through B repetitions gave the ingredients needed for `bcapar`: $\hat{\boldsymbol{\theta}}^* = (\hat{\theta}^*(1), \dots, \hat{\theta}^*(B))$ and

$$z^* = \begin{pmatrix} \vdots \\ z^*(i)^\top \\ \vdots \end{pmatrix} \tag{5.42}$$

(playing the role of $\boldsymbol{Y}^*$ in (5.35)).

Homework 5.13 Suppose $\hat{\theta}$ is the bodyweight coefficient bwei $= -0.295$ in Table 5.5; use `bcapar`$(\hat{\theta}, \hat{\boldsymbol{\theta}}^*, \boldsymbol{z}^*)$ to calculate bca confidence intervals for θ. Use the plotting program `bcaplot` to compare them with the standard intervals. How would you describe the comparison?

The main goal of the study was to accurately predict from the baseline measurements whether or not the baby would die, with the intention of intensified treatment for those with poor prognosis. Figure 5.5 shows the estimated death probabilities $\hat{\pi}_j$, separated into the 605 babies who lived (solid blue histogram) and the 207 who died (line histogram).

We see a fair degree of separation. Using $\hat{\pi}_j$ greater or less than 0.25 as the prediction threshold, 112 of the 605 babies who lived were predicted to die (those having $\hat{\pi}_j > 0.25$), while 41 of the 207 who died were predicted to live ($\hat{\pi}_j < 0.25$). Estimated error rates of the first and second kind, $\widehat{\text{err}}_1$ and $\widehat{\text{err}}_2$, were

$$\widehat{\text{err}}_1 = \frac{112}{605} = 0.182 \quad \text{and} \quad \widehat{\text{err}}_2 = \frac{81}{207} = 0.198,$$

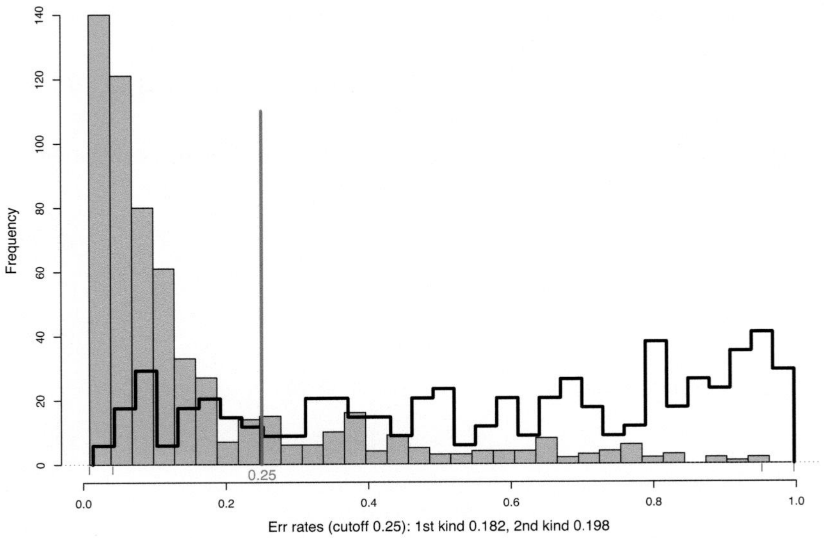

Figure 5.5 Predicted probability of death for neonate data after logistic regression; solid histogram $y = 0$, line histogram $y = 1$.

for an error ratio $\widehat{\text{erat}}$,

$$\widehat{\text{erat}} = \frac{\widehat{\text{err}}_2}{\widehat{\text{err}}_1} = 1.09.$$

How accurate is this estimate?

$B = 2000$ bootstrap replications gave the histogram of the $\widehat{\text{erat}}^*$ values shown in Figure 5.6. A large upward bias is evident, with only 32% of the 2000 $\widehat{\text{erat}}^*$s being less than $\widehat{\text{erat}}$. The histogram is also mildly long-tailed to the right. We can expect trouble for the standard intervals.

Homework 5.14 Give an explicit description of how the $\widehat{\text{erat}}^*$ values were calculated.

Following steps (5.38)–(5.41) $B = 2000$ times gave $\hat{\boldsymbol{\theta}}^*$, the 2000-vector of bootstrap replications $\widehat{\text{erat}}^*(i)$, and also z^* (5.42). (Notice that z^*, once calculated, can then be applied to any desired statistic $\hat{\theta}$ and $\hat{\theta}^*$.) Table 5.7 shows the bca and standard interval endpoints for erat, e.g., $\hat{\theta}_{\text{bca}}[0.975] = 1.303$ and $\hat{\theta}_{\text{stand}}[0.975] = 1.375$. Figure 5.7, from the `bcaboot` program `bcaplot`, graphs both sets of confidence limits versus their claimed two-sided coverages. The bca limits lie substantially below the standard limits,

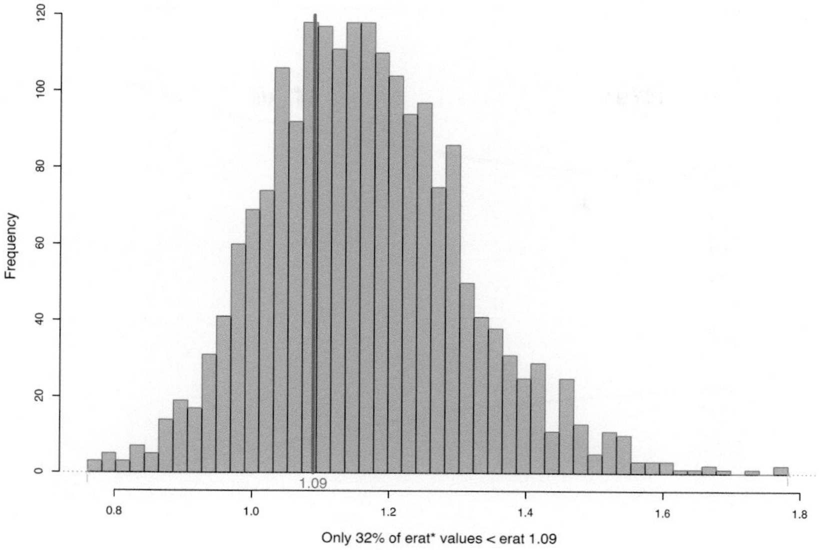

Figure 5.6 $B = 2000$ bootstrap replications of prediction error ratio $\widehat{\text{err}}_2/\widehat{\text{err}}_1$.

this being mainly due to the large bias corrector $\hat{z}_0 = \Phi^{-1}(0.32) = -0.470$ in formula (5.32).

Table 5.7 *Bca and standard limits for neonate error ratio,* $\alpha = 0.025, \ldots, 0.975$, $B = 2000$*; jacksd is estimated Monte Carlo stdevs for bca limits; pct shows percentage points of bootstrap replications corresponding to bca endpoints, e.g.,* $\alpha = 0.025$ *endpoint* 0.782 *corresponds to fourth smallest bootstrap replication,* $4/2000 = 0.002$.

α	Bca	Standard	Jacksd	Pct
0.025	0.782	0.803	0.015	0.002
0.05	0.813	0.849	0.024	0.005
0.1	0.873	0.902	0.011	0.013
0.16	0.905	0.944	0.009	0.027
0.5	1.031	1.089	0.006	0.174
0.84	1.163	1.234	0.006	0.522
0.9	1.204	1.276	0.006	0.634
0.95	1.258	1.329	0.009	0.760
0.975	1.303	1.375	0.013	0.847

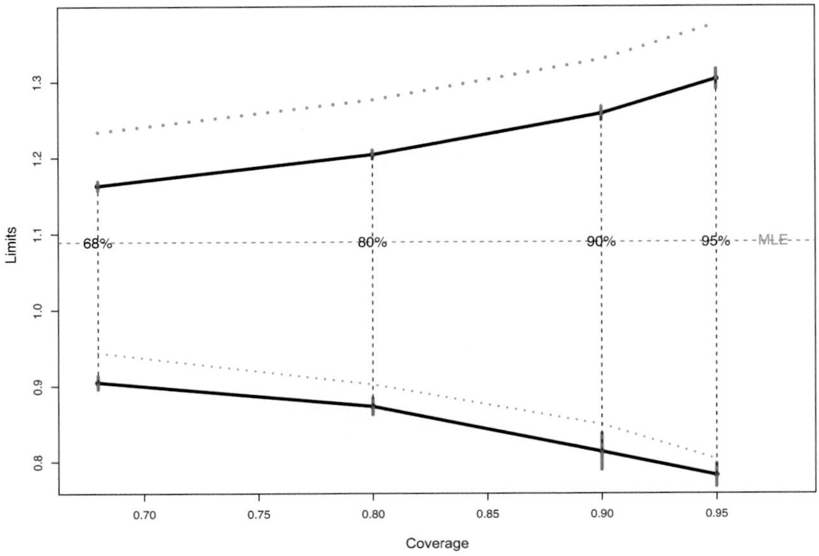

Figure 5.7 Approximate two-sided intervals for error ratio versus claimed coverage; black solid line is bca, green dotted line is standard, red bars show Monte Carlo error.

Homework 5.15 Use `bcapar` to recreate Table 5.7 for $B = 1000, 2000$, and 4000. Why would it be most efficient to do the $B = 4000$ case first?

Acceleration Computations

The fact that `bcapar` requires $\boldsymbol{Y}^*$, the matrix of sufficient vectors (5.35), in addition to $\hat{\theta}$ and $\hat{\boldsymbol{\theta}}^*$, detracts from its claim to automatic application, even though $\boldsymbol{Y}^*$ is often easily available. The sufficient vectors are needed to let `bcapar` know which exponential family has generated the data.

Neyman's construction for the student score correlation example in Figure 5.1 suggests why $\widehat{G}$, the bootstrap CDF, by itself might not be enough for the bca algorithm. There the black curve $f_{\hat{\theta}}(\cdot)$, which is equivalent to $\widehat{G}$, plays no role in Neyman's calculations, which depend on densities $f_\theta(\cdot)$ with θ far away from $\hat{\theta}$. Bootstrap methods, by definition, work entirely from the MLE distribution, the black curve $f_{\hat{\theta}}(\cdot)$ in Figure 5.1. In some way, `bcapar` has to learn about $f_\theta(\cdot)$ for values of θ to the left and right of $\hat{\theta}$. At the very least, it must infer how the standard deviation of $f_\theta(\cdot)$ changes as a function of θ. This is exactly the role of the acceleration a.

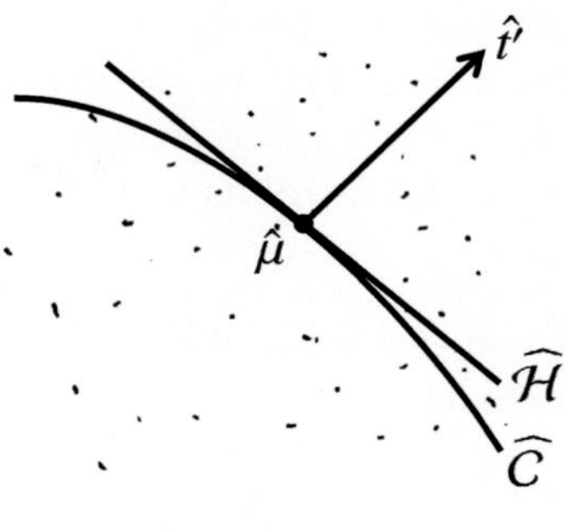

Figure 5.8 Heuristic representation of difference between $\hat{z}_0$ and $\hat{a}$ as explained in the text. Small dots represent bootstrap sufficient vectors $y^*(i)$, $i = 1, \ldots, B$.

In one-parameter exponential families, $\hat{a} \doteq \hat{z}_0$ (5.25) and $\hat{a}$ *is* calculable directly from the bootstrap distribution, being equal to $\hat{z}_0 = \Phi^{-1}\widehat{G}(\hat{\theta})$ (5.22). It would be nice if $\hat{a} \doteq \hat{z}_0$ in multiparameter exponential families, but that isn't the case. Figure 5.8 gives a heuristic picture of the difference, as seen in the space B and $\mathcal{Y}$ of Figure 2.5: $\hat{\mu} = y$, the MLE of the expectation parameter μ, gives $\hat{\theta} = t(\hat{\mu})$ as the MLE of θ; it determines the level surface

$$\widehat{C} = \left\{\mu : t(\mu) = \hat{\theta}\right\}$$

of vectors μ having the same estimate of θ as does $\hat{\mu}$. The gradient vector $\hat{t}'$ for Stein's least favorable family (5.29) is orthogonal to $\widehat{C}$ at $\mu = \hat{\mu}$.

The small dots in Figure 5.8 represent the bootstrap sufficient vectors $y^*(i)$, $i = 1, \ldots, B$. Let $\hat{p}_z$ be the proportion of them lying below $\widehat{C}$, that is, in the direction opposite $\hat{t}'$,

$$\hat{p}_z = \#\left\{y^*(i) \text{ below } \widehat{C}\right\}\Big/B.$$

Then, under monotonicity conditions, $\widehat{G}(\hat{\theta}) = \hat{p}_z$ and

$$\hat{z}_0 = \Phi^{-1}(\hat{p}_z). \tag{5.43}$$

Also let $\widehat{\mathcal{H}}$ be the hyperplane passing through $\hat{\mu}$ orthogonally to $\hat{t}'$, and define

$$\hat{p}_a = \#\left\{y^*(i) \text{ below } \widehat{\mathcal{H}}\right\}\Big/B.$$

Notice that in general $\hat{p}_a \neq \hat{p}_z$, the difference being the proportion of points lying between $\widehat{C}$ and $\widehat{\mathcal{H}}$. It turns out that

$$\hat{a} \doteq \Phi^{-1}(\hat{p}_a), \tag{5.44}$$

this following from (5.25) applied to the one-parameter exponential family (5.29). Comparing (5.43) with (5.44), we see that $\hat{a}$ will exceed $\hat{z}_0$ if $\widehat{C}$ curves away from $\hat{t}'$, as in Figure 5.7, and conversely if the curvature is in the opposite direction. Section 5.8 will have more to say about curvature calculations.

Homework 5.16 Equation (5.44) provides one way to estimate a from $\boldsymbol{Y}^*$. Give another way based on (5.23).

Monte Carlo Error

How accurate are the bca endpoints $\hat{\theta}_{\text{bca}}[\alpha]$? There are two sources of error:

- *Sampling error*, arising from the fact that the data set $\boldsymbol{x}$ is a random sample, as in most statistical error calculations.
- *Monte Carlo error*, arising from the bootstrap sampling process. This can be eliminated by letting the number of bootstrap replicates go to infinity, but in practice we have to stop at some finite value B (as with $B = 2000$ in the neonate `bcapar` calculations); `bcapar` provides an estimate of the Monte Carlo standard deviation of $\hat{\theta}_{\text{bca}}[\alpha]$, indicated by the red bars in Figure 5.7. These depend on *jackknife* calculations, as described next.

Suppose that $\boldsymbol{x} = (x_1, \ldots, x_n)$ is an i.i.d. sample from some unknown distribution F on a space $\mathcal{X}$, and that we wish to assess the standard deviation of a statistic $\hat{\theta} = T(\boldsymbol{x})$; $\mathcal{X}$ can be anything at all – one-dimensional, multidimensional, functional – while $T(\boldsymbol{x})$ is any function defined exchangeably in the n components.

Define $\boldsymbol{x}_{(i)}$ to be $\boldsymbol{x}$ with the ith component x_i removed and let

$$\hat{\theta}_{(i)} = T(\boldsymbol{x}_{(i)}), \qquad \text{for } i = 1, \ldots, n. \tag{5.45}$$

Then the jackknife estimate of standard deviation for $\hat{\theta}$ is

$$\hat{\sigma}_{\text{jack}} = \left[\frac{n-1}{n}\sum_{i=1}^{n}\left(\hat{\theta}_{(i)} - \hat{\theta}_{(\cdot)}\right)^2\right]^{1/2} \qquad \left(\hat{\theta}_{(\cdot)} = \sum_{i=1}^{n}\frac{\hat{\theta}_{(i)}}{n}\right). \tag{5.46}$$

Efron and Stein (1981) shows $\hat{\sigma}^2_{\text{jack}}$ to be conservative (upwardly biased) as an estimate of the variance of $\hat{\sigma}$.

Reference Efron and Hastie (2016), *Computer Age Statistical Inference: Algorithms, Evidence, and Data Science*, Section 10.1.

REFERENCE Efron and Stein (1981), "The jackknife estimate of variance", *Ann. Stat.* 586–596.

With the data $\boldsymbol{x}$ (and hence $\hat{\eta}$) considered fixed, the bootstrap process produces an i.i.d. sequence of resamples $\boldsymbol{x}^*(i)$, replications $\hat{\theta}^*(i) = T(\boldsymbol{x}^*(i))$, and sufficient vectors $\boldsymbol{y}^*(i) = S(bx^*(i))$,

$$g_{\hat{\eta}}(\cdot) \longrightarrow \boldsymbol{x}^*(i) \longrightarrow \left[\hat{\theta}^*(i), \boldsymbol{y}^*(i)\right], \qquad \text{for } i = 1, \ldots, B.$$

Moreover, $\hat{\theta}_{\text{bca}}[\alpha]$ is a symmetric function of the B pairs $[\hat{\theta}^*(i), \boldsymbol{y}^*(i)]$ so we are set up to use the jackknife for the Monte Carlo standard deviation of $\hat{\theta}_{\text{bca}}[\alpha]$. To reduce the number of jackknife recomputations (5.45), `bcapar` randomly partitions the integers $\{1, \ldots, B\}$ into J sets of size B/J (default $J = 10$) and lets $\boldsymbol{S}_j$ be the set of pairs $[\hat{\theta}^*(i), \boldsymbol{y}^*(i)]$ for i in the jth set. Now $\hat{\theta}_{\text{bca}}[\alpha]$ can be thought of as a symmetric function of $(\boldsymbol{S}_1, \ldots, \boldsymbol{S}_J)$, that is, as exchangeable, and we can apply the jackknife formula (5.45)–(5.46) with $\boldsymbol{S}_i$, $i = 1, \ldots, J$ playing the role of $\boldsymbol{x}_i$. The red bars in Figure 5.7 indicate $\pm\hat{\sigma}_{\text{jack}}$ for $\hat{\theta}_{\text{bca}}[\alpha]$.

Homework 5.17 Continuing from Homework 5.15, compute the column Jacksd from Table 5.7 for $B = 1000, 2000, 4000$.

Theoretical calculations in Section 19.3 of Efron and Tibshirani (1996) suggest $B \geq 1000$ as a minimum for bootstrap confidence interval calculations. The jackknife estimate of standard deviation for $\hat{\theta}_{\text{bca}}[\alpha]$ are far from perfect, but they give a data-based idea of whether the choice of B is adequate. In Figure 5.7 for example, we see that $B = 2000$ was more than adequate for upper values of α, but perhaps borderline for the lower values.

REFERENCE Efron and Tibshirani (1996), "Using specially designed exponential families for density estimation", *Ann. Stat.* 2431–2461.

The `bcapar` algorithm also returns jackknife Monte Carlo standard errors for the bias corrector $\hat{z}_0$ and the acceleration $\hat{a}$, as well as for the usual bootstrap standard deviation estimate of $\hat{\theta}$. Table 5.8 shows the output for the neonate example: $\hat{a}$ is nearly zero, and would evidently stay that way as $B \to \infty$; the large negative value of $\hat{z}_0$ has about 5% Monte Carlo coefficient of variation, so would not change much under increased bootstrap sampling.

Given the intricacies of the bca formula and its implementation, one can legitimately ask "Does all this really work?" The following example, which is unusual in offering an exact gold standard comparison formula, is

Table 5.8 *Some estimates and their jackknife Monte Carlo standard errors for neonate error ratio, B = 2000.*

	$\hat{\theta}$	Bootstrap stdev	Bias-corrector $\hat{z}_0$	Acceleration $\hat{a}$
Estimate	1.09	0.146	−0.470	0.003
Jacksd	0.00	0.002	0.026	0.010

reassuring: we observe two independent estimates of variance,

$$\hat{\sigma}_1 \sim \frac{\sigma_1^2 G_{N_1}}{N_1}, \qquad \hat{\sigma}_2 \sim \frac{\sigma_2^2 G_{N_2}}{N_2}, \tag{5.47}$$

where G_N is gamma with N degrees of freedom as in Section 1.5, and are interested in the ratio

$$\theta = \frac{\sigma_1^2}{\sigma_2^2};$$

the estimates of (5.47) form a two-parameter exponential family with sufficient vector $y = (\hat{\sigma}_1^2, \hat{\sigma}_2^2)$.

Homework 5.18

(a) Apply `bcapar` with $B = 10,000$ to obtain bca confidence intervals for

$$\hat{\theta} = \frac{\hat{\sigma}_1^2}{\hat{\sigma}_2^2},$$

taking $N_1 = 5$, $N_2 = 21$, and $\hat{\sigma}_1^2 = \hat{\sigma}_2^2 = 1$.

(b) Show that the exact α-level confidence limit for θ is

$$\hat{\theta}_{\text{exact}}[\alpha] = \frac{\hat{\theta}}{F_{2N_1,2N_2}^{(1-\alpha)}},$$

where the notation indicates the $(1-\alpha)$ quantile of an F distribution with degrees of freedom $(2N_1, 2N_2)$.

(c) Compare exact and bca limits. Why would the comparison be essentially the same for any choice of $(\hat{\sigma}_1^2, \hat{\sigma}_2^2)$?

An Empirical Bayes Example

The insurance data of Table 1.3 provided estimates of

$$E\{\mu \mid x\}, \tag{5.48}$$

the expected number of claims made in a new year by a driver making x claims during the previous year. Egam(x), the third column of Table 5.9, is an estimate of $E\{\mu \mid x\}$ obtained from a gamma prior (1.54), the bottom row in Table 1.3.

Egam(x) is based on *g-modeling*, Section 4.8. In terms of the notation (4.80)–(4.84), $g(\mu)$ was assumed to be supported on the $K = 56$ points

$$\mathcal{T} = \exp(-8.0, -7.8, \ldots, 1.8, 2.0). \tag{5.49}$$

Matrix $\boldsymbol{Q}$ (4.83) had jth row

$$q_j^\top = \left(\log \mu_{(j)}, \mu_{(j)}\right), \tag{5.50}$$

for $j = 1, \ldots, 56$. Penalized maximum likelihood estimation (penalty $0.1 \cdot (\sum_1^2 \beta_j^2)^{1/2}$ rather than (4.86)) gave $\hat{g}(\mu)$ and, by Bayes rule, the conditioned expectation $\widehat{E}\{\mu \mid x\}$, listed as Egam($x$) in Table 5.9.

Table 5.9 *Empirical Bayes analysis of insurance data, Table 1.3;* Egam(x) *is estimate of* $E\{\mu \mid x\}$ *using Fisher's gamma prior, Table 1.3 row 3;* Ens(x) *is estimate of* $E\{\mu \mid x\}$ *starting from a natural spline model for prior* $g(\mu)$*, 4df. Last two columns are lower and upper limits of approximate 95% confidence intervals for* $E\{\mu \mid x\}$ *from* `bcapar`*, as explained in the text.*

x	Count	Egam(x)	Ens(x)	0.025	0.975
0	7840	0.16	0.17	0.16	0.18
1	1317	0.40	0.35	0.33	0.38
2	239	0.64	0.67	0.60	0.75
3	42	0.87	1.08	0.93	1.29
4	14	1.10	1.58	1.33	1.94
5	4	1.34	2.57	1.92	4.64
6	4	1.57	6.36	2.56	14.1
7	1	1.61	14.2	3.94	19.1

Homework 5.19 Ignoring penalization and the discreteness of μ, why is g-modeling with $\boldsymbol{Q}$ as in (5.50) equivalent to the MLE choice of a gamma prior (1.54) for μ?

Ens(x), column 4 of Table 5.9, is the g-modeling estimate of $E\{\mu \mid x\}$ starting from a natural spline model with four degrees of fredom: $\boldsymbol{Q} = \text{ns}(\mathcal{T}, 4)$ in R notation like that of (4.88). It gave an estimate $\hat{\boldsymbol{f}}$ of the marginal density $\boldsymbol{f}$, (4.81)–(4.82) of Section 4.8, closely matching the observed count proportions in Table 5.9. Column "yns" of Table 5.10 shows the match in terms of the estimated counts,

$$N \cdot \hat{f} \qquad (N = 9461). \tag{5.51}$$

"CSns" is the chi-square goodness-of-fit statistic

$$\text{CSns}(x) = (y(x) - \text{yns}(x))^2 \,/\, \text{yns}(x); \tag{5.52}$$

$\sum_0^7 \text{CSns}(x) = 4.63$, which shows the ns model passing a goodness-of-fit test to the observed counts (null distribution χ^2_ν with $\nu = 4 = 8 - 4$). The fit is better than that for the gamma model, columns "ygam" and "CSgam".

Table 5.10 *Counts y and estimated counts* ygam *(gamma model) and* yns *(natural spline model, 4df);* CSgam *and* CSns *are chi-square goodness-of-fit statistics, as in* (2.76).

x	y	ygam	yns	CSgam	CSns
0	7840	7847.7	7844.0	0.0	0.0
1	1317	1287.1	1314.0	0.7	0.0
2	239	256.6	230.1	1.2	0.3
3	42	54.3	51.2	2.8	1.6
4	14	11.8	13.8	0.4	0.0
5	4	2.6	4.3	0.8	0.0
6	4	0.6	1.9	19.3	2.3
7	1	0.1	1.7	8.1	0.3

What prior density $g(\mu)$, in the Bayesian setup

$$\mu_i \sim g(\cdot) \quad \text{and} \quad z_i \mid \mu_i \sim \text{Poi}(\mu_i), \qquad \text{for } i = 1, \dots, 9461, \tag{5.53}$$

could produce the counts $y(x)$ in Table 5.10? Figure 5.9 shows the estimated density $\hat{\boldsymbol{g}}$ (4.83) from our natural spline model, plotted continuously. It is no surprise that $\hat{\boldsymbol{g}}(\mu)$ is concentrated near $\mu = 0$ given that $y(0) = 7840$ is 83% of the total 9461.

In the Poisson setup (5.53) there is a simple formula for the posterior expectation of μ given x,

$$E\{\mu \mid x\} = (x+1)f(x+1)/f(x), \tag{5.54}$$

as in (1.53), saving the trouble of numerically calculating $E\{\mu \mid x\}$ from Bayes rule.

Having found the point estimate

$$\text{Ens}(x) = (x+1)\hat{f}(x+1)/\hat{f}(x), \tag{5.55}$$

$\hat{f}(x)$ from the natural spline model, we would like to compute a confidence

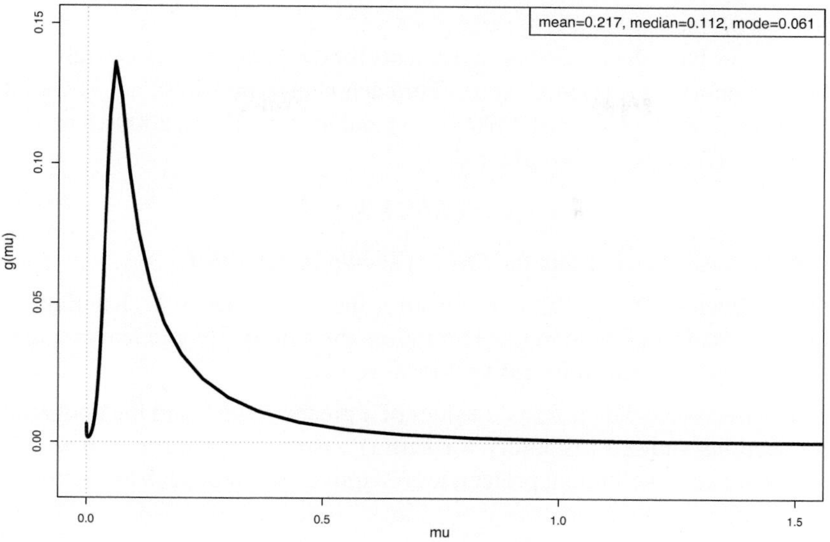

Figure 5.9 Estimated prior density $g(\mu)$ for insurance data, Table 1.3.

interval for $E\{\mu \mid x\}$. The last two columns of Table 5.9 give approximate 95% bca confidence limits for $E\{\mu \mid x\}$, $x = 0, \ldots, 7$, calculated as follows.

The count vector $\boldsymbol{y} = (7840, 1317, \ldots, 1)$ in Table 5.10 is assumed to be generated from a four-parameter natural spline model for $\boldsymbol{Q}$ in (4.83),

$$\beta \longrightarrow \boldsymbol{g} \longrightarrow \boldsymbol{f} \longrightarrow \boldsymbol{y} \sim \text{Mult}_8(N, \boldsymbol{f}), \tag{5.56}$$

for $N = 9461$, as in (4.80). Penalized maximum likelihood has produced estimates $\hat{\beta}$, $\hat{\boldsymbol{g}}$, and $\hat{\boldsymbol{f}}$,

$$\boldsymbol{y} \longrightarrow \hat{\beta} \longrightarrow \hat{\boldsymbol{g}} \longrightarrow \hat{\boldsymbol{f}}, \tag{5.57}$$

from which we obtain parametric bootstrap samples

$$\boldsymbol{y}^*(i) \sim \text{Mult}_8(N, \hat{\boldsymbol{f}}), \qquad i = 1, \ldots, B \tag{5.58}$$

($B = 2000$ for Table 5.9), and bootstrap replications of $\hat{\boldsymbol{f}}$,

$$\boldsymbol{y}^* \longrightarrow \hat{\beta}^* \longrightarrow \hat{\boldsymbol{g}}^* \longrightarrow \hat{\boldsymbol{f}}^*. \tag{5.59}$$

Finally, each replication $\hat{\boldsymbol{f}}^*(i)$ gives a bootstrap replication of $\text{Ens}(x)$,

$$\text{Ens}(x)^* = (x+1)\hat{\boldsymbol{f}}^*_{x+1}/\hat{\boldsymbol{f}}^*(x), \tag{5.60}$$

for $x = 0, \ldots, 7$.

We now have the necessary ingredients for the parametric bootstrap confidence interval program `bcapar`. For each choice of x in $0, \ldots, 7$, we let $\hat{\theta} = \mathrm{Ens}(x)$, $\hat{\boldsymbol{\theta}}^*$ the vector of 2000 $\mathrm{Ens}(x)^*$ values, and $\boldsymbol{Y}^*$ the 2000×8 matrix having $\boldsymbol{y}^*(i)$ as its ith row. The call

$$\texttt{bcapar}\left(\hat{\theta}, \hat{\boldsymbol{\theta}}^*, \boldsymbol{Y}^*[, 1:7]\right) \tag{5.61}$$

gave the confidence limits for $E\{\mu \mid x\}$ shown in Table 5.9.

Note Because $\sum_{j=1}^{8} y_j^*(i) = N$ for all i, the $B \times 8$ matrix $\boldsymbol{Y}^*$ has rank 7. Deleting its last column in (5.61) removes the singularity and leaves a matrix having minimal sufficient vectors as rows.

The insurance data is thin on values of x greater than 4, and the bootstrap replications showed instability for $\mathrm{Ens}(x)^*$, particularly for $x = 6$ and 7, where we very well might prefer a less adaptive estimator such as $\mathrm{Egam}(x)$.

5.7 Nonparametric Bootstrap Confidence Intervals

The bootstrap was originally introduced as a nonparametric device for assigning standard errors to estimates. In the same spirit, there is also a nonparametric version of the bca method for constructing accurate approximate confidence intervals. The nonparametric venue turns out to be particularly amenable to the bca approach, permitting a fully automated computer algorithm applicable to almost any nonparametric estimation problem. And of course there are many situations where not having to choose a parametric model is quite a relief.

One-sample nonparametric estimation problems can be described in the following framework: sampling from unknown probability distribution F on a space $\mathcal{X}$ yields an observed sample $\boldsymbol{x} = (x_1, \ldots, x_n)$,

$$x_1, \ldots, x_n \overset{\text{iid}}{\sim} F, \tag{5.62}$$

where $\mathcal{X}$ can be anything – $\mathcal{R}^1$, $\mathcal{R}^p$, categorical, an image space, time series – and F can be any distribution on $\mathcal{X}$. The sample $\boldsymbol{x}$ defines an empirical distribution $\widehat{F}$ that puts probability $1/n$ on each x_i.

We are interested in a real-valued parameter

$$\theta = \tau(F),$$

its nonparametric maximum likelihood estimate being

$$\hat{\theta} = \tau\left(\widehat{F}\right).$$

Having observed $\boldsymbol{x}$, or equivalently $\widehat{F}$, we wish to set confidence intervals for θ.

As an example, $\mathcal{X}$ could be $\mathcal{R}^5$, $n = 22$, and $\boldsymbol{x}$ the 22×5 matrix of Table 5.1, with $\hat{\theta} = \tau(\widehat{F})$ the sample correlation between the first two columns of $\boldsymbol{x}$. In this case, $\theta = \tau(F)$ would be the correlation between the first two coordinates of the five-dimensional distribution F. Here F could be any distribution on $\mathcal{R}^5$, not necessarily the multivariate normal distributions used previously.

A nonparametric bootstrap sample $\boldsymbol{x}^*$ is a random sample of size n drawn from $\widehat{F}$,

$$x_1^*, \ldots, x_n^* \overset{\text{iid}}{\sim} \widehat{F}.$$

That is, $\boldsymbol{x}^* = (x_1^*, \ldots, x_n^*)$ is obtained by independently drawing n times *with replacement* from the original sample $\boldsymbol{x}$. Let s_i be the number of times x_i occurs in $\boldsymbol{x}^*$. The count vector $s = (s_1, \ldots, s_n)$ follows a multinomial distribution of n draws on n categories, each of which has probability $1/n$,

$$s \sim \text{Mult}_n(n, \pi_0) \qquad (\pi_0 = (1, \ldots, 1)/n).$$

The vector of proportions $p = s/n$ has distribution

$$p \sim \text{Mult}_n(n, \pi_0)/n, \tag{5.63}$$

as in Section 2.9. Let $\widehat{F}(p)$ be the distribution that puts probability p_i on x_i, so $\widehat{F}(\pi_0) = \widehat{F}$, the empirical distribution.[8] A *bootstrap sample* $\boldsymbol{x}^*$ determines a *bootstrap replication* $\hat{\theta}^*$ of $\hat{\theta}$, namely the nonparametric MLE of θ based on $\boldsymbol{x}^*$. We can write $\hat{\theta}^*$ as a function of p,

$$\hat{\theta}^* = \tau\left(\widehat{F}(p)\right) = t(p), \tag{5.64}$$

where p has distribution (5.63). The notation $\hat{\theta}^* = t(p)$ makes sense because the original data $\boldsymbol{x}$ is held fixed in bootstrap computations, and does not need to be denoted. For instance, if the x_i are real-valued and the parameter of interest θ is the expectation of F, then

$$\tau\left(\widehat{F}\right) = \sum_{i=1}^{n} x_i/n \quad \text{and} \quad t(p) = \sum_{i=1}^{n} p_i x_i.$$

The nonparametric bootstrap estimate of standard error for $\hat{\theta}$, $\hat{\sigma}(\hat{\theta})$, is obtained by sampling B times from (5.63),

$$p(i) \overset{\text{iid}}{\sim} \text{Mult}_n(n, \pi_0)/n, \qquad i = 1, \ldots, B, \tag{5.65}$$

[8] Notice that $\widehat{F}(p)$ is the empirical distribution $\widehat{F}^*$ for the bootstrap sample $\boldsymbol{x}^*$.

computing the bootstrap replications

$$\hat{\theta}^*(i) = t\left(p(i)\right), \tag{5.66}$$

and setting

$$\hat{\sigma}\left(\hat{\theta}\right) = \left\{\sum_{i=1}^{B} \frac{\left(\hat{\theta}^*(i) - \hat{\theta}^*(\cdot)\right)^2}{B-1}\right\}^{1/2},$$

where $\hat{\theta}^*(\cdot) = \sum_1^B \hat{\theta}^*(i)/B$. This gives the nonparametric standard interval endpoint

$$\hat{\theta}_{\text{stand}}[\alpha] = \hat{\theta} + z^{(\alpha)}\hat{\sigma}\left(\hat{\theta}\right),$$

having the same virtues and defects as noted previously.

Improved nonparametric intervals $\hat{\theta}_{\text{bca}}[\alpha]$ will be introduced next, along with a computer program `bcajack` for their calculation; $\hat{\theta}_{\text{bca}}[\alpha]$ is just as "automatic" to calculate as $\hat{\theta}_{\text{stand}}[\alpha]$. All that is required is to carry out steps (5.65)–(5.66) B times: B needs to be perhaps 10 times larger for $\hat{\theta}_{\text{bca}}[\alpha]$ as for $\hat{\sigma}(\hat{\theta})$, but that is time spent by the machine and not by the statistician, most often a good bargain these days.

Homework 5.20 Calculate the nonparametric bootstrap standard error $\hat{\sigma}(\hat{\theta})$ for $\hat{\theta}$ the student score correlation coefficient cor(mechanics, vectors), Table 5.1, using $B = 100, 200, 400, 800$, and 1600. (Notice that it isn't necessary to express $\hat{\theta}^*$ in the form $t(p)$; if $\boldsymbol{x}^*$ in the 22×5 bootstrap student score data matrix then $\hat{\theta}^*$ is directly calculated as the sample correlation between the first two columns of $\boldsymbol{x}^*$.)

We wish to implement the bca endpoints $\hat{\theta}_{\text{bca}}[\alpha]$ (5.32) nonparametrically. Steps (5.65)–(5.66) yield the vector of bootstrap replications

$$\hat{\boldsymbol{\theta}}^* = \left(\hat{\theta}^*(1), \ldots, \hat{\theta}^*(B)\right)^\top,$$

which gives bootstrap CDF $\widehat{G}(t) = \#\{\hat{\theta}^*(i) \leq t\}/B$, and bias-corrector $\hat{z}_0 = \Phi^{-1}(\widehat{G}(\hat{\theta}))$ (5.22), leaving us the acceleration a to estimate. As discussed next, this turns out to be easy to do in the nonparametric framework, or at least easier to automate into a single program than were the parametric calculations.

The formula[9] for the estimated acceleration $\hat{a}$ involves the *directional*

[9] The description here is taken from the more detailed discussion in Sections 7 and 8 of Efron (1987).

derivatives of the function $\hat{\theta}^* = t(p)$ (5.64), evaluated at $p = \pi_0 = (1, \dots, 1)^\top/n$. As in Section 2.9, let e_i be the ith indicator vector,

$$e_i = (0, \dots, 1, 0, \dots, 0)^\top \qquad \text{(1 in } i\text{th place)},$$

for $i = 1, \dots, n$, and set

$$U_i = \lim_{\epsilon \to 0} \frac{t(\pi_0 + \epsilon(e_i - \pi_0)) - t(\pi_0)}{\epsilon}. \tag{5.67}$$

U_i is the rate of change of $t(p)$ as p moves from π_0 toward the ith corner of the simplex $\mathcal{S}_n$. This assumes $t(p)$ is smoothly defined near π_0, but we will relax things later.

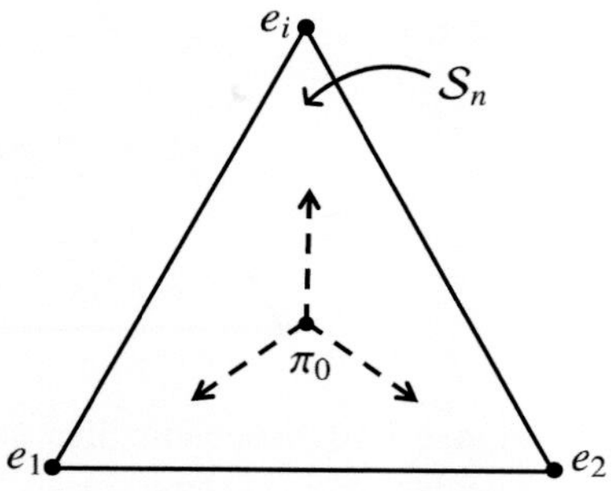

Homework 5.21 Show that $\sum_1^n U_i = 0$.

Lemma 5.5 *In the nonparametric framework, the least favorable family estimate of acceleration is*

$$\hat{a} = \frac{1}{6} \frac{\sum_1^n U_i^3}{\left(\sum_1^n U_i^2\right)^{3/2}}. \tag{5.68}$$

Proof The multinomial family

$$p \sim \text{Mult}_n(n, \pi_0)/n \qquad (\pi \in \mathcal{S}_n)$$

is an $(n-1)$-parameter exponential family, with (π, p) playing the role of (μ, y) in our general notation, and $\eta = \log \pi +$ constant (2.73). We have observed $\hat{\pi} = \pi_0$ and wish to set confidence intervals for a real-valued parameter $\theta = \tau(F)$. The proof of Lemma 5.5 consists of showing that (5.68) is (5.31), $\hat{a} = \hat{\gamma}/6$, applied to the multinomial situation.

Let $\widehat{C}$ be those values of p giving the same estimate of θ as π_0, i.e., $t(p) = \hat{\theta}$,

$$\widehat{C} = \{p : t(p) = t(\pi_0)\}.$$

Here p is thought of as defined continuously on $\mathcal{S}_n$, not restricted to the integer$/n$ values in (5.63).

Homework 5.22 Show that $\boldsymbol{U} = (U_1, \dots, U_n)^\top$ is orthogonal to the surface $\widehat{C}$ at $\pi = \pi_0$. See Figure 5.10.

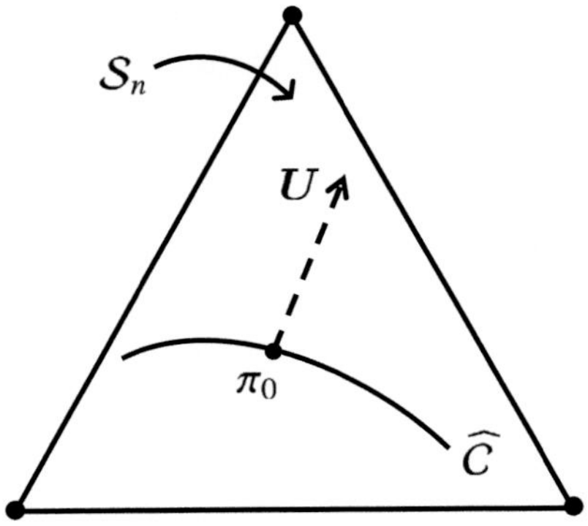

Figure 5.10 Schematic illustration of elements for nonparametric calculation of acceleration: surface $\widehat{C}$ of points p having $t(p)$ equal to MLE $\hat{\theta} = t(\pi_0)$; $\boldsymbol{U}$ the vector of directional derivatives (5.67) is orthogonal to $\widehat{C}$ at point π_0; $\boldsymbol{U}$ gives $\hat{a}$ (5.68).

Looking at Figure 2.5 in Section 2.6 we see that $\boldsymbol{U}$ must lie along the least favorable direction $\boldsymbol{t}'$; $\hat{\eta} = \log \pi_0$ is a vector of constants, say c_0, so the least favorable family $\hat{\eta} + \lambda \boldsymbol{U}$ has expectation vectors $\pi(\lambda)$ with coordinates

$$\pi_i(\lambda) = \frac{e^{c_0 + \lambda U_i}}{\sum_{j=1}^n e^{c_0 + \lambda U_j}} = \frac{e^{\lambda U_i}}{\sum_{j=1}^n e^{\lambda U_j}}. \tag{5.69}$$

Homework 5.23 (a) Show that (5.69) represents a one-parameter exponential family, as in Section 2.6, with sufficient statistic $x = \boldsymbol{U}^\top p$ and CGF

$$\hat{\phi}(\lambda) = \log\left(\sum_{j=1}^n e^{\lambda U_j}\right).$$

(b) Show that family (5.69) has skewness

$$\hat{\gamma} = \frac{\sum_1^n U_i^3}{\left(\sum_1^n U_i^2\right)^{3/2}},$$

for $\lambda = 0$.

This shows that (5.31) is given by (5.68), verifying Lemma 5.5. ■

There is one drawback to estimating acceleration by formula (5.68): evaluating U_i in (5.67) requires knowing $\hat{\theta}$ as a function of p, $\hat{\theta} = t(p)$. For the automatic calculation of $\hat{\theta}_{\text{bca}}[\alpha]$ it is better only to require $\hat{\theta} = T(\boldsymbol{x})$,

$T(\cdot)$ being the original function the statistician used. Program `bcajack`, our nonparametric bca algorithm, estimates a using the jackknife values $\hat{\theta}_{(i)}$ (5.45),

$$\hat{a} = \frac{1}{6} \frac{\sum_1^n \left(\hat{\theta}_{(i)} - \hat{\theta}_{(\cdot)}\right)^3}{\left[\sum_1^n \left(\hat{\theta}_{(i)} - \hat{\theta}_{(\cdot)}\right)^2\right]^{3/2}}, \tag{5.70}$$

which needs only $T(\cdot)$.

Homework 5.24 Formula (5.70) amounts to evaluating U_i in (5.67) with a certain nonzero value of ϵ. What is the value?

Program `bcajack` assumes that the data set $\boldsymbol{x}$ is an $n \times m$ matrix, whose ith row is the data for observation x_i. Then $\boldsymbol{x}_{(i)}$ (5.45) is the $(n-1) \times m$ matrix $\boldsymbol{x}$ with its ith row removed, and $\hat{\theta}_{(i)} = T(\boldsymbol{x}_{(i)})$.

A call to `bcajack` takes the form

$$\texttt{bcajack}\left(\hat{\theta}, B, T\right), \tag{5.71}$$

where $\hat{\theta}$ is the numerical value of the estimate, B the desired number of bootstrap replications, and T an R function that evaluates $T(\boldsymbol{x}^*)$ for bootstrap samples $\boldsymbol{x}^*$. (Because of (5.70), $T(\cdot)$ must also apply to matrices with $n-1$ rows.) The `bcajack` algorithm forms bootstrap matrices $\boldsymbol{x}^*$ by resampling the rows of $\boldsymbol{x}$ and then evaluates $\hat{\theta}^* = T(\boldsymbol{x}^*)$, doing all this B times.

As an example we reconsider the student score maxeig problem (5.11). Here $\boldsymbol{x}$ is the 22×5 matrix in Table 5.1. The R function $T(\boldsymbol{x})$ computes the empirical covariance matrix of $\boldsymbol{x}$, takes its eigenvalues and returns the ratio of the largest eigenvalue to their sum, giving $\hat{\theta} = T(\boldsymbol{x}) = 0.699$.

Table 5.11 shows part of the output of the call `bcajack(`$\hat{\theta}$`,B=10,000,T)` (5.1 seconds execution time); jacksd is the jackknife estimate of Monte Carlo error, as in Section 5.6; pct is the percentile point of the bca endpoint, so for instance, the 0.025 endpoint 0.533 was the 290th smallest value of the 10,000 bootstrap replications, percentile 0.029. The resulting nonparametric confidence intervals are shown as the black curves in Figure 5.11 which also includes the parametric bca curves in red (based on $B = 10,000$ parametric bootstrap samples) and the standard intervals $\hat{\theta} + z^{(\alpha)}\hat{\sigma}$, $\hat{\sigma}$ the nonparametric bootstrap standard error.

Table 5.12 compares $\hat{z}_0$, $\hat{a}$, and bootstrap standard error $\hat{\sigma}$ from the parametric and nonparametric analyses: $\hat{z}_0$ is parametrically more than twice as negative, accounting for a substantial amount of the parametric/nonparametric differences seen in Figure 5.11.

Table 5.11 *Nonparametric bca and standard confidence interval endpoints for student score maxeig parameter, using* `bcajack` *with* $B = 10,000$.

α	Nonpar bca	Standard	Jacksd	Pct
0.025	0.533	0.540	0.003	0.029
0.05	0.555	0.565	0.002	0.052
0.1	0.580	0.595	0.002	0.092
0.16	0.605	0.618	0.002	0.141
0.5	0.693	0.699	0.001	0.442
0.84	0.774	0.780	0.001	0.818
0.9	0.795	0.804	0.002	0.890
0.95	0.821	0.833	0.002	0.952
0.975	0.843	0.859	0.002	0.979

The very small jacksd values of Table 5.11 show that $B = 10,000$ was much greater than required. (It was intended to make the parametric/nonparametric comparison sharp.) Jacksd is roughly proportional to $B^{-1/2}$, so the values in Table 5.11 suggest that $B = 1,000$ would still give jacksd < 0.01.

Table 5.12 *Bca parameter estimates for student score maxeig analysis, using* `bcapar` *and* `bcajack`*, each with* $B = 10,000$.

	$\hat{\theta}$	Bootsd	$\hat{z}_0$	$\hat{a}$
Parametric	0.699	0.072	−0.184	0.055
Jacksd		0.001	0.015	0.005
Nonparametric	0.699	0.081	−0.075	0.060
Jacksd		0.001	0.011	0.000

Note There is an intermediate case between the parametric and nonparametric bca methods: categorical data. Suppose the observations x_i in (5.62) can take on only L distinct values, e.g., "beer, wine, spirits, soda" in Table 2.2. This constitutes an $(L - 1)$-parameter exponential family, Section 2.9, that we could analyze using `bcapar`. Equivalently, we could use the nonparametric program `bcajack`, taking $\boldsymbol{x}$ to be an n-vector with components x_i equaling 1 through L according to category. This can be interpreted to say that nonparametric bca analysis automatically groups similar observations together, so even if n is large, `bcajack` still may effectively be working with a small number of parameters.

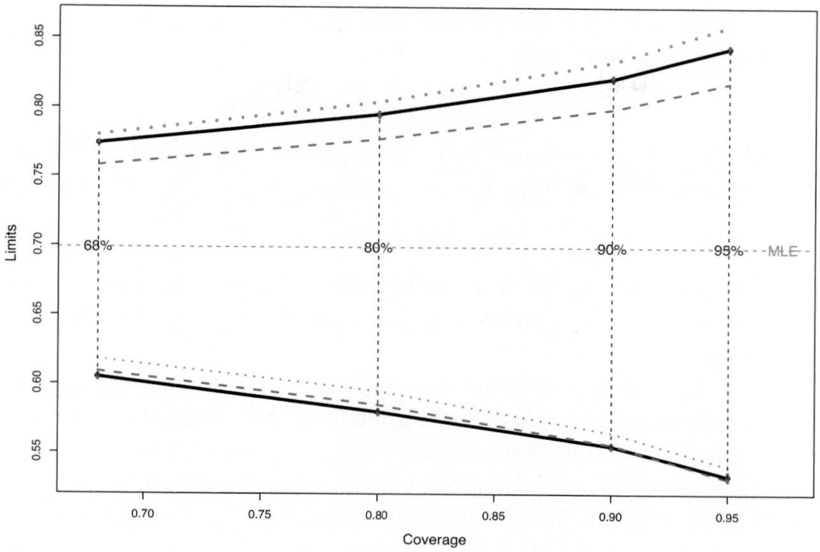

Figure 5.11 Student score maxeig confidence intervals, plotted vertically, versus nominal two-sided coverage; green standard, black nonparametric bca, red parametric bca.

Homework 5.25 For the situation just described, show that the parametric and nonparametric methods give the same estimates $\hat{\theta}_{\text{bca}}[\alpha]$.

5.8 The Abc Algorithm

The bca method is a pure child of the computer age, lavishly spending thousands of times more calculations than the standard intervals. Surprisingly, there is an alternate approach that approximates the bca intervals using local Taylor series in place of bootstrap replications. This is the *approximate bootstrap confidence* (abc) algorithm.[10] We will see that it provides quite satisfactory bca intervals at a small fraction of the computational cost, at least for smoothly defined statistics $\hat{\theta} = t(\boldsymbol{x})$. The catch is that, besides being less general than `bcapar`, it is also less automatic in requiring a certain amount of theoretical setup from the statistician. Here we will discuss only the *parametric* abc algorithm, although there is a nonparametric version. The main reference is DiCiccio and Efron (1996).

[10] Not to be confused with the more recent ABC method, or Approximate Bayesian Computation.

As in Section 5.5, we assume that data $\boldsymbol{x}$ has been observed from a p-parameter exponential family

$$g_\eta(\boldsymbol{x}) = e^{\eta^\top y - \psi(\eta)} g_0(\boldsymbol{x}) \qquad (y = Y(\boldsymbol{x})),$$

and that we wish to estimate a real-valued parameter θ, which can be expressed as a function of either η or $\mu = E_\eta\{y\} = \dot\psi(\eta)$,

$$\theta = s(\eta) = t(\mu),$$

now with $s(\eta)$ and $t(\mu)$ being smoothly differentiable. The data $\boldsymbol{x}$ gives MLE $\hat\mu = y$, $\hat\eta = \dot\psi^{-1}(\hat\mu)$, and $\hat\theta = s(\hat\eta) = t(\hat\mu)$. We wish to set confidence intervals for θ.

The standard interval endpoint $\hat\theta_{\text{stand}}[\alpha] = \hat\theta + z^{(\alpha)}\hat\sigma$ requires the specification of two quantities, $\hat\theta$ and $\hat\sigma$. We already have $\hat\theta$. Applied statisticians used to spend lots of time deriving tedious Taylor series approximations for $\hat\sigma$. The `abc` program instead begins by using numerical differentiation[11] to estimate $\hat\sigma$. Let $mu(\eta)$ stand for the vector function $\dot\psi(\eta)$. The abc estimate of covariance matrix $\widehat{V} = V_{\hat\eta}$ has jth column

$$\left[mu(\hat\eta + e_j d) - mu(\hat\eta - e_j d)\right](2d)^{-1}, \qquad (5.72)$$

for d small and e_j the p-dimensional coordinate vector $(0, \dots, 1, 0, \dots, 0)^\top$, 1 in the jth place.

Homework 5.26 Why does (5.72) give a good approximation to $\widehat{V}$?

Unlike `bcapar`, `abc` requires $\hat\theta$ to be specified in the form $\hat\theta = t(y)$; this is one of the ways it is less convenient. The least favorable direction $\hat t'$ (5.29) is obtained as

$$\hat t'_j = \left[t(\hat\mu + e_j d) - t(\hat\mu - e_j d)\right](2d)^{-1}, \qquad \text{for } j = 1, \dots, p,$$

d sufficiently small. The first-order Taylor expansion

$$t(y) \doteq t(\mu) + t'^\top(y - \mu) \qquad \left(t' = (\cdots \partial t/\partial\mu_j \cdots)^\top\right)$$

suggests the standard error estimate $\sigma = (t^\top V t)^{1/2}$ for $\hat\theta$, estimated by

$$\hat\sigma = \left(\hat t'^\top \widehat{V} \hat t'\right)^{1/2} \qquad \left(\hat t' = (\cdots \hat t'_j \cdots)^\top\right); \qquad (5.73)$$

$\hat\sigma$ is the traditional "delta method" estimate of standard error, derived here numerically rather than by formula.

[11] If $h(z)$ is a real-valued smooth function of a real-valued argument z, then the derivative $\dot h(z)$ is approximated by $[h(z+d) - h(z-d)](2d)^{-1}$, with d small. The second derivative approximation is $[h(z+d) - 2h(z) + h(z-d)]d^{-2}$. Some experimentation in the choice of d is usually needed.

Homework 5.27 Justify (5.73).

The standard intervals require just two parameter estimates, $\hat{\mu}$ and $\hat{\sigma}$; the abc intervals require three more:

- $\hat{a}$ for acceleration;
- $\hat{b}$ for bias;
- $\hat{c}$ for curvature, discussed next.

The five quantities together, $(\hat{\theta}, \hat{\sigma}, \hat{a}, \hat{b}, \hat{c})$, are enough to produce good approximations to the bca endpoints (5.32). "Good" here means accurate enough to preserve the second-order accuracy of $\hat{\theta}_{\text{bca}}[\alpha]$.

The abc acceleration estimate is

$$\hat{a} = \frac{\partial^2}{\partial \lambda^2}\left[\hat{t'}^\top mu\left(\hat{\eta} + \lambda\hat{t'}\right)\right]_{\lambda=0}\left(6\hat{\sigma}^2\right)^{-1}, \tag{5.74}$$

where $mu(\cdot)$ is the function $\dot{\psi}(\cdot)$ and the second derivative is approximated numerically, as in the previous footnote.

Homework 5.28 Show that (5.74) follows from $\hat{a} = \hat{\gamma}/6$ (5.23) applied within the least favorable family $\hat{f}_\lambda(\boldsymbol{x}) = g_{\hat{\eta}+\hat{t'}\lambda}(\boldsymbol{x})$ (5.29).

The quantity $\hat{b}$ concerns the expectation bias of $\hat{\theta} = t(y)$ as an estimate of $\theta = t(\mu)$. The second-order Taylor expansion

$$t(y) \doteq t(\mu) + t'^\top(y - \mu) + 1/2(y - \mu)^\top t''(y - \mu),$$

where t'' is the $p \times p$ matrix $(\partial^2 t(y)/\partial y_i \partial y_j)|_{y=\mu}$, gives

$$\begin{aligned}
\text{bias} &\doteq (1/2)E_\mu\left\{(y - \mu)^\top t''(y - \mu)\right\} \\
&= (1/2)E_\mu\left\{\text{trace}\left[t''(y - mu)(y - mu)^\top\right]\right\} \\
&= \text{trace}\left(t''V/2\right),
\end{aligned}$$

where $V = \text{Cov}_\mu\{y\}$; $\hat{b}$ is *bias* evaluated at $\mu = \hat{\mu}$. It is efficiently calculated by making use of the eigen decomposition of $\widehat{V}$,

$$\widehat{V} = \Gamma D \Gamma',$$

with D the diagonal matrix of eigenvalues d_j, and $\Gamma = (\gamma_1, \dots, \gamma_p)$ the matrix of eigenvectors.

Homework 5.29 Show that

$$\hat{b} = \frac{1}{2}\sum_{j=1}^{p}\frac{\partial^2}{\partial\lambda^2} t\left(\hat{\mu} + \lambda d_j^{1/2}\gamma_j\right)\Bigg|_{\lambda=0}$$

(which is approximated numerically in abc, as with $\hat{a}$).

The least favorable family $\hat{\eta} + \lambda\hat{t}'$ is a straight line through the η space but a curve in the μ space, as suggested in Figure 2.3 of Section 2.6. The curve passes through $\hat{\mu}$ in the direction

$$\dot{s} = \widehat{V}\hat{t}',$$

this following from $d\mu/d\eta = V$. The *curvature* $\hat{c}$ measures the nonlinearity of the function $\theta = t(\mu)$ as it moves away from $\hat{\mu}$ in direction $\dot{s}$,

$$\hat{c} = \frac{1}{2\hat{\sigma}} \frac{\partial^2}{\partial\lambda^2} t\left(\hat{\mu} + \frac{\dot{s}}{\hat{\sigma}}\lambda\right)\bigg|_{\lambda=0}. \tag{5.75}$$

The quantity $\hat{c}$ is better called "naming curvature": all monotone transformations $\phi = m(\theta)$ have the same level surfaces $\widehat{C}$ and $\widehat{\mathcal{H}}$ (Figure 5.8 of Section 5.6) and the same $\hat{z}_0$ and $\hat{a}$, with confidence limits transforming correctly,

$$\hat{\phi}_{\text{abc}}[\alpha] = m\left(\hat{\theta}_{\text{abc}}[\alpha]\right).$$

All parameters $\phi = m(\theta)$, $m(\cdot)$ monotone, are essentially the same from a confidence-interval point of view, differing only in the name attached to them as we move along the least favorable family; $\hat{c}$ (5.75) gives abc enough information to compensate for different namings.

The acceleration estimate $\hat{a}$ does double duty in abc, also helping to approximate the bootstrap CDF $\widehat{G}$ in formula (5.32) for $\hat{\theta}_{\text{bca}}[\alpha]$. Suppose that $g(x)$ is a one-dimensional probability density having mean 0 and variance 1, and that we have reason to believe $g(x)$ is roughly $\mathcal{N}(0,1)$. A two-term Edgeworth expansion for the CDF $G(x)$ is

$$G(x) \doteq \Phi(x) - \varphi(x)\frac{\gamma}{6}(x^2 - 1), \tag{5.76}$$

with $\varphi(x)$ the standard $\mathcal{N}(0,1)$ density and γ the skewness of x. But $\hat{a} = \hat{\gamma}/6$, showing a connection of acceleration with $\widehat{G}$.

Homework 5.30 Use (5.76) to derive

$$\Phi^{-1}(G(0)) \doteq \frac{\gamma}{6}.$$

A more elaborate Edgeworth analysis shows that a bootstrap replication $\hat{\theta}^* \sim g_{\hat{\eta}}(\cdot)$ from a p-parameter exponential family has approximate mean, standard deviation, and skewness

$$\hat{\theta}^* \mathrel{\dot{\sim}} \left[\hat{\theta} + \hat{b}, \hat{\sigma}, 6(\hat{a} + \hat{c})\right], \tag{5.77}$$

the approximation being good enough to maintain second-order accuracy (Diciccio and Efron, 1996).

Homework 5.31 Use Homework 5.30 and expression (5.77) to derive the abc approximation

$$\hat{z}_0 \doteq \hat{a} + \hat{c} - \frac{\hat{b}}{\hat{\sigma}}. \tag{5.78}$$

Putting them all together (and ignoring terms too small to effect second-order accuracy) eventually produces the abc confidence intervals endpoint

$$\hat{\theta}_{\text{abc}}[\alpha] = t\left[\hat{\mu} + \widehat{V}\left(\frac{\hat{t}'}{\hat{\sigma}}\right)c_{1\alpha}\right],$$

where

$$c_{1\alpha} = c_{0\alpha} + \hat{a}c_{0\alpha}^2 \quad \text{and} \quad c_{0\alpha} = (\hat{z}_0 + z^{(\alpha)})\left[1 - \hat{a}(\hat{z}_0 + z^{(\alpha)})\right]^{-1}.$$

A slightly cruder approximation gives the *abcq* endpoint

$$\hat{\theta}_{\text{abcq}}[\alpha] = \hat{\theta} + \hat{\sigma}(c_{1\alpha} + \hat{c}z_{1\alpha}^2).$$

The advantage of abcq is that it is entirely *local*, that is, it only requires $t(\cdot)$ recalculated arbitrarily close to $\hat{\mu}$, which can be advantageous for some functions $t(\cdot)$. Keeping count of numerical derivates, $\hat{\theta}_{\text{abc}}[\alpha]$ and $\hat{\theta}_{\text{abcq}}[\alpha]$ require about three times as many recomputations of $t(\cdot)$ as $\hat{\theta}_{\text{stand}}[\alpha]$, which compares to factors of thousands for `bcapar`.

As a simple example, Figure 5.12 shows the results of applying abc to the ulcer data 2×2 table of Figure 1.1, Section 1.5. Here we can take

$$y = (9, 12, 7, 17) \quad \text{and} \quad \hat{\eta} = \log y.$$

The multinomial mapping $\mu = mu(\eta)$ from η to μ, with $\mu = n\pi$ in the notation of Section 2.9, is

$$mu(\eta) = \frac{ne^{\eta}}{\sum_1^4 e^{\eta_j}} \qquad (n = 45).$$

Two parameters $\theta = t(\mu)$ are considered in Figure 5.12: the cross-product ratio $\theta_1 = (\mu_1/\mu_2)/(\mu_3/\mu_4)$, and also $\theta_2 = \log\theta_1$. The left panel shows confidence limits obtained from the call

$$\texttt{abc}(\hat{\theta}_1, \texttt{y}, \hat{\eta}, \texttt{mu}),$$

where the standard limits are seen to be extremely inaccurate. On the right,

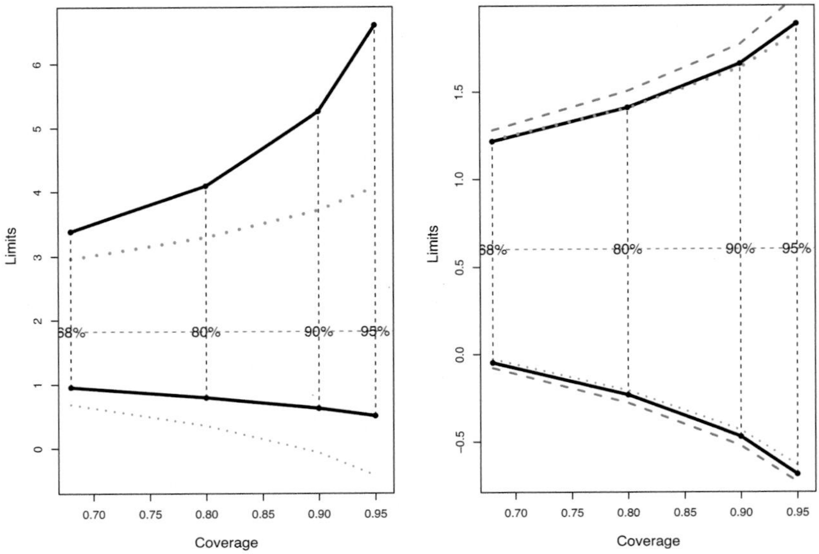

Figure 5.12 Abc (black) and standard (green) confidence limits for 2×2 ulcer table cross-product ratio (left) and its logarithm (right). Abc endpoints transform correctly but standard endpoints do not. Red dashed curves in right panel show parametric bca limits, using `bcapar` with $B = 10,000$.

the abc and standard limits for θ_2 are in near-perfect agreement. This illustrates the value of transformation invariance,

$$\hat{\theta}_{2\,\mathrm{abc}}[\alpha] = \log\left(\hat{\theta}_{1\,\mathrm{abc}}[\alpha]\right),$$

which isn't true for the standard limits. Applied statisticians always knew to compute these standard intervals on the log scale, which is the kind of trick `abc` and `bcapar` automate.

Homework 5.32 Compute the abc limits for θ_1 and θ_2 now assuming that the components of y are independent Poissons (so $mu(\eta) = e^{\eta}$).

Table 5.13 shows estimates for the abc parameters. Notably, the bias $\hat{b}$ and curvature $\hat{c}$ are huge for θ_1. The dashed red curves in the right panel of Figure 5.12 show $\hat{\theta}_{\mathrm{bca}}[\alpha]$ for $\hat{\theta}_2$, obtained using `bcapar` with $B = 10,000$ (again, taken so large for the sake of accurate comparisons). The bca limits are about 9% longer. This was mainly due to the bootstrap $\hat{\sigma}$ estimate 0.71

being 13% larger than the abc estimate 0.63.[12] Note that bcapar was run using the Poisson trick (Section 2.8), that is, taking $y^* \sim \text{Poi}(y)$.

The maxeig example, (5.11) and Figure 5.4, was based on modeling the rows of the 22×5 student score matrix as independent $\mathcal{N}_5(\lambda, \Gamma)$ vectors. Here the sufficient vector y is $(p = 20)$-dimensional. The mapping $mu(\cdot)$ from η to μ is somewhat intricate (Section 2.8), as is the calculation of $\hat{\eta}$ from y. (Though if you save your programs, they will apply to all future multivariate normal situations.)

Table 5.13 *Ulcer data, Figure 1.1: abc estimates.*

	θ	St. err	z_0	a	b	c
Crossprod	1.82	1.15	−0.029	−0.008	0.412	0.338
Log.crossprod	0.60	0.63	−0.029	−0.008	0.028	0.023

Homework 5.33 Perform the intricate calculations and apply the abc algorithm.

Figure 5.13 shows the abc confidence limits for the maxeig parameter. These are compared with the bcapar results from the $B = 10,000$ bootstrap replications in Figure 5.4, which they closely match. Both the abc and bcapar intervals are shifted downwards from the standard intervals.

Note It can be shown that

$$\hat{\nu} = \frac{\hat{b}}{\hat{\sigma}} - \hat{c}$$

is a measure of the total curvature of the level surface $\widehat{C} = \{\mu : t(\mu) = \hat{\theta}\}$ at $\hat{\mu}$; see Figure 5.8 in Section 5.6. That is, $\hat{\nu}$ measures how bowl-shaped $\widehat{C}$ is at $\hat{\mu}$. Relation (5.78) gives

$$\hat{z}_0 = \hat{a} - \hat{\nu}. \tag{5.79}$$

If $\widehat{C}$ if flat then $\hat{\nu} = 0$ and $\hat{z}_0 = \hat{a}$. This is always the case in one-dimensional exponential families (5.25) but most often not so for multiparameter families. Looking at Figure 5.8, the $\hat{a}$ part of $\hat{z}_0$ in (5.79) concerns the skewness of the one-parameter least favorable family (5.29), while $\hat{\nu}$ measures the global effects of curvature on the proportion of y^* values lying below $\widehat{C}$.

In practice, the curvature effect $\hat{\nu}$ can be much larger than the acceleration effect, especially in high-dimension situations where the total curva-

[12] Delta-method standard errors, pervasive in classical applications, tend to be a little downwardly biased, though there is no theorem to this effect.

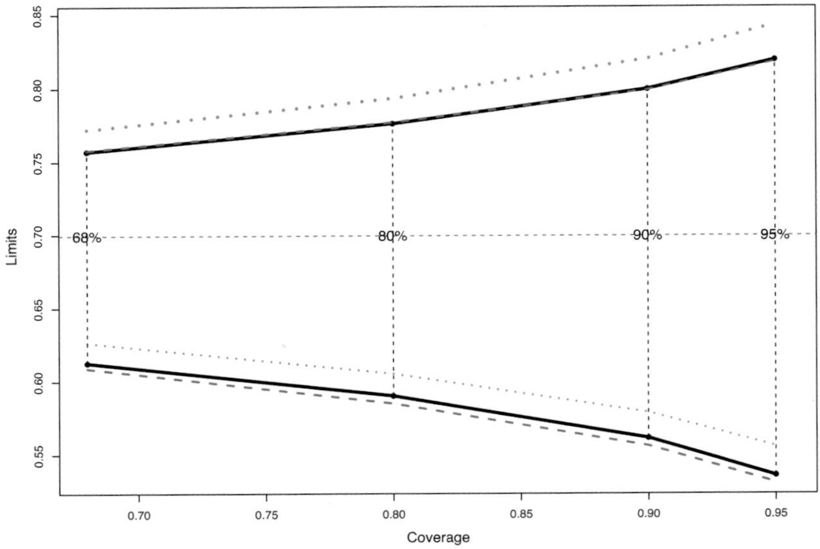

Figure 5.13 Student score maxeig confidence intervals; solid black curves are abc, dashed red are bca with $B = 10,000$, dotted green are standard.

ture can build up from individual parameter causes. As an example, suppose that $x \sim \mathcal{N}_p(\mu, I)$ and $\theta = ||\mu||^2$, in which case $\hat{\theta} = ||x||^2$ has a noncentral chi-squared distribution with p degrees of freedom and noncentrality parameter $\theta : \hat{\theta} \sim \chi^2_p(\theta)$. Then $\hat{\theta}^* \sim \chi^2_p(\hat{\theta})$ and

$$\hat{z}_0 = \Phi^{-1}\left\{\Pr\left(\chi^2_p\left(\hat{\theta}\right) \le \hat{\theta}\right)\right\}.$$

If for instance $p = 10$ and $\hat{\theta} = 20$ then $\Pr\{\chi^2_{10}(20) \le 20\} = 0.156$, giving $\hat{z}_0 = -1.01$. Bca methods tend to fail for such highly curved situations, the corrections to the standard intervals being too large for stable estimation.

5.9 Confidence Densities and Implied Likelihoods

Reference Efron and Hastie (2016), Section 11.6.

Reference Efron (1993), "Bayes and likelihood calculations from confidence intervals", *Biometrika* 3–26.

Reference Xie and Singh (2013), "Confidence distribution, the frequentist distribution estimator of a parameter: A review", *Int. Stat. Rev.* 3–39.

Among Fisher's major ideas, only one met with general disapproval and disdain–fiducial inference. Confidence densities and confidence distributions are direct descendents of fiducial inference, and they speak to the same goal: to isolate an appropriate likelihood or posterior density for a *single* parameter of interest, given data from a multiparameter family of densities. Our subject here, the bca theory of confidence densities, clarifies and to a certain extent justifies fiducial ideas, and makes them computationally accessible even in very complicated situations.

Confidence densities are derived from confidence intervals. We begin with observed data $\boldsymbol{x}$ having confidence limits $\theta_{\boldsymbol{x}}[\alpha]$ for a real-valued parameter θ (now including $\boldsymbol{x}$ in the notation and eliminating the hat), and define $\alpha_{\boldsymbol{x}}(\theta)$ as the inverse function of $\theta_{\boldsymbol{x}}[\alpha]$. That is, $\alpha_{\boldsymbol{x}}(\theta)$ is the confidence level α corresponding to confidence limit θ. It is assumed that α is a differentiable increasing function of θ.

Definition The confidence density of θ given $\boldsymbol{x}$ is

$$\pi_{\boldsymbol{x}}(\theta) = d\alpha_{\boldsymbol{x}}(\theta)/d\theta. \qquad (5.80)$$

For any two values $\theta_1 < \theta_2$ we have

$$\int_{\theta_1}^{\theta_2} \pi_{\boldsymbol{x}}(\theta)\, d\theta = \alpha_{\boldsymbol{x}}(\theta_2) - \alpha_{\boldsymbol{x}}(\theta_1).$$

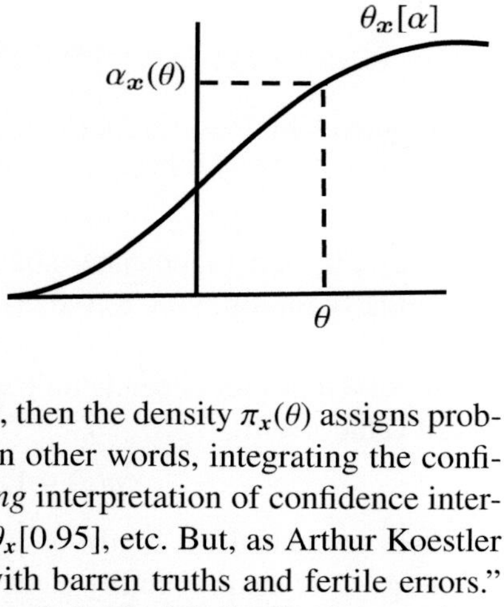

If say $\alpha_{\boldsymbol{x}}(\theta_1) = 0.95$ and $\alpha_{\boldsymbol{x}}(\theta_2) = 1.00$, then the density $\pi_{\boldsymbol{x}}(\theta)$ assigns probability 0.05 to the event $\theta \in (\theta_1, \theta_2)$. In other words, integrating the confidence density gives the famously *wrong* interpretation of confidence intervals, 0.05 probability of θ exceeding $\theta_{\boldsymbol{x}}[0.95]$, etc. But, as Arthur Koestler said, "The history of ideas is filled with barren truths and fertile errors." Confidence densities are what Fisher called fiducial distributions, at least in the special cases Fisher considered, and if they are errors they are useful and evocative ones.

In the example of Figure 5.14, we have observed $x = 10$ from the Poisson model $x \sim \text{Poi}(\theta)$, and wish to infer the possible values of θ. The heavy black curve is the confidence density for θ, computed from exact Poisson confidence intervals; $\pi_x(\theta)$ stretches far right, nicely illustrating increased uncertainty to the right side of $x = 10$. The percentiles of the distribution $\pi_x(\theta)$ correspond to confidence limits for θ; for instance $\theta = 17.82$, the 0.975 quantile of the confidence distribution, is the 0.975 confidence limit

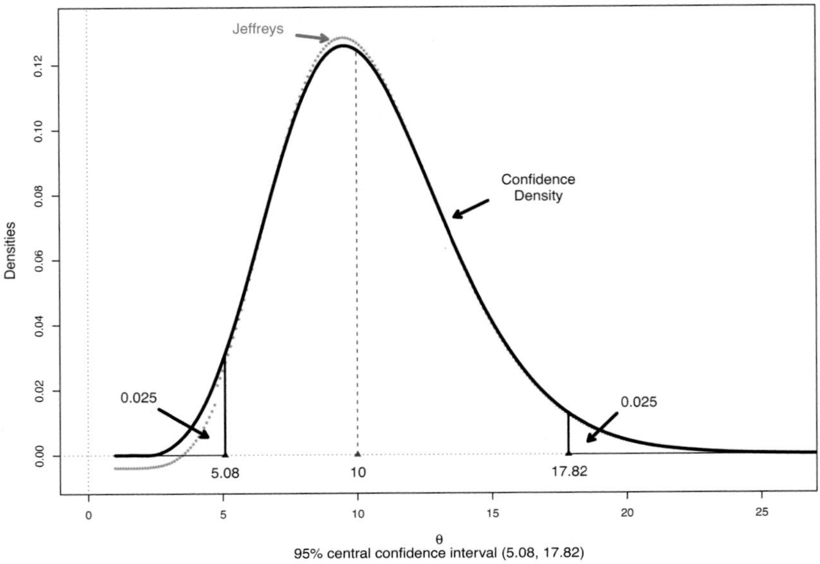

Figure 5.14 Poisson(10) example of bca confidence density; dotted red curve is posterior from Jeffreys matching prior.

$\theta_x[0.975]$. Figure 5.14 indicates 0.025 confidence probability for $\theta \geq 17.82$, and similarly for $\theta \leq 5.08$, $\theta_x[0.025]$.

The Poisson confidence density depends on the definition of $\alpha_x(\theta)$ in (5.80), which in turn depends on the exceedence probabilities for the observed value $x = 10$,

$$p(\theta) = \Pr_\theta\{x \geq 10\}.$$

This was computed for a grid of θ values ranging from 0 to 25, always counting 1/2 of the probability atom[13] at $x = 10$.

Homework 5.34 Compute the Poisson confidence density having observed $x = 8$.

The red curve in Figure 5.14 is *Jeffreys uninformative prior* density

$$\pi_{\text{Jeff}}(\theta) = c\theta^{-1/2} \qquad (\theta > 0),$$

with c chosen to make $\int_0^\infty \pi_{\text{Jeff}}(\theta)\, d\theta = 1$. An enormous amount of intellectual effort has gone into the definition of uninformative priors, by which

[13] Some such specification is necessary in defining confidence limits for discrete distributions. Equal splitting at the observed point is the most natural choice.

is meant Bayesian prior distributions that correspond to a complete lack of prior knowledge (for θ in our situation). One approach is through *matching priors*: that is, prior distributions whose posterior intervals match confidence limits. Since by definition confidence density percentiles perfectly match the confidence limits, and $\pi_{\text{Jeff}}(\theta)$ is close to $\pi_x(\theta)$, the former must be close to a matching prior in Figure 5.14.

For one-parameter families $f_\theta(x)$, x real-valued, Jeffreys uninformative prior can be defined to be

$$\pi_{\text{Jeff}}(\theta) = c/\sigma(\theta), \tag{5.81}$$

with $\sigma(\theta)$ the standard deviation of the MLE $\hat{\theta}$ for parameter value θ. This yields approximate matchingness, but doesn't extend to more complicated situations. The real challenge is to produce uninformative priors for a real-valued parameter of interest $\theta = t(\mu)$ when the data comes from a multi-parameter family $g_\mu(\theta)$; in other words, to get rid of nuisance parameters in a neutral way. This is where bca intervals and their confidence densities come into play.

Homework 5.35 Create a graph comparing $\pi_{\text{Jeff}}(\theta)$ for $x \sim \text{Poi}(\theta)$, $x = 10$, with the posterior density of θ starting from a flat prior, $\pi(\theta) = 1$.

There turns out to be a convenient computational formula for the confidence density $\pi_x(\theta)$ (5.80) based on the bca endpoints $\hat{\theta}_{\text{bca}}[\alpha]$. Following the notation in (5.32) of Section 5.5, define

$$u_\theta = \Phi^{-1}\left(\widehat{G}(\theta)\right) - \hat{z}_0$$

and

$$w(\theta) = \frac{\varphi\left[u_\theta/(1+\hat{a}u_\theta) - \hat{z}_0\right]}{\varphi(u_\theta + \hat{z}_0)(1+\hat{a}u_\theta)^2}, \tag{5.82}$$

with $\varphi(\cdot)$ the standard normal density. For a parametric bootstrap sample $\hat{\theta}_1^*, \ldots, \hat{\theta}_B^*$, let $\hat{w}_i = w(\hat{\theta}_i^*)$. The weights $\hat{w}_i$ turn out to produce the confidence density based on bca confidence intervals; program `bcapar` in package `bcaboot` returns the weights $\{w_i\}$ under the prompt `cd=1`.

Theorem 5.6 *The bca confidence density is estimated by the empirical distribution that puts weight* $\hat{w}_i(\sum_1^B \hat{w}_j)^{-1}$ *on* $\hat{\theta}_i^*$.

Homework 5.36 Show that the unweighted bootstrap histogram is the confidence density based on the percentile method (5.9) of Section 5.3.

The left panel of Figure 5.15 graphs the weights $\hat{w}_i$ against the bootstrap

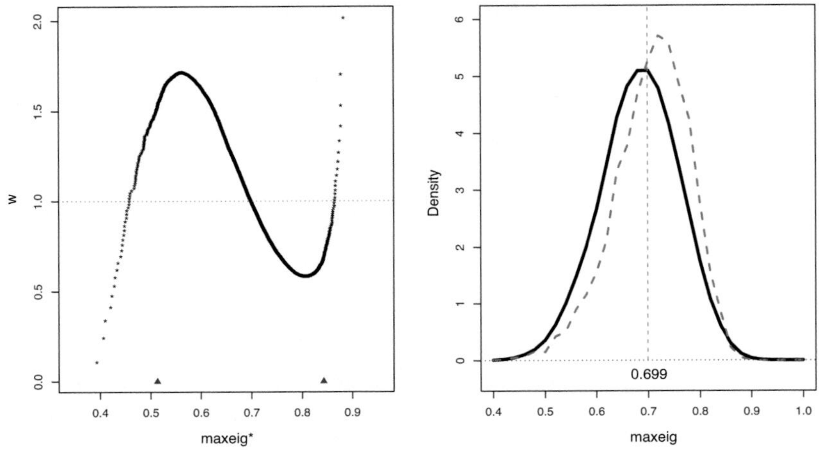

Figure 5.15 Maxeig example, Figure 5.4. *Left*: bca confidence density weights $\hat{w}_i$ plotted versus bootstrap replications $\hat{\theta}_i^*$; triangles show 1st and 99th percentiles of 10,000 $\hat{\theta}^*$s. *Right*: bca confidence density (black) compared with percentile confidence density (red dashed).

replications $\hat{\theta}_i^*$ for the maxeig example of Figure 5.4 (with the $\hat{w}_i$ standardized to $\bar{w} = 1$). Smaller values of $\hat{\theta}_i^*$ mostly have bigger values of w_i. This shifts the bca confidence density leftwards from the raw histogram of the $\hat{\theta}_i^*$s (red dashed curve). In this case the bca confidence density is less skewed than the percentile confidence density, skewness -0.17 compared to -0.54.

The Student-t Intervals

Fisher's motivating example for fiducial inference was the student-t intervals. Having observed $x_1, \dots, x_n \overset{\text{iid}}{\sim} \mathcal{N}(\theta, \sigma^2)$, the random variable

$$T = \frac{\hat{\theta} - \theta}{s} \qquad \left(\hat{\theta} = \bar{x}, s^2 = \sum \frac{(x_i - \bar{x})^2}{n(n-1)}\right) \tag{5.83}$$

has a student-t distribution with $n-1$ degrees of freedom, $T \sim t_{n-1}$, yielding the confidence limits

$$\hat{\theta}_t[\alpha] = \hat{\theta} - s t_{n-1}^{(1-\alpha)}, \tag{5.84}$$

where $t_{n-1}^{(1-\alpha)}$ indicates the $(1-\alpha)$ quantile of t_{n-1}.

Homework 5.37 Show that the confidence distribution based on $\hat{\theta}_t[\alpha]$ has quantiles matching (5.84).

Fisher rewrote (5.83) as

$$\theta = \hat{\theta} - sT,$$

and then argued as follows: the sufficient statistics $(\hat{\theta}, s)$ exhaust the information about θ available in $\boldsymbol{x}$, leaving the randomness in $T \sim t_{n-1}$ as irreducible variability for θ; therefore, given $\boldsymbol{x}$ we have θ distributed as

$$\theta \sim \hat{\theta} - st_{n-1}, \tag{5.85}$$

with $(\hat{\theta}, s)$ fixed as observed. This is the *fiducial distribution* of θ, which coincides with the confidence distribution argument in Homework 5.37.

All of this seems harmless enough since it gives the usual student-t endpoints (5.84) as the corresponding fiducial quantiles. But Fisher had bigger game in mind. The notorious Behrens–Fisher problem concerns two independent normal samples,

$$x_{11}, x_{12}, \dots, x_{1n_1} \overset{\text{iid}}{\sim} \mathcal{N}(\mu_1, \sigma_1^2) \quad \text{and} \quad x_{21}, x_{22}, \dots, x_{2n_2} \overset{\text{iid}}{\sim} \mathcal{N}(\mu_2, \sigma_2^2).$$

A confidence interval is desired for $\theta = \mu_2 - \mu_1$. In this situation there is no pivotal quantity available to provide confidence intervals for θ.

Let

$$\bar{x}_1 = \sum_{i=1}^{n_1} \frac{x_{1i}}{n_1} \quad \text{and} \quad s_1^2 = \sum_{i=1}^{n_1} \frac{(x_{1i} - \bar{x}_1)^2}{n_1(n_1 - 1)},$$

and similarly for $\bar{x}_2$ and s_2. Then

$$\frac{\bar{x}_1 - \mu_1}{s_1} \sim t_{n_1-1} \quad \text{independent of} \quad \frac{\bar{x}_2 - \mu_2}{s_2} \sim t_{n_2-1},$$

the notation denoting independent t variables with the indicated degrees of freedom. Fisher suggested

$$\theta \sim \bar{x}_2 - \bar{x}_1 + s_1 t_{n_1-1} - s_2 t_{n_2-1} \tag{5.86}$$

as the fiducial distribution of θ, with $(\bar{x}_1, s_1, \bar{x}_2, s_2)$ fixed as observed and t_{n_1-1} and t_{n_2-1} as independent t variables.

Homework 5.38 What was Fisher's argument for (5.86)?

Fisher's solution to the Behrens–Fisher problem seemed either magical or wrong to the 1930s statistical community. From a modern point of view it looks like an attempt to get a posterior distribution without the need for a

Bayes prior: "making the Bayesian omelette without breaking the Bayesian eggs," in the words of L. J. Savage.

The theory of confidence densities refers the argument back to confidence intervals, albeit approximate ones, perhaps giving results like (5.86) some of the legitimacy of (5.85). In what follows we will pursue the Bayesian roots of confidence density inferences.

Implied Likelihood

A confidence density $\pi_{\boldsymbol{x}}(\theta)$ is intended to answer this question: given observed data $\boldsymbol{x}$ from a multiparameter family $g_{\mu}(\boldsymbol{x})$, what is a posterior density for a real-valued parameter of interest $\theta = t(\mu)$, based on an uninformative prior distribution? "Uninformative" means that the prior is not supposed to add any outside information to the estimation of θ. Confidence densities are thought to fill the bill since their quantiles match confidence limit endpoints, and confidence intervals have no Bayesian input. All of this sounds vague, but we will make it less so in what follows.

Uninformative inference for a parameter $\theta = t(\mu)$ can be approached through another question: given data $\boldsymbol{x}$ from a p-parameter family $g_{\mu}(\boldsymbol{x})$, what is an individual likelihood function $L_{\boldsymbol{x}}(\theta)$ for θ alone, stripped of the $p - 1$ nuisance parameters in μ? An answer would be invaluable for applied Bayesian inference. Suppose an outside expert provides a prior density $\pi_{\text{expert}}(\theta)$ (an easier task than providing a p-dimensional prior for μ); then Bayes rule would give posterior density,

$$\pi_{\text{expert}}(\theta \mid \boldsymbol{x}) = c_{\boldsymbol{x}}\pi_{\text{expert}}(\theta)L_{\boldsymbol{x}}(\theta),$$

$c_{\boldsymbol{x}}$ the constant making $\pi_{\text{expert}}(\theta \mid \boldsymbol{x})$ integrate to 1. This would be true for any choice of $\pi_{\text{expert}}(\theta)$, demonstrating the uninformative nature of $L_{\boldsymbol{x}}(\theta)$.

Once again, there has been an enormous amount of theoretical effort directed at the production of individual likelihoods $L_{\boldsymbol{x}}(\theta)$. Confidence densities offer a computational approach that avoids the theory. The basic idea is motivated by the case where there are no nuisance parameters, say with $g_{\theta}(\boldsymbol{x})$ as the probability family.

Starting from prior density $\pi_0(\theta)$, Bayes rule gives posterior density

$$\pi_{\boldsymbol{x}}(\theta) = c_{\boldsymbol{x}}\pi_0(\theta)g_{\theta}(\boldsymbol{x}) = c_{\boldsymbol{x}}\pi_0(\theta)L_{\boldsymbol{x}}(\theta),$$

where here $L_{\boldsymbol{x}}(\theta)$ is the usual likelihood $g_{\theta}(\boldsymbol{x})$ thought of as a function of θ. Now imagine having observed a second realization from $g_{\theta}(\boldsymbol{x})$ that happens to exactly equal the original observation $\boldsymbol{x}$. Using hopefully obvious notation, likelihood $L_{xx}(\theta) = L_x(\theta)^2$, and the posterior density is

$$\pi_{xx}(\theta) = c_{xx}\pi_0(\theta)g_\theta(\boldsymbol{x})^2 = c_{xx}\pi_0(\theta)L_x(\theta)^2.$$

The ratio $\pi_{xx}(\theta)/\pi_x(\theta)$ equals $L_x(\theta)$,

$$L_x(\theta) = \frac{\pi_{xx}(\theta)}{\pi_x(\theta)} \tag{5.87}$$

(ignoring c_{xx}/c_x since constant multipliers don't affect likelihoods). Formula (5.87) can just as well be applied to multiparameter families $g_\mu(\boldsymbol{x})$. Efron (1993) offers evidence in favor of using (5.87) with bca confidence intervals in exponential families. There it is called the *implied likelihood*.

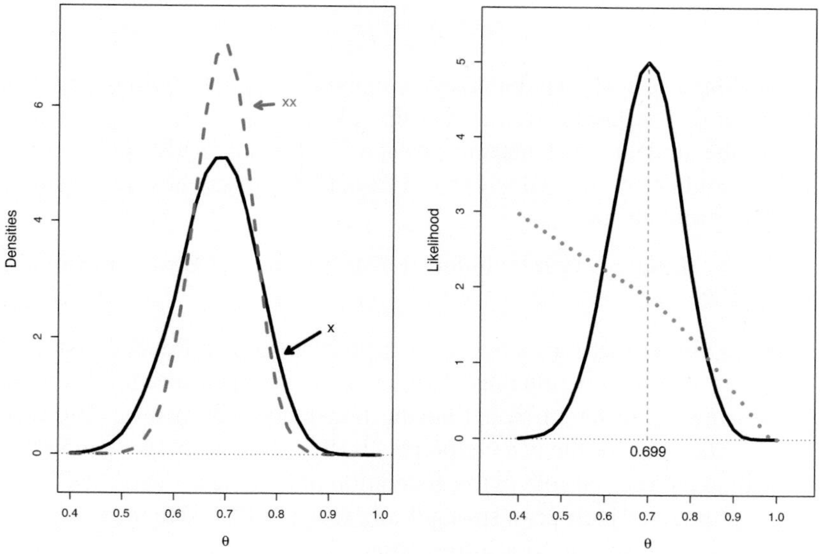

Figure 5.16 *Left*: $\pi_x(\theta)$ and $\pi_{xx}(\theta)$ for student score maxeig problem. *Right*: Implied likelihood (5.87) and implied prior density (5.88), green dots.

Figure 5.16 shows this in action for the maxeig problem of Figure 5.4. The solid curve in the left panel is the confidence density $\pi_x(\theta)$ (as in Figure 5.15). The red dashed curve is $\pi_{xx}(\theta)$: two copies of the student score data matrix $\boldsymbol{X}$ in Table 5.1 were concatenated, giving the 44×5 data matrix

$$\boldsymbol{X}_{(2)} = \begin{pmatrix} \boldsymbol{X} \\ \boldsymbol{X} \end{pmatrix};$$

the mean $\hat{\lambda}_{(2)}$ and maximum likelihood covariance estimate $\widehat{\Gamma}_{(2)}$ based on $\boldsymbol{X}_{(2)}$ were calculated; 10,000 bootstrap data matrices $\boldsymbol{X}^*_{(2)}$ were generated as 44 independent draws from $\mathcal{N}_5(\hat{\lambda}_{(2)}, \widehat{\Gamma}_{(2)})$; and `bcapar` was applied to give the confidence density $\pi_{\boldsymbol{xx}}(\theta)$, which appears as the red dashed curve.

Homework 5.39 How do $\hat{\lambda}_{(2)}$ and $\widehat{\Gamma}_{(2)}$ compare to the original estimates $\hat{\lambda}$ and $\widehat{\Gamma}$?

The solid curve in the right panel is the implied likelihood $L_{\boldsymbol{x}}(\theta)$ (5.87). Perhaps surprisingly, it's quite close to a normal density function, $\mathcal{N}(0.696, 0.079^2)$, with skewness only -0.11 and kurtosis -0.15. From Bayes rule $\pi_{\boldsymbol{x}}(\theta) = c_{\boldsymbol{x}}\pi_0(\theta)L_{\boldsymbol{x}}(\theta)$, one can also infer an *implied prior*

$$\pi_0(\theta) = \frac{\pi_{\boldsymbol{x}}(\theta)^2}{\pi_{\boldsymbol{xx}}(\theta)}. \tag{5.88}$$

This is plotted as the green dotted curve in the right panel, declining almost linearly in θ. The implication is that this plays the role of uninformative prior in the student score maxeig problem. A hypothetical expert's prior $\pi_{\text{expert}}(\theta)$ could be compared with $\pi_0(\theta)$ from (5.88) to say how informative that expert actually was.

Reference Lindley (1958), "Fiducial distributions and Bayes' theorem", *JRSS-B* 102–107.

Lindley shows that for one-parameter problems $g_\theta(x)$, fiducial distributions agree with Bayes rule only if (θ, x) can be mapped into (ζ, W), with W following a translation model having translation parameter ζ. But this is the justification for the bca endpoints (5.15) shown at (5.16), giving the bca method a preferred role in the generation of confidence densities.

Current practice includes a strong desire to apply Bayesian ideas even in the absence of Bayesian prior information. Confidence densities, like their fiducial forebears, are speculative in nature but offer a halfway house between frequentist and Bayesian methodology that has considerable allure, particularly now that they can be routinely computed.

References

Barndorff-Nielsen, O. 1980. Conditionality resolutions. *Biometrika*, **67**(2), 293–310.

Bradley, Ralph Allan, and Terry, Milton E. 1952. Rank analysis of incomplete block designs I. The method of paired comparisons. *Biometrika*, **39**, 324–345.

Brown, Lawrence D. 1986. *Fundamentals of Statistical Exponential Families with Applications in Statistical Decision Theory*. Vol. 9. Institute of Mathematical Statistics, Hayward, CA.

Cox, D. R. 1972. Regression models and life-tables. *J. Roy. Statist. Soc. Ser. B*, **34**, 187–220.

Cox, D. R. 1975. Partial likelihood. *Biometrika*, **62**(2), 269–276.

Daniels, H. E. 1983. Saddlepoint approximations for estimating equations. *Biometrika*, **70**(1), 89–96.

Dempster, A. P., Laird, N. M., and Rubin, D. B. 1977. Maximum likelihood from incomplete data via the EM algorithm. *J. Roy. Statist. Soc. Ser. B*, **39**(1), 1–38. with discussion.

Diaconis, Persi, and Ylvisaker, Donald. 1979. Conjugate priors for exponential families. *Ann. Stat.*, **7**(2), 269–281.

DiCiccio, T. J. 1984. On parameter transformations and interval estimation. *Biometrika*, **71**(3), 477–485.

DiCiccio, Thomas, and Efron, Bradley. 1992. More accurate confidence intervals in exponential families. *Biometrika*, **79**(2), 231–245.

DiCiccio, Thomas J., and Efron, Bradley. 1996. Bootstrap confidence intervals. *Statist. Sci.*, **11**(3), 189–228. With comments and a rejoinder by the authors.

Efron, Bradley. 1975. Defining the curvature of a statistical problem (with applications to second order efficiency). *Ann. Stat.*, **3**(6), 1189–1242. With discussion and a reply by the author.

Efron, Bradley. 1977. The efficiency of Cox's likelihood function for censored data. *J. Amer. Statist. Assoc.*, **72**(359), 557–565.

Efron, Bradley. 1978. The geometry of exponential families. *Ann. Stat.*, **6**(2), 362–376.

Efron, Bradley. 1982. Transformation theory: How normal is a family of distributions? *Ann. Stat.*, **10**(2), 323–339.

Efron, Bradley. 1986a. Double exponential families and their use in generalized linear regression. *J. Amer. Statist. Assoc.*, **81**(395), 709–721.

Efron, Bradley. 1986b. How biased is the apparent error rate of a prediction rule? *J. Amer. Statist. Assoc.*, **81**(394), 461–470.

Efron, Bradley. 1987. Better bootstrap confidence intervals. *J. Amer. Statist. Assoc.*, **82**(397), 171–200. With comments and a rejoinder by the author.

Efron, Bradley. 1993. Bayes and likelihood calculations from confidence intervals. *Biometrika*, **80**(1), 3–26.

Efron, Bradley. 2011. Tweedie's formula and selection bias. *J. Amer. Statist. Assoc.*, **106**(496), 1602–1614.

Efron, Bradley. 2018. Curvature and inference for maximum likelihood estimates. *Ann. Stat.*, **46**(4), 1664–1692.

Efron, Bradley, and Hastie, Trevor. 2016. *Computer Age Statistical Inference: Algorithms, Evidence, and Data Science*. Cambridge: Cambridge University Press.

Efron, Bradley, and Hinkley, David V. 1978. Assessing the accuracy of the maximum likelihood estimator: Observed versus expected Fisher information. *Biometrika*, **65**(3), 457–487. With comments and a reply by the authors.

Efron, Bradley, and Stein, Charles. 1981. The jackknife estimate of variance. *Ann. Stat.*, **9**(3), 586–596.

Efron, Bradley, and Tibshirani, Robert. 1996. Using specially designed exponential families for density estimation. *Ann. Stat.*, **24**(6), 2431–2461.

Fisher, R.A. 1925. *Statistical Methods for Research Workers*. Oliver and Boyd.

Hoeffding, Wassily. 1965. Asymptotically optimal tests for multinomial distributions. *Ann. Math. Stat.*, **36**, 369–408.

Hougaard, Philip. 1982. Parametrizations of nonlinear models. *J. Roy. Statist. Soc. Ser. B*, **44**(2), 244–252.

Johnson, Norman L., and Kotz, Samuel. 1970a. *Distributions in Statistics. Continuous Univariate Distributions. 1*. Houghton Mifflin Co., Boston.

Johnson, Norman L., and Kotz, Samuel. 1970b. *Distributions in Statistics. Continuous Univariate Distributions. 2*. Boston, Mass.: Houghton Mifflin Co.

Jørgensen, Bent. 1987. Exponential dispersion models. *J. Roy. Statist. Soc. Ser. B*, **49**(2), 127–162. With discussion and a reply by the author.

Kendall, Maurice G., and Stuart, Alan. 1958. *The Advanced Theory of Statistics. Vol. 1. Distribution Theory*. Hafner Publishing Co., New York.

Kotz, Samuel, Balakrishnan, N., and Johnson, Norman L. 2000. *Continuous Multivariate Distributions Vol. 1*. 2nd edn. Wiley-Interscience, New York.

Lehmann, Erich L., and Casella, George. 1998. *Theory of Point Estimation*. 2nd edn. Springer-Verlag, New York.

Lehmann, Erich L., and Romano, Joseph P. 2005. *Testing Statistical Hypotheses*. 3rd edn. Springer, New York.

Lindley, D. V. 1958. Fiducial distributions and Bayes' theorem. *J. Roy. Statist. Soc. Ser. B*, **20**, 102–107.

Love, Michael I, Huber, Wolfgang, and Anders, Simon. 2014. Moderated estimation of fold change and dispersion for RNA-seq data with DESeq2. bioRxiv 002832.

Mardia, Kantilal Varichand, Kent, John T., and Bibby, John M. 1979. *Multivariate Analysis*. London: Academic Press [Harcourt Brace Jovanovich Publishers].

McCullagh, P., and Nelder, J. A. 1983. *Generalized Linear Models*. Chapman & Hall, London.

McCullagh, P., and Nelder, J. A. 1989. *Generalized Linear Models*. 2nd edn. Chapman & Hall, London.

Reid, N. 1988. Saddlepoint methods and statistical inference. *Statist. Sci.*, **3**(2), 213–238. With comments and a rejoinder by the author.

Robbins, Herbert. 1956. An empirical Bayes approach to statistics. Pages 157–163 of: *Proceedings of the Third Berkeley Symposium on Mathematical Statistics and Probability, 1954–1955, Vol. I.* University of California Press, Berkeley and Los Angeles.

Stein, Charles. 1956. Efficient nonparametric testing and estimation. Pages 187–195 of: *Proceedings of the Third Berkeley Symposium on Mathematical Statistics and Probability, 1954–1955, Vol. I.* University of California Press, Berkeley and Los Angeles.

Sundberg, Rolf. 2019. *Statistical Modelling by Exponential Families.* Vol. 12. Cambridge University Press, Cambridge.

Thisted, Ronald, and Efron, Bradley. 1987. Did Shakespeare write a newly-discovered poem? *Biometrika*, **74**(3), 445–455.

Xie, Min-ge, and Singh, Kesar. 2013. Confidence distribution, the frequentist distribution estimator of a parameter: A review. *Int. Stat. Rev.*, **81**(1), 3–39. with discussion.

Index

Printed in the United States
by Baker & Taylor Publisher Services